VECTOR AND TENSOR ANALYSIS WITH APPLICATIONS

VECTOR AND TENSOR ANALYSIS WITH APPLICATIONS

by

A. I. BORISENKO

and

I. E. TARAPOV

Revised English Edition

Translated and Edited by

Richard A. Silverman

Dover Publications, Inc.

New York

This Dover edition, first published in 1979, is an unabridged and corrected republication of the work originally published in 1968 by Prentice-Hall, Inc.

International Standard Book Numbers

ISBN-13: 978-0-486-63833-1
ISBN-10: 0-486-63833-2

Library of Congress Catalog Card Number: 79-87809

Manufactured in the United States by LSC Communications
63833223 2019
www.doverpublications.com

EDITOR'S PREFACE

The present book is a freely revised and restyled version of the third edition of the Russian original (Moscow, 1966). As in other volumes of this series, I have not hesitated to introduce a number of pedagogical and mathematical improvements that occurred to me in the course of doing the translation. I have also added a brief Bibliography, confined to books in English dealing with approximately the same topics, at about the same level.

In their prefaces to the three Russian editions, the authors acknowledge the help of the following colleagues: Professors Y. P. Blank, V. L. German, G. I. Drinfeld, A. D. Myshkis, N. I. Akhiezer, P. K. Rashevsky, V. A. Marchenko and N. V. Yefimov.

R. A. S.

CONTENTS

3 TENSOR ALGEBRA, Page 103.

4 VECTOR AND TENSOR ANALYSIS: RUDIMENTS, Page 134.

VECTOR AND TENSOR ANALYSIS WITH APPLICATIONS

1

VECTOR ALGEBRA

In this chapter we define vectors and then discuss algebraic operations on vectors. The vector concept will be generalized in a natural way in Chapter 2, leading to the concept of a tensor. Then in Chapter 3 we will consider algebraic operations on tensors.

1.1. Vectors and Scalars

Quantities which can be specified by giving just one number (positive, negative or zero) are called *scalars*. For example, temperature, density, mass and work are all scalars. Scalars can be compared only if they have the same physical dimensions. Two such scalars measured in the same system of units are said to be equal if they have the same magnitude (absolute value) and sign.

However, one must often deal with quantities, called *vectors*, whose specification requires a direction as well as a numerical value. For example, displacement, velocity, acceleration, force, the moment of a force, electric field strength and dielectric polarization are all vectors. Operations on vectors obey the rules of vector algebra, to be considered below.

Scalars and vectors hardly exhaust the class of quantities of interest in applied mathematics and physics. In fact, there are quantities of a more complicated structure than scalars or vectors, called *tensors* (of order 2 or higher), whose specification requires more than knowledge of a magnitude and a direction. For example, examination of the set of all stress vectors acting on all elements of area which can be drawn through a given point of an elastic body leads to the concept of the *stress tensor* (of order 2), while examination of the deformation of an arbitrary elementary volume of an elastic body leads to the concept of the *deformation* (or *strain*) *tensor*. We will defer further discussion of tensors until Chapter 2, concentrating our attention for now on vectors.

A vector **A** is represented by a directed line segment, whose direction and length coincide with the direction and magnitude (measured in the chosen system of units) of the quantity under consideration. Vectors are denoted by boldface letters, **A**, **B**, . . . and their magnitudes by $|\mathbf{A}|$, $|\mathbf{B}|$, . . . or by the corresponding lightface letters A, B, . . . When working at the blackboard, it is customary to indicate vectors by the presence of little arrows, as in $\vec{A}$, $\vec{B}$, ...

The vector of magnitude zero is called the *zero vector*, denoted by 0 (ordinary lightface zero). This vector cannot be assigned a definite direction, or alternatively can be regarded as having any direction at all.

Vectors can be compared only if they have the same physical or geometrical meaning, and hence the same dimensions. Two such vectors **A** and **B** measured in the same system of units are said to be equal, written $\mathbf{A} = \mathbf{B}$, if they have the same magnitude (length) and direction.

1.1.1. Free, sliding and bound vectors. A vector like the velocity of a body undergoing uniform translational motion, which can be displaced parallel to itself and applied at any point, is called a *free vector* [see Fig. 1.1(a)]. In

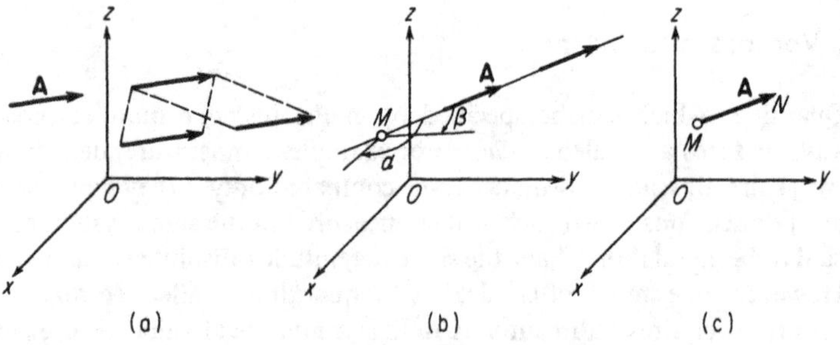

FIG. 1.1. (a) A free vector, which can be displaced parallel to itself; (b) A sliding vector, which can be displaced along its line of action; (c) A bound vector.

three dimensions a free vector is uniquely determined by three numbers, e.g., by its three projections on the axes of a rectangular coordinate system. It can also be specified by giving its magnitude (the length of the line segment representing the vector) and two independent angles α and β (any two of the angles between the vector and the coordinates axes).

A vector like the force applied to a rigid body fastened at a fixed point, which can only be displaced along the line containing the vector, is called a *sliding vector* [see Fig. 1.1(b)]. In three dimensions a sliding vector is determined by five numbers, e.g., by the coordinates of the point of intersection M of one of the coordinate planes and the line containing the vector (two numbers), by the magnitude of the vector (one number) and by two independent angles α and β between the vector and two of the coordinate axes (two numbers).

A vector like the wind velocity at a given point of space, which is referred to a fixed point, is called a *bound vector* [see Fig. 1.1(c)]. In three dimensions a bound vector is determined by six numbers, e.g., the coordinates of the initial and final points of the vector (M and N in the figure).

Free vectors are the most general kind of quantity specified by giving a magnitude and a direction, and the study of sliding and bound vectors can always be reduced to that of free vectors. Therefore we shall henceforth consider only free vectors.

1.2. Operations on Vectors

1.2.1. Addition of vectors. Given two vectors $\mathbf{A}$ and $\mathbf{B}$, suppose we put the initial point of $\mathbf{B}$ at the final point of $\mathbf{A}$. Then by the sum $\mathbf{A} + \mathbf{B}$ we mean the vector joining the initial point of $\mathbf{A}$ to the final point of $\mathbf{B}$. This is also the diagonal of the parallelogram constructed on $\mathbf{A}$ and $\mathbf{B}$, in the way shown in Fig. 1.2(a). It follows that the sum $\mathbf{A} + \mathbf{B} + \mathbf{C} + \cdots$ of several vectors

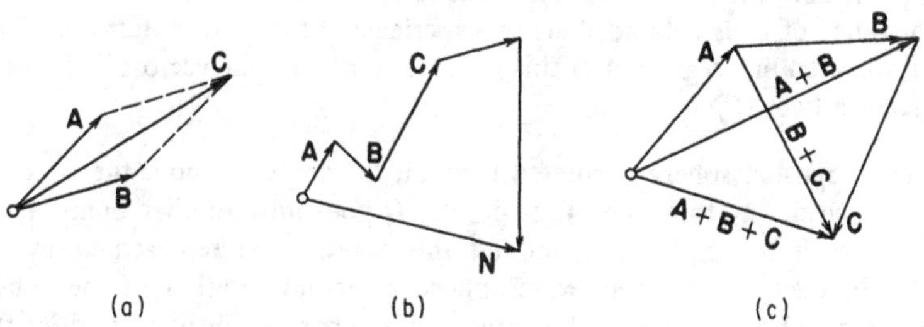

FIG. 1.2. (a) The sum of two vectors $\mathbf{A} + \mathbf{B} = \mathbf{C}$;
(b) The sum of several vectors $\mathbf{A} + \mathbf{B} + \mathbf{C} + \cdots = \mathbf{N}$;
(c) Associativity of vector addition: $(\mathbf{A} + \mathbf{B}) + \mathbf{C} = \mathbf{A} + (\mathbf{B} + \mathbf{C}) = \mathbf{A} + \mathbf{B} + \mathbf{C}$.

$\mathbf{A}, \mathbf{B}, \mathbf{C}, \ldots$ is the vector closing the polygon obtained by putting the initial point of $\mathbf{B}$ at the final point of $\mathbf{A}$, the initial point of $\mathbf{C}$ at the final point of $\mathbf{B}$, and so on, as in Fig. 1.2(b). The physical meaning of vector addition is clear if we interpret $\mathbf{A}, \mathbf{B}, \mathbf{C}, \ldots$ as consecutive displacements of a point in space.

Then the sum $\mathbf{A} + \mathbf{B} + \mathbf{C} + \cdots$ is just the resultant of the individual displacements, equal to the total displacement of the point.

The definition just given implies that vector addition has the characteristic properties of ordinary algebraic addition, i.e.,

1) Vector addition is *commutative:*

$$\mathbf{A} + \mathbf{B} = \mathbf{B} + \mathbf{A};$$

2) Vector addition is *associative:*

$$(\mathbf{A} + \mathbf{B}) + \mathbf{C} = \mathbf{A} + (\mathbf{B} + \mathbf{C}) = \mathbf{A} + \mathbf{B} + \mathbf{C}.$$

In other words, a sum of two vectors does not depend on the order of the terms, and there is no need to write parentheses in a sum involving three or more vectors [see Fig. 1.2(c)].

Since the zero vector has no length at all, its initial and final points coincide, and hence

$$\mathbf{A} + 0 = \mathbf{A}$$

for any vector $\mathbf{A}$.

The fact that a quantity is characterized by a magnitude and a direction is a necessary but not sufficient condition for it to be a vector. The quantity must also obey the laws of vector algebra, in particular the law of vector addition. To illustrate this remark, suppose a rigid body is rotated about some axis. Then the rotation can be represented by a line segment of length equal to the angle of rotation directed along the axis of rotation, for example in the direction from which the rotation seems to be counterclockwise (the direction of advance of a right-handed screw experiencing the given rotation). However, directed line segments of this kind do not "add like vectors." This can be seen in two ways:

1) Suppose a sphere is rotated through the angle α_1 about the y-axis in such a way that one of its points A_1 goes into another point A_2, as shown in Fig. 1.3(a), and let this rotation be represented by the directed line segment $\boldsymbol{\alpha}_1$. Suppose a second rotation of the sphere through the angle α_2 about the z-axis carries A_2 into A_3, and let this rotation be represented by the directed line segment $\boldsymbol{\alpha}_2$. Then if $\boldsymbol{\alpha}_1$ and $\boldsymbol{\alpha}_2$ add like vectors, the rotation carrying A_1 into A_3 must equal $\boldsymbol{\alpha}_1 + \boldsymbol{\alpha}_2$ on the one hand, while on the other hand it must be represented by a directed line segment $\boldsymbol{\alpha}_3$ perpendicular to the plane A_1OA_3. But since $\boldsymbol{\alpha}_3$ does not lie in the yz-plane, it cannot be the sum of $\boldsymbol{\alpha}_1$ and $\boldsymbol{\alpha}_2$. This is particularly apparent in the case where

$$\alpha_1 = \alpha_2 = \frac{\pi}{2}.$$

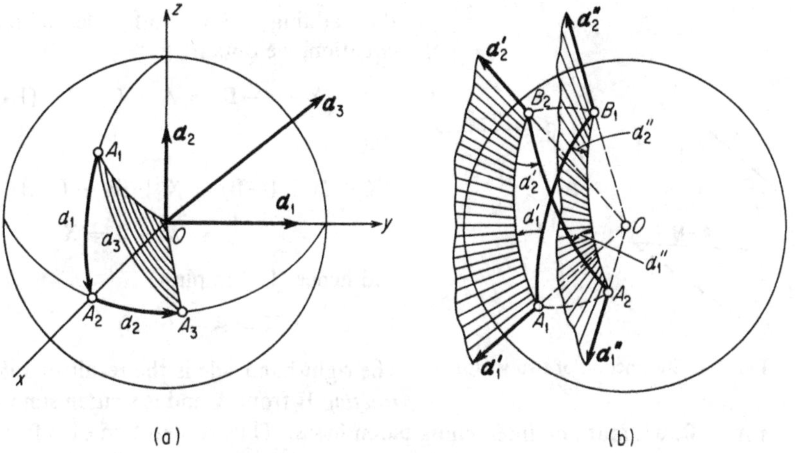

FIG. 1.3. Rotation of a sphere about its axis.

(a) The directed line segment $\boldsymbol{\alpha}_3$ does not lie in the plane of the segments $\boldsymbol{\alpha}_1$ and $\boldsymbol{\alpha}_2$;

(b) Two ways of carrying a circular arc from one position (A_1B_1) to another (A_2B_2).

2) Both of the following operations carry a circular arc A_1B_1 on a sphere into a new position A_2B_2 [see Fig. 1.3(b)]:

   a) A rotation through α_1' radians about the axis OA_1 (the directed line segment $\boldsymbol{\alpha}_1'$) followed by a rotation through α_2' radians about the axis OB_2 (the directed line segment $\boldsymbol{\alpha}_2'$);

   b) A rotation through α_2'' radians about the axis OB_1 (the directed line segment $\boldsymbol{\alpha}_2''$) followed by a rotation through α_1'' radians about the axis OA_2 (the directed line segment $\boldsymbol{\alpha}_1''$).[1]

Hence $\boldsymbol{\alpha}_1' + \boldsymbol{\alpha}_2'$ must equal $\boldsymbol{\alpha}_2'' + \boldsymbol{\alpha}_1''$ if rotations add like vectors. But this is impossible, since $\boldsymbol{\alpha}_1' + \boldsymbol{\alpha}_2'$ and $\boldsymbol{\alpha}_2'' + \boldsymbol{\alpha}_1''$ lie in nonparallel planes.

Remark. Unlike the case of finite rotations of a rigid body, it turns out that infinitesimal rotations are vectors (see Prob. 10, p. 44).

1.2.2. Subtraction of vectors. By the vector $-\mathbf{A}$ ("minus $\mathbf{A}$") we mean the vector with the same magnitude as $\mathbf{A}$ but with the opposite direction. Each of the vectors $\mathbf{A}$ and $-\mathbf{A}$ is called the *opposite* or *negative* of the other. Obviously, the only vector equal to its own negative is the zero vector, and the sum of a vector and its negative is the zero vector.

If

$$\mathbf{X} + \mathbf{B} = \mathbf{A},$$

[1] Note that $\alpha_1' = \alpha_1''$, $\alpha_2' = \alpha_2''$.

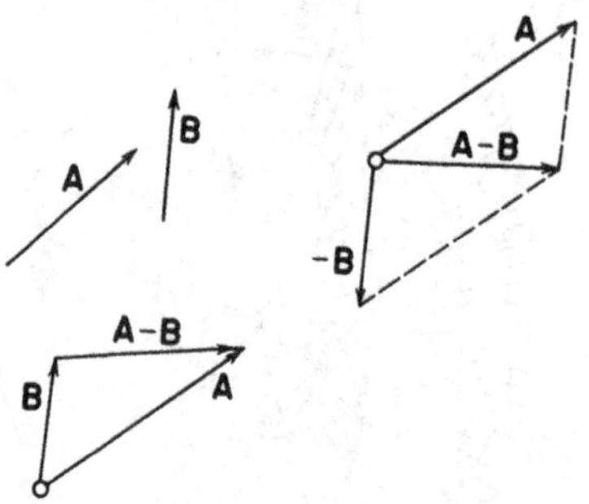

FIG. 1.4. Subtraction of two vectors.

then adding $-\mathbf{B}$ to both sides of the equation, we obtain

$$\mathbf{X} + \mathbf{B} + (-\mathbf{B}) = \mathbf{A} + (-\mathbf{B}). \quad (1.1)$$

But

$$\mathbf{X} + \mathbf{B} + (-\mathbf{B}) = \mathbf{X} + [\mathbf{B} + (-\mathbf{B})]$$
$$= \mathbf{X} + 0 = \mathbf{X},$$

and hence (1.1) implies

$$\mathbf{X} = \mathbf{A} + (-\mathbf{B}).$$

The right-hand side is the result of *subtracting* **B** from **A** and is written simply as $\mathbf{A} - \mathbf{B}$, without any intervening parentheses. Thus subtraction of **B** from **A** reduces to addition of **A** and the negative of **B**, as shown in Fig. 1.4.

1.2.3. Projection of a vector onto an axis. Given a vector **A** and an axis u (see Fig. 1.5), by the *projection* of **A** onto u, denoted by A_u, we mean the

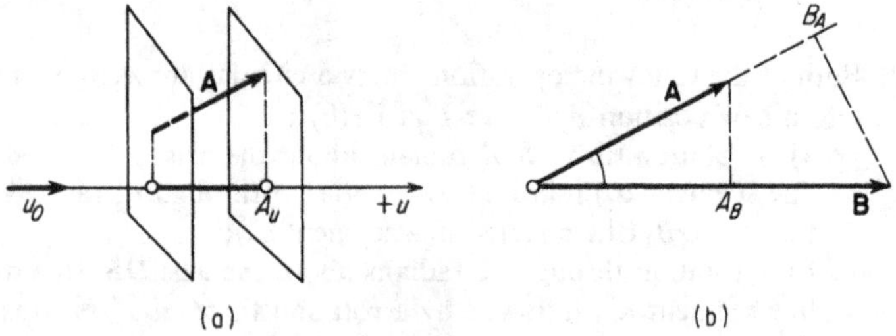

FIG. 1.5. (a) Projection of a vector onto an axis;
(b) Projection of one vector onto the direction of another.

length of the segment cut from u by the planes drawn through the end points of **A** perpendicular to u, taken with the plus sign if the direction from the projection (onto u) of the initial point of **A** to the projection of the end point of **A** coincides with the positive direction of u, and with the minus sign otherwise. By the *unit vector* corresponding to the axis u we mean the vector $\mathbf{u}_0$ of unit length whose direction coincides with that of u (thus $u_0 = |\mathbf{u}_0| = 1$). If $\varphi = (\mathbf{A}, \mathbf{u}_0)$ denotes the angle between **A** and $\mathbf{u}_0$,[2] then

$$A_u = A \cos \varphi = A \cos (\mathbf{A}, \mathbf{u}_0). \quad (1.2)$$

[2] Given two vectors **A** and **B**, the angle between **A** and **B**, denoted by (**A**, **B**), will always be chosen to lie between 0 and π.

In fact, it is clear that $|A \cos \varphi|$ always gives the length of the segment of the axis u between the planes drawn through the end points of **A** perpendicular to u. Moreover, the direction from the projection (onto u) of the initial point of **A** to the projection of the end point of **A** is the same as the positive direction of u if $\varphi < \pi/2$ and the opposite if $\varphi > \pi/2$. But then (1.2) automatically gives A_u the correct sign, since $\varphi < \pi/2$ implies $\cos \varphi > 0$ and hence $A_u > 0$, while $\varphi > \pi/2$ implies $\cos \varphi < 0$ and hence $A_u < 0$.

Thus the projection of a vector **A** onto an axis u equals the product of the length of **A** and the cosine of the angle between **A** and the positive direction of u.[3]

1.2.4. Multiplication of a vector by a scalar. By the product of a vector **A** and a scalar m we mean the vector of magnitude $|m|$ times that of **A**, with the same direction as **A** if $m > 0$ and the opposite direction if $m < 0$. Thus

$$\mathbf{B} = m\mathbf{A}$$

implies

$$|\mathbf{B}| = B = |m|\,|\mathbf{A}| = |m|\,A.$$

If $m = -1$, then **B** and **A** are opposite vectors. In any events, the vectors **B** and **A** are parallel or for that matter *collinear* (there is no distinction between parallel and collinear *free* vectors).

It is clear that multiplication of a vector by a scalar obeys the following rules:

$$m(n\mathbf{A}) = (mn)\mathbf{A},$$

$$m(\mathbf{A} + \mathbf{B}) = m\mathbf{A} + m\mathbf{B},$$

$$(m + n)\mathbf{A} = m\mathbf{A} + n\mathbf{A}.$$

1.3. Bases and Transformations

1.3.1. Linear dependence and linear independence of vectors. We say that n vectors $\mathbf{A}_1, \mathbf{A}_2, \ldots, \mathbf{A}_n$ are *linearly dependent* if there exist n scalars $c_1, c_2, \ldots, c_n$ not all equal to zero such that

$$c_1\mathbf{A}_1 + c_2\mathbf{A}_2 + \cdots + c_n\mathbf{A}_n = 0, \tag{1.3}$$

i.e., if some (nontrivial) linear combination of the vectors equals zero.[4] Vectors which are not linearly dependent are said to be *linearly independent*.

---

[3] Before reading further, solve Exercise 1, p. 54.

[4] By a "trivial" linear combination of vectors we mean a combination all of whose coefficients vanish.

In other words, n vectors $\mathbf{A}_1, \mathbf{A}_2, \ldots, \mathbf{A}_n$ are said to be linearly independent if (1.3) implies

$$c_1 = c_2 = \cdots = c_n = 0.$$

Two linearly dependent vectors are *collinear*. This follows from Sec. 1.2.4 and the fact that

$$c_1\mathbf{A} + c_2\mathbf{B} = 0$$

implies

$$\mathbf{A} = -\frac{c_2}{c_1}\mathbf{B}$$

if $c_1 \neq 0$ or

$$\mathbf{B} = -\frac{c_1}{c_2}\mathbf{A}$$

if $c_2 \neq 0$.

Three linearly dependent vectors are *coplanar*, i.e., lie in the same plane (or are parallel to the same plane). In fact, if

$$c_1\mathbf{A} + c_2\mathbf{B} + c_3\mathbf{C} = 0 \tag{1.4}$$

where at least one of the numbers c_1, c_2, c_3 is nonzero, say c_3, then

$$\mathbf{C} = m\mathbf{A} + n\mathbf{B}$$

where

$$m = -\frac{c_1}{c_3}, \qquad n = -\frac{c_2}{c_3}, \tag{1.5}$$

i.e., $\mathbf{C}$ lies in the same plane as $\mathbf{A}$ and $\mathbf{B}$ (being the sum of the vector $m\mathbf{A}$ collinear with $\mathbf{A}$ and the vector $n\mathbf{B}$ collinear with $\mathbf{B}$).

1.3.2. Expansion of a vector with respect to other vectors. *Let* $\mathbf{A}$ *and* $\mathbf{B}$ *be two linearly independent* (*i.e., noncollinear*) *vectors. Then any vector* $\mathbf{C}$ *coplanar with* $\mathbf{A}$ *and* $\mathbf{B}$ *has a unique expansion*

$$\mathbf{C} = m\mathbf{A} + n\mathbf{B} \tag{1.6}$$

with respect to $\mathbf{A}$ *and* $\mathbf{B}$. In fact, since $\mathbf{A}$, $\mathbf{B}$ and $\mathbf{C}$ are coplanar, (1.4) holds with at least one nonzero coefficient, say c_3. Dividing (1.4) by c_3, we get (1.6), where m and n are the same as in (1.5). To prove the uniqueness of the expansion (1.6), suppose there is another expansion

$$C = m'\mathbf{A} + n'\mathbf{B}. \tag{1.7}$$

Subtracting (1.7) from (1.6), we obtain

$$(m - m')\mathbf{A} + (n - n')\mathbf{B} = 0.$$

But then $m = m'$, $n = n'$ since $\mathbf{A}$ and $\mathbf{B}$ are linearly independent. In other words, the coefficients m and n of the expansion (1.6) are uniquely determined.

Let **A**, **B** *and* **C** *be three linearly independent (i.e., noncoplanar) vectors. Then any vector* **D** *has a unique expansion*

$$\mathbf{D} = m\mathbf{A} + n\mathbf{B} + p\mathbf{C} \tag{1.8}$$

with respect to **A**, **B** *and* **C**. To see this, draw the vectors **A**, **B**, **C** and **D** from a common origin O (see Fig. 1.6). Then through the end point of **D** draw the three planes parallel to the plane of the vectors **A** and **B**, **A** and **C**, **B** and **C**. These planes, together with the planes of the vectors **A** and **B**, **A** and **C**, **B** and **C** form a parallelepiped with the vector **D** as one of its diagonals and the vectors **A**, **B** and **C** (drawn from the origin O) along three of its edges. If the numbers m, n and p are such that $m\mathbf{A}$, $n\mathbf{B}$ and $p\mathbf{C}$ have magnitudes equal to the lengths of the corresponding edges of the parallelepiped, then clearly

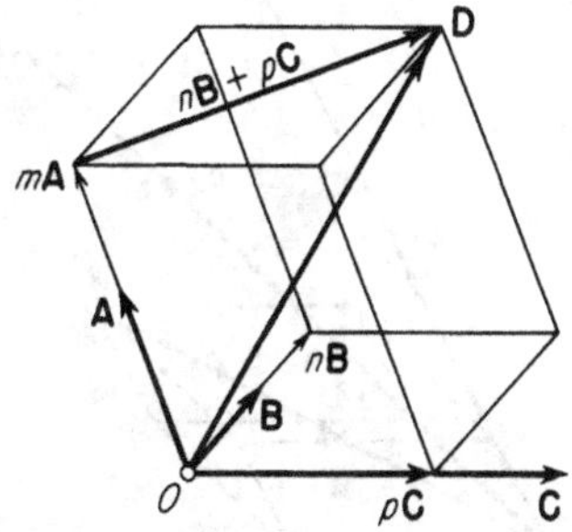

FIG. 1.6. An arbitrary vector **D** has a unique expansion with respect to three noncoplanar vectors **A**, **B** and **C**.

$$\mathbf{D} = m\mathbf{A} + (n\mathbf{B} + p\mathbf{C}) = m\mathbf{A} + n\mathbf{B} + p\mathbf{C}$$

as shown in Fig. 1.6.

To prove the uniqueness of the expansion (1.8), suppose there is another expansion

$$\mathbf{D} = m'\mathbf{A} + n'\mathbf{B} + p'\mathbf{C}. \tag{1.9}$$

Subtracting (1.9) from (1.8), we obtain

$$(m - m')\mathbf{A} + (n - n')\mathbf{B} + (p - p')\mathbf{C} = 0.$$

But then $m = m'$, $n = n'$, $p' = p$ since **A**, **B** and **C** are linearly independent by hypothesis.

Remark. It follows from the above considerations that any four vectors in three-dimensional space are linearly dependent.

1.3.3. Bases and basis vectors. By a *basis* for three-dimensional space we mean any set of three linearly independent vectors $\mathbf{e}_1$, $\mathbf{e}_2$, $\mathbf{e}_3$. Each of the vectors $\mathbf{e}_1$, $\mathbf{e}_2$, $\mathbf{e}_3$ is called a *basis vector*. Given a basis $\mathbf{e}_1$, $\mathbf{e}_2$, $\mathbf{e}_3$, it follows from the above remark that every vector **A** has a unique expansion of the form

$$A = m\mathbf{e}_1 + n\mathbf{e}_2 + p\mathbf{e}_3.$$

It should be emphasized that *any* triple of noncoplanar vectors can serve as a basis (in three dimensions).

Suppose the vectors of a basis $\mathbf{e}_1, \mathbf{e}_2, \mathbf{e}_3$ are all drawn from a common origin O, and let Ox^k denote the line containing the vector $\mathbf{e}_k$, where k ranges from 1 to 3.[5] This gives an *oblique* coordinate system with axes Ox^1, Ox^2, Ox^3 and origin O (see Fig. 1.7). The case of greatest importance is where the basis vectors $\mathbf{e}_1, \mathbf{e}_2, \mathbf{e}_3$ are orthogonal to one another and are all of unit length (such vectors are said to be *orthonormal* and so is the corresponding basis). The coordinate system is then called *rectangular* instead of oblique, and we write x_1, x_2, x_3 instead of x^1, x^2, x^3 (for reasons that will become clear in Sec. 1.6) and $\mathbf{i}_1, \mathbf{i}_2, \mathbf{i}_3$ instead of $\mathbf{e}_1, \mathbf{e}_2, \mathbf{e}_3$.

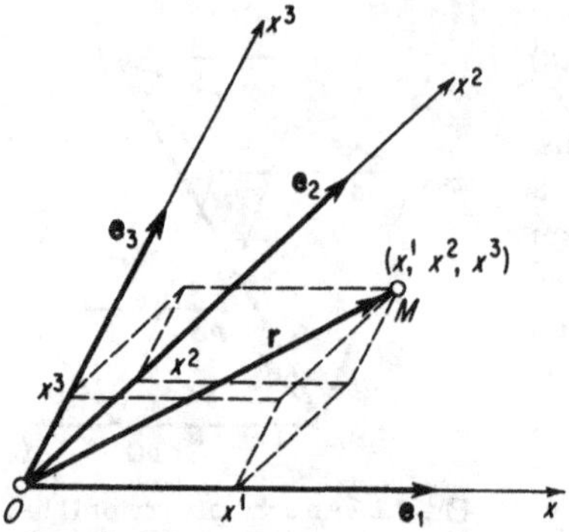

FIG. 1.7. An oblique coordinate system with basis $\mathbf{e}_1, \mathbf{e}_2, \mathbf{e}_3$.

The position of a point M relative to a given coordinate system (rectangular or not) is uniquely determined by its *radius vector* $\mathbf{r} = \mathbf{r}(M)$, i.e., by the vector drawn from the origin of the coordinate system to the point M (see Figs. 1.7 and 1.8). Suppose M has coordinates x_1, x_2, x_3 in a rectangular coordinate system. Then x_1, x_2, x_3 are the signed distances between M and

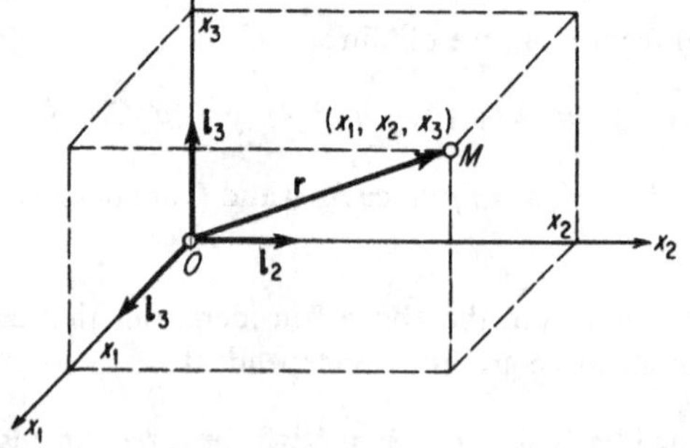

FIG. 1.8. A rectangular coordinate system with basis $\mathbf{i}_1, \mathbf{i}_2, \mathbf{i}_3$.

---

[5] The reason for writing k as a subscript in $\mathbf{e}_k$ and as a superscript in x^k will soon be apparent. Do not think of x^k as x raised to the kth power!

the planes x_2Ox_3, x_3Ox_1 and x_1Ox_2, and hence

$$\mathbf{r} = x_1\mathbf{i}_1 + x_2\mathbf{i}_2 + x_3\mathbf{i}_3.$$

The great merit of vectors in applied problems is that equations describing physical phenomena can be formulated without reference to any particular coordinate system. However, in actually carrying out the calculations needed to solve a given problem, one must eventually cast the problem into a form involving scalars. This is done by introducing a suitable coordinate system, and then replacing the given vector (or tensor) equations by an equivalent system of scalar equations involving only numbers obeying the ordinary rules of arithmetic. The key step is to expand the vectors (or tensors) with respect to a suitable basis, corresponding to the chosen system of coordinates.

Consider, for example, the case of two dimensions. The position of a point M in the plane is uniquely determined by its radius vector $\mathbf{r}$ relative to some fixed point O which can be chosen arbitrarily and is independent of any coordinate system. However, before making any calculations, we must introduce a coordinate system. Then the position of the point M is given by two numbers p and q (called its *coordinates*), which now depend both on the coordinate system and on the units of measurement. In a rectangular system, these coordinates are just the (signed) distances $p \equiv x_1$ and $q \equiv x_2$ between M and two perpendicular lines going through the origin of coordinates. Holding one coordinate fixed, say $p \equiv$ const, and continuously varying the other coordinate, we obtain a *coordinate curve*. Thus there are two coordinate curves passing through every point of the plane. In rectangular coordinates these curves are simply the lines parallel to the coordinate axes. As the basis vectors corresponding to the coordinates p and q, we choose the *unit vectors* (i.e., the vectors of unit length) tangent to the coordinate curves at the point M. In rectangular coordinates these are just the unit vectors $\mathbf{i}_1$ and $\mathbf{i}_2$ parallel to the coordinate axes.

Clearly, the basis vectors $\mathbf{i}_1$ and $\mathbf{i}_2$ of a rectangular coordinate system are independent of the point M and always intersect at right angles. Suppose, however, that the position of M is specified in *polar coordinates*, i.e., by giving the distance R between M and a fixed point O (called the *pole*) and the angle φ between the line joining O to M and a fixed ray (called the *polar axis*) drawn from O. The coordinate curves are then the circles of radius R and the rays of inclination φ, and the corresponding basis vectors are the unit vectors $\mathbf{e}_R$ and $\mathbf{e}_\varphi$ shown in Fig. 1.9(a). Note that although $\mathbf{e}_R$ and $\mathbf{e}_\varphi$ vary from point to point, they always intersect at right angles (compare the basis at M with that at N). Coordinate systems whose basis vectors intersect at right angles are called *orthogonal systems*, and are the systems of greatest importance in the applications. Coordinate systems whose coordinate curves are not straight lines are called systems of *curvilinear coordinates* (as opposed to

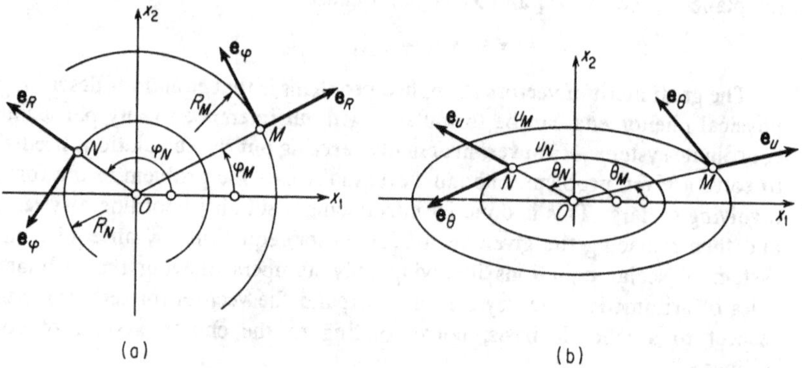

FIG. 1.9. Curvilinear coordinates in the plane.
(a) Polar coordinates;
(b) "Generalized" polar coordinates.

rectangular or oblique coordinates). Three-dimensional curvilinear coordinates will be considered in detail in Sec. 2.8.

Given a system of polar coordinates R and θ with pole O, consider a system of rectangular coordinates x_1 and x_2 with origin O, and suppose the x_1-axis coincides with the polar axis of the system of polar coordinates. Then the relation between the rectangular coordinates x_1, x_2 and the polar coordinates R, φ is given by the formula

$$x_1 = R\cos\varphi, \qquad x_2 = R\sin\varphi \tag{1.10}$$

and

$$R = \sqrt{x_1^2 + x_2^2} \qquad (0 \leqslant R < \infty),$$

$$\tan\varphi = \frac{x_2}{x_1} \qquad (0 \leqslant \varphi < 2\pi).$$

More generally, consider the system of "generalized" polar coordinates u and θ whose coordinate curves are the ellipses

$$u = \sqrt{\frac{x_1^2}{a^2} + \frac{x_2^2}{b^2}} \qquad (0 \leqslant u < \infty)$$

$(a > 0,\ b > 0,\ a \neq b)$ and the rays

$$\tan\theta = \frac{ax_2}{bx_1} \qquad (0 \leqslant \theta < 2\pi),$$

where the analogue of (1.10) is now

$$x_1 = au\cos\theta, \qquad x_2 = bu\sin\theta.$$

The corresponding basis vectors are the unit vectors $\mathbf{e}_u$ and $\mathbf{e}_\theta$ shown in Fig. 1.9(b). Note that not only do $\mathbf{e}_u$ and $\mathbf{e}_\theta$ vary from point to point as in the case of ordinary polar coordinates, but the angle between them also varies (compare the basis at M with that at N) and is in general not a right angle. For this reason, the system of coordinates u, θ is said to be *nonorthogonal.*

Remark. Orthogonal coordinates are the preferred coordinates for solving physical problems. However, many properties of orthogonal coordinates become clearer when they are regarded as a limiting case of nonorthogonal curvilinear coordinates. Thus, in this book, vectors and tensors will be considered in orthogonal coordinates for the most part, but nonorthogonal coordinates will also be used occasionally.

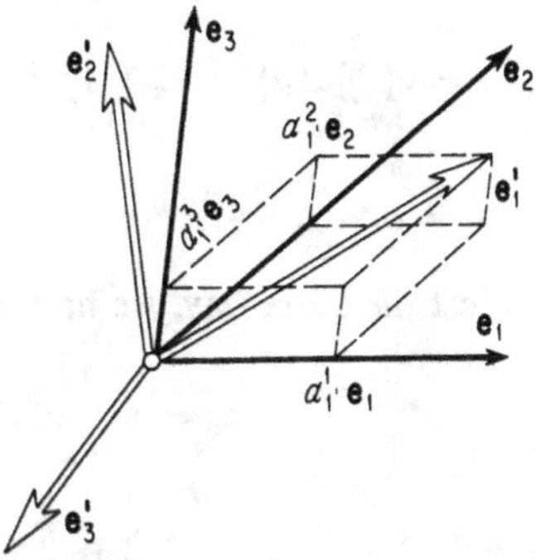

FIG. 1.10. Expansion of the vector $\mathbf{e}'_1$ with respect to the vectors $\mathbf{e}_1$, $\mathbf{e}_2$, $\mathbf{e}_3$.

1.3.4. Direct and inverse transformations of basis vectors. Consider two bases $\mathbf{e}_1, \mathbf{e}_2, \mathbf{e}_3$ and $\mathbf{e}'_1, \mathbf{e}'_2, \mathbf{e}'_3$ drawn from the same point O. Then any vector of the first basis can be expanded with respect to the vectors of the second basis and conversely (see Fig. 1.10). Let $\alpha^1_{i'}, \alpha^2_{i'}, \alpha^3_{i'}$ be the coefficients of the expansion of $\mathbf{e}'_i$ with respect to the vectors $\mathbf{e}_1, \mathbf{e}_2, \mathbf{e}_3$, respectively. Then

$$\mathbf{e}'_1 = \alpha^1_{1'}\mathbf{e}_1 + \alpha^2_{1'}\mathbf{e}_2 + \alpha^3_{1'}\mathbf{e}_3 = \sum_{k=1}^{3}\alpha^k_{1'}\mathbf{e}_k,$$

$$\mathbf{e}'_2 = \alpha^1_{2'}\mathbf{e}_1 + \alpha^2_{2'}\mathbf{e}_2 + \alpha^3_{2'}\mathbf{e}_3 = \sum_{k=1}^{3}\alpha^k_{2'}\mathbf{e}_k,$$

$$\mathbf{e}'_3 = \alpha^1_{3'}\mathbf{e}_1 + \alpha^2_{3'}\mathbf{e}_2 + \alpha^3_{3'}\mathbf{e}_3 = \sum_{k=1}^{3}\alpha^k_{3'}\mathbf{e}_k,$$

or more concisely,

$$\mathbf{e}'_i = \sum_{k=1}^{3}\alpha^k_{i'}\mathbf{e}_k \qquad (i = 1, 2, 3). \tag{1.11}$$

The nine numbers $\alpha^k_{i'}$ ($i, k = 1, 2, 3$) are called the *coefficients of the direct transformation* (from the unprimed basis to the primed basis).

Similarly, let $\alpha^{1'}_i, \alpha^{2'}_i, \alpha^{3'}_i$ be the coefficients of the expansion of $\mathbf{e}_i$ with respect to the vectors $\mathbf{e}'_1, \mathbf{e}'_2, \mathbf{e}'_3$, respectively, so that

$$\mathbf{e}_i = \sum_{k=1}^{3}\alpha^{k'}_i\mathbf{e}'_k \qquad (i = 1, 2, 3). \tag{1.12}$$

Then the nine numbers $\alpha^{k'}_i$ ($i, k = 1, 2, 3$) are called the *coefficients of the inverse transformation* (from the primed basis back to the unprimed basis).

There are simple relations between the coefficients of the direct and the inverse transformations. In fact, substituting the expansions of the vectors $\mathbf{e}_1, \mathbf{e}_2, \mathbf{e}_3$ from (1.12) into (1.11) and regrouping terms, we obtain

$$\begin{aligned}
\mathbf{e}'_i &= \alpha^1_{i'}\mathbf{e}_1 + \alpha^2_{i'}\mathbf{e}_2 + \alpha^3_{i'}\mathbf{e}_3 \\
&= \alpha^1_{i'}(\alpha^{1'}_1\mathbf{e}'_1 + \alpha^{2'}_1\mathbf{e}'_2 + \alpha^{3'}_1\mathbf{e}'_3) + \alpha^2_{i'}(\alpha^{1'}_2\mathbf{e}'_1 + \cdots) + \alpha^3_{i'}(\alpha^{1'}_3\mathbf{e}'_1 + \cdots) \\
&= (\alpha^1_{i'}\alpha^{1'}_1 + \alpha^2_{i'}\alpha^{1'}_2 + \alpha^3_{i'}\alpha^{1'}_3)\mathbf{e}'_1 + (\alpha^1_{i'}\alpha^{2'}_1 + \cdots)\mathbf{e}'_2 + (\alpha^1_{i'}\alpha^{3'}_1 + \cdots)\mathbf{e}'_3 \\
&= \mathbf{e}'_1\sum_{l=1}^{3}\alpha^l_{i'}\alpha^{1'}_l + \mathbf{e}'_2\sum_{l=1}^{3}\alpha^l_{i'}\alpha^{2'}_l + \mathbf{e}'_3\sum_{l=1}^{3}\alpha^l_{i'}\alpha^{3'}_l \\
&= \sum_{k=1}^{3}\mathbf{e}'_k\sum_{l=1}^{3}\alpha^l_{i'}\alpha^{k'}_l.
\end{aligned} \tag{1.13}$$

In just the same way, we find that

$$\begin{aligned}
\mathbf{e}_i &= \mathbf{e}_1\sum_{l=1}^{3}\alpha^{l'}_i\alpha^1_{l'} + \mathbf{e}_2\sum_{l=1}^{3}\alpha^{l'}_i\alpha^2_{l'} + \mathbf{e}_3\sum_{l=1}^{3}\alpha^{l'}_i\alpha^3_{l'} \\
&= \sum_{k=1}^{3}\mathbf{e}_k\sum_{l=1}^{3}\alpha^{l'}_i\alpha^k_{l'}.
\end{aligned}$$

Together this equation and (1.13) imply the following eighteen relations:

$$\begin{aligned}
\sum_{l=1}^{3}\alpha^l_{i'}\alpha^{j'}_l &= \begin{cases} 0 & \text{if } i \neq j, \\ 1 & \text{if } i = j, \end{cases} \\
\sum_{l=1}^{3}\alpha^{l'}_i\alpha^j_{l'} &= \begin{cases} 0 & \text{if } i \neq j, \\ 1 & \text{if } i = j. \end{cases}
\end{aligned} \tag{1.14}$$

1.4. Products of Two Vectors

1.4.1. The scalar product. By the *scalar product* (synonymously, the *dot product*) of two vectors $\mathbf{A}$ and $\mathbf{B}$, denoted by $\mathbf{A} \cdot \mathbf{B}$, we mean the quantity

$$\mathbf{A} \cdot \mathbf{B} = |\mathbf{A}|\,|\mathbf{B}| \cos(\mathbf{A}, \mathbf{B}), \tag{1.15}$$

i.e., the product of the magnitudes of the vectors times the cosine of the angle between them. It follows from (1.15) and the considerations of Sec. 1.2.3 that the scalar product of $\mathbf{A}$ and $\mathbf{B}$ equals the magnitude of $\mathbf{A}$ times the projection of $\mathbf{B}$ onto the direction of $\mathbf{A}$ [see Fig. 1.5(b)] or vice versa, i.e.,

$$\mathbf{A} \cdot \mathbf{B} = A_B B = A B_A.$$

Scalar multiplication of two vectors is obviously *commutative:*

$$\mathbf{A} \cdot \mathbf{B} = \mathbf{B} \cdot \mathbf{A}.$$

It is also *distributive* (with respect to addition), i.e.,

$$\mathbf{A} \cdot (\mathbf{B} + \mathbf{C}) = \mathbf{A} \cdot \mathbf{B} + \mathbf{B} \cdot \mathbf{C},$$

as can be seen at once from Fig. 1.11.

A necessary and sufficient condition for two vectors **A** and **B** to be perpendicular is that

$$\mathbf{A} \cdot \mathbf{B} = 0.$$

In fact, if $\mathbf{A} \perp \mathbf{B}$, then $\cos(\mathbf{A}, \mathbf{B}) = 0$ and hence $\mathbf{A} \cdot \mathbf{B} = 0$, while if $\mathbf{A} \cdot \mathbf{B} = 0$ and $\mathbf{A} \neq 0$, $\mathbf{B} \neq 0$, then $\cos(\mathbf{A}, \mathbf{B}) = 0$ and hence $\mathbf{A} \perp \mathbf{B}$.[6]

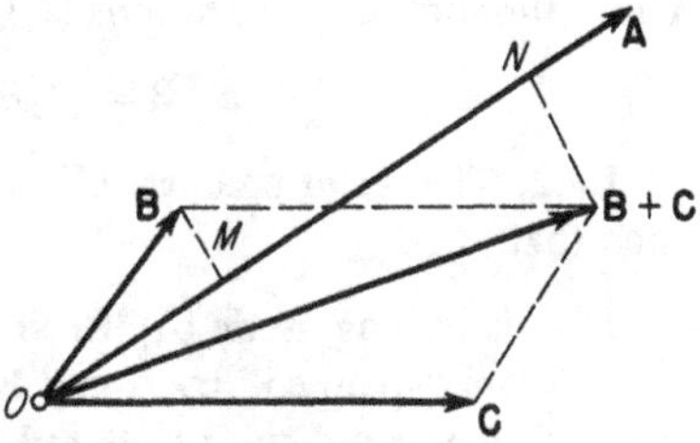

FIG. 1.11. Distributivity of the scalar product:
Since $ON = OM + MN$,

$$|\mathbf{A}|\, ON = |\mathbf{A}|\, OM + |\mathbf{A}|\, MN,$$

i.e.,

$$\mathbf{A} \cdot (\mathbf{B} + \mathbf{C}) = \mathbf{A} \cdot \mathbf{B} + \mathbf{B} \cdot \mathbf{C}.$$

The projection of a vector **A** onto an axis u equals the scalar product of **A** and the unit vector $\mathbf{u}_0$ corresponding to u:

$$A_u = \mathbf{A} \cdot \mathbf{u}_0 = A \cos(\mathbf{A}, \mathbf{u}_0).$$

This is an immediate consequence of formula (1.2).

Given a system of rectangular coordinates x^1, x^2, x^3, let $\mathbf{i}_1, \mathbf{i}_2, \mathbf{i}_3$ be the corresponding basis vectors. Then any vector **A** can be represented in the form

$$\mathbf{A} = A_1\mathbf{i}_1 + A_2\mathbf{i}_2 + A_3\mathbf{i}_3 \tag{1.16}$$

[recall (1.8)]. Since the vectors $\mathbf{i}_1, \mathbf{i}_2, \mathbf{i}_3$ are orthonormal (see p. 10), we have the *orthonormality conditions*

$$\mathbf{i}_j \cdot \mathbf{i}_k = \begin{cases} 0 & \text{if } j \neq k, \\ 1 & \text{if } j = k. \end{cases} \tag{1.17}$$

But then

$$\begin{aligned} \mathbf{A} \cdot \mathbf{i}_1 &= (A_1\mathbf{i}_1 + A_2\mathbf{i}_2 + A_3\mathbf{i}_3) \cdot \mathbf{i}_1 = A_1\mathbf{i}_1 \cdot \mathbf{i}_1 + A_2\mathbf{i}_1 \cdot \mathbf{i}_2 + A_3\mathbf{i}_1 \cdot \mathbf{i}_3 \\ &= A_1 \cdot 1 + A_2 \cdot 0 + A_3 \cdot 0 = A_1, \end{aligned} \tag{1.18}$$

and similarly

$$\mathbf{A} \cdot \mathbf{i}_2 = A_2, \qquad \mathbf{A} \cdot \mathbf{i}_3 = A_3. \tag{1.19}$$

In other words, A_1, A_2 and A_3 are the projections of the vector **A** onto the coordinate axes. The numbers A_1, A_2 and A_3 are called the *components* of **A** with respect to the given coordinate system. It follows from (1.17)–(1.19) that

$$\mathbf{A} = (\mathbf{A} \cdot \mathbf{i}_1)\mathbf{i}_1 + (\mathbf{A} \cdot \mathbf{i}_2)\mathbf{i}_2 + (\mathbf{A} \cdot \mathbf{i}_3)\mathbf{i}_3. \tag{1.20}$$

---

[6] If $\mathbf{A} = 0$, say, then **A** can be regarded as having any direction at all, in particular the direction perpendicular to **B** or parallel to **B**.

The scalar product of two vectors **A** and **B** can easily be expressed in terms of their components:

$$\mathbf{A} \cdot \mathbf{B} = (A_1\mathbf{i}_1 + A_2\mathbf{i}_2 + A_3\mathbf{i}_3) \cdot (B_1\mathbf{i}_1 + B_2\mathbf{i}_2 + B_3\mathbf{i}_3)$$
$$= A_1B_1\mathbf{i}_1 \cdot \mathbf{i}_1 + A_2B_1\mathbf{i}_2 \cdot \mathbf{i}_1 + A_3B_1\mathbf{i}_3 \cdot \mathbf{i}_1 + A_1B_2\mathbf{i}_1 \cdot \mathbf{i}_2 + \cdots + A_3B_3\mathbf{i}_3 \cdot \mathbf{i}_3.$$

The orthonormality conditions (1.17) then imply the important formula

$$\mathbf{A} \cdot \mathbf{B} = A_1B_1 + A_2B_2 + A_3B_3. \tag{1.21}$$

1.4.2. The vector product. Given two vectors **A** and **B**, let **C** be the vector such that

1) **C** is of magnitude $|\mathbf{A}|\,|\mathbf{B}| \sin(\mathbf{A}, \mathbf{B})$ equal to the area of the parallelogram "spanned" by **A** and **B** (see Fig. 1.12);
2) **C** is perpendicular to the plane of **A** and **B**;
3) **C** points in the direction from which the rotation from **A** and **B** (through the smaller of the two possible angles) appears to be counterclockwise.

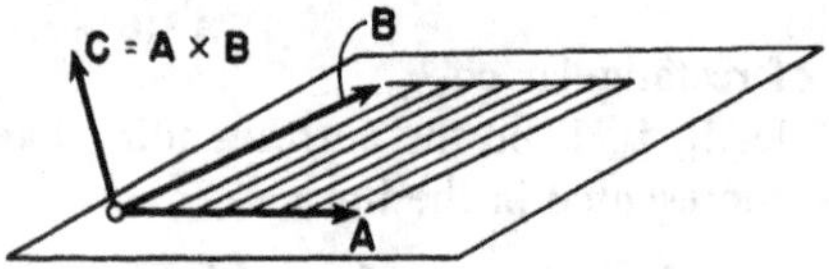

FIG. 1.12. The vector product.

Then **C** is called the *vector product* (synonymously, the *cross product*) of **A** and **B**, denoted by **A** × **B**. Note that the vector product **A** × **B** points in the direction of advance of a right-handed screw turned from **A** to **B**.

Unlike the scalar product, the vector product is not commutative, and in fact

$$\mathbf{A} \times \mathbf{B} = -\mathbf{B} \times \mathbf{A}$$

(why?), but it does obey the distributive law

$$\mathbf{A} \times (\mathbf{B} + \mathbf{C}) = \mathbf{A} \times \mathbf{B} + \mathbf{A} \times \mathbf{C}$$

as shown in Fig. 1.13.

A necessary and sufficient condition for two vectors **A** and **B** to be parallel is that

$$\mathbf{A} \times \mathbf{B} = 0.$$

In fact, if $\mathbf{A} \parallel \mathbf{B}$, then $\sin(\mathbf{A}, \mathbf{B}) = 0$ and hence $\mathbf{A} \times \mathbf{B} = 0$, while if $\mathbf{A} \times \mathbf{B} = 0$ and $\mathbf{A} \neq 0$, $\mathbf{B} \neq 0$, then $\sin(\mathbf{A}, \mathbf{B}) = 0$ and hence $\mathbf{A} \parallel \mathbf{B}$ (see footnote 6, p. 15).

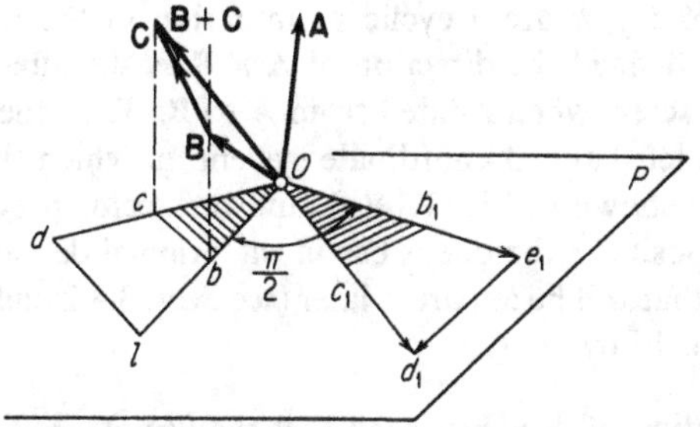

FIG. 1.13. Distributivity of the vector product: $\mathbf{A} \times (\mathbf{B} + \mathbf{C}) = \mathbf{A} \times \mathbf{B} + \mathbf{A} \times \mathbf{C}$:

Project the vectors **B**, **C** and **B** + **C** onto the plane P perpendicular to **A**. Then enlarge the sides of the resulting triangle Ocb $|\mathbf{A}|$ times, obtaining the triangle Ode. Rotate Ode through the angle $\pi/2$ in a way that would cause a right-handed screw to advance in the direction of **A**. This carries Ode into Od_1e_1 (and the points c, b into c_1, b_1). Draw the vectors $\overrightarrow{Od_1}$, $\overrightarrow{Oe_1}$ and $\overrightarrow{e_1d_1}$. Then the distributivity of the vector product follows from

$$\overrightarrow{Od_1} = \mathbf{A} \times (\mathbf{B} + \mathbf{C}), \quad \overrightarrow{e_1d_1} = \mathbf{A} \times \mathbf{C}, \quad \overrightarrow{Oe_1} = \mathbf{A} \times \mathbf{B},$$

together with

$$\overrightarrow{Od_1} = \overrightarrow{Oe_1} + \overrightarrow{e_1d_1}.$$

Let $\mathbf{i}_1$, $\mathbf{i}_2$ and $\mathbf{i}_3$ be the basis vectors of a right-handed coordinate system like that shown in Fig. 1.14(a). Then

$$\mathbf{i}_1 \times \mathbf{i}_2 = \mathbf{i}_3, \quad \mathbf{i}_2 \times \mathbf{i}_3 = \mathbf{i}_1, \quad \mathbf{i}_3 \times \mathbf{i}_1 = \mathbf{i}_2,$$
$$\mathbf{i}_1 \times \mathbf{i}_1 = 0, \quad \mathbf{i}_2 \times \mathbf{i}_2 = 0, \quad \mathbf{i}_3 \times \mathbf{i}_3 = 0,$$

or more concisely

$$\mathbf{i}_j \times \mathbf{i}_k = \mathbf{i}_l, \quad \mathbf{i}_m \times \mathbf{i}_m = 0, \tag{1.22}$$

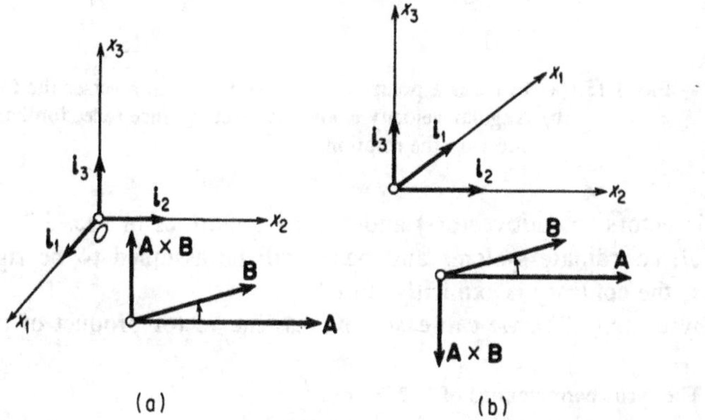

FIG. 1.14. The vector product **A** × **B** in right-handed and left-handed coordinate systems.

where the indices i, j, k are a cyclic permutation of the numbers 1, 2, 3.[7] Suppose we had defined the direction of $\mathbf{A} \times \mathbf{B}$ as the direction of advance of a left-handed screw when rotated from **A** to **B**. Then the formulas (1.22) would hold in a left-handed coordinate system, in which the axes have the relative positions shown in Fig. 1.14(b), and the vector product would have the direction opposite to the one it has in the original definition. From now on, for a reason that will be apparent later (see Secs. 3.4.2 and 3.7), we replace condition 3 on p. 16 by

3′) The direction of $\mathbf{A} \times \mathbf{B}$ is such that the vectors **A**, **B**, $\mathbf{A} \times \mathbf{B}$ (in that order) have the same "handedness" as the underlying basis $\mathbf{i}_1$, $\mathbf{i}_2$, $\mathbf{i}_3$ itself (see Fig. 1.14).

Vectors (like the moment of a force, the angular velocity, etc.) whose direction is established by convention, and which therefore change direction when the "handedness" of the coordinate system is changed (from right-handed to left-handed, say) are called *axial vectors*. Vectors (like force, velocity, etc.) whose direction depends only on their physical meaning, and which therefore do not change direction when the "handedness" of the coordinate system is changed, are called *polar vectors*. To determine the nature of a vector, imagine it reflected in a mirror perpendicular to itself. If the reflection preserves the direction of the quantity describing the physical phenomenon, then the vector is axial (see Fig. 1.15). More will be said about

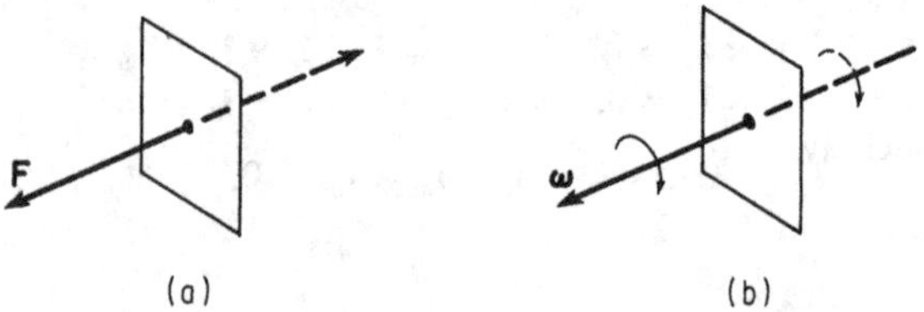

FIG. 1.15. (a) Force is a polar vector, since reflection reverses the force; (b) Angular velocity is an axial vector, since reflection has no effect on the rotation.

axial vectors (pseudovectors) and related quantities in Sec. 3.7. From now on, all coordinate systems and bases will be assumed to be right-handed unless the contrary is explicitly stated.

By using (1.22), we can easily express the vector product of two vectors

---

[7] The cyclic permutations of 1, 2, 3 are

$$1, 2, 3, \quad 2, 3, 1, \quad 3, 1, 2.$$

They differ from 1, 2, 3 by an even number of transpositions.

A and B in terms of their components:

$$\mathbf{C} = \mathbf{A} \times \mathbf{B} = (A_1\mathbf{i}_1 + A_2\mathbf{i}_2 + A_3\mathbf{i}_3) \times (B_1\mathbf{i}_1 + B_2\mathbf{i}_2 + B_3\mathbf{i}_3)$$
$$= (A_2B_3 - A_3B_2)\mathbf{i}_1 + (A_3B_1 - A_1B_3)\mathbf{i}_2 + (A_1B_2 - A_2B_1)\mathbf{i}_3.$$

Thus

$$C_1 = (\mathbf{A} \times \mathbf{B})_1 = A_2B_3 - A_3B_2,$$
$$C_2 = (\mathbf{A} \times \mathbf{B})_2 = A_3B_1 - A_1B_3,$$
$$C_3 = (\mathbf{A} \times \mathbf{B})_3 = A_1B_2 - A_2B_1,$$

or more concisely

$$C_i = A_jB_k - A_kB_j, \tag{1.23}$$

where i, j, k is a cyclic permutation of the numbers 1, 2, 3. Formula (1.23) leads to the following simple way of writing the vector product as a determinant:

$$\mathbf{C} = \mathbf{A} \times \mathbf{B} = \begin{vmatrix} \mathbf{i}_1 & \mathbf{i}_2 & \mathbf{i}_3 \\ A_1 & A_2 & A_3 \\ B_1 & B_2 & B_3 \end{vmatrix}. \tag{1.24}$$

1.4.3. Physical examples. Vector and scalar products are intimately associated with a variety of physical concepts. For example, the work done by a force applied at a point is defined as the product of the displacement and the component of the force in the direction of displacement.[8] Thus the component of the force perpendicular to the displacement "does no work." If $\mathbf{F}$ is the force and $\mathbf{s}$ the displacement, then the work W is by definition equal to

$$W = F_s s = Fs \cos(\mathbf{F}, \mathbf{s}) = \mathbf{F} \cdot \mathbf{s}. \tag{1.25}$$

Suppose the force makes an obtuse angle with the displacement, so that the force is "resistive." Then the work is regarded as negative, in keeping with formula (1.25).

The simplest physical model of a vector product is the moment $\mathbf{M}$ of a force $\mathbf{F}$ about a point O, defined as

$$\mathbf{M} = \mathbf{r} \times \mathbf{F},$$

where $\mathbf{r}$ is the vector joining O to the initial point of $\mathbf{F}$. The positive direction of the vector $\mathbf{M}$ depends on the choice of the coordinate system, like the direction of the angular velocity vector due to the action of the moment $\mathbf{M}$.

An electric charge e in an electric field $\mathbf{E}$ experiences a force

$$\mathbf{F}_e = e\mathbf{E},$$

while an electric charge e moving with velocity $\mathbf{v}$ in a magnetic field $\mathbf{H}$ experiences a force

$$\mathbf{F}_m = \frac{e}{c}(\mathbf{v} \times \mathbf{H}),$$

[8] I.e., the projection of the force onto the direction of the displacement.

where c is the velocity of light.[9] Hence the total force experienced by an electric charge e moving with velocity $\mathbf{v}$ in an electromagnetic field with electric field $\mathbf{E}$ and magnetic field $\mathbf{H}$ is given by

$$\mathbf{F} = e\mathbf{E} + \frac{e}{c}(\mathbf{v} \times \mathbf{H}).$$

If the charge moves a distance $d\mathbf{r}$ in time dt, the work done by the electromagnetic field is

$$\mathbf{F} \cdot d\mathbf{r} = \mathbf{F} \cdot \frac{d\mathbf{r}}{dt}\,dt = \mathbf{F} \cdot \mathbf{v}\,dt = \left[e\mathbf{E} + \frac{e}{c}(\mathbf{v} \times \mathbf{H})\right] \cdot \mathbf{v}\,dt.$$

This work goes into changing the kinetic energy U of the moving charge. The magnetic field does no work, since the force $\mathbf{F}_m$ due to the magnetic field is perpendicular to $\mathbf{v}$ at every instant. Hence the work done by the electromagnetic field is due to the electric field alone and is done at the rate

$$\frac{dU}{dt} = e\mathbf{E} \cdot \mathbf{v}.$$

The magnetic field changes the direction of the charge's velocity, but not the magnitude of the velocity.

1.5. Products of Three Vectors

1.5.1. The scalar triple product. By the *scalar triple product* of three vectors $\mathbf{A}$, $\mathbf{B}$ and $\mathbf{C}$, we mean the scalar

$$V = (\mathbf{A} \times \mathbf{B}) \cdot \mathbf{C} = |\mathbf{A} \times \mathbf{B}|\, C_{\mathbf{A}\times\mathbf{B}} = |\mathbf{A} \times \mathbf{B}|\, h,$$

where $C_{\mathbf{A}\times\mathbf{B}} = h$ is the projection of $\mathbf{C}$ onto the direction of $\mathbf{A} \times \mathbf{B}$ (see Fig. 1.16), i.e., the height of the parallelepiped "spanned" by $\mathbf{A}$, $\mathbf{B}$ and $\mathbf{C}$. Since $|\mathbf{A} \times \mathbf{B}|$ is the area of the base of this parallelepiped, the scalar triple product is just the volume of the parallelepiped taken with the plus sign or the minus sign depending on whether the angle between the vectors $\mathbf{C}$ and $\mathbf{A} \times \mathbf{B}$ is acute or obtuse.

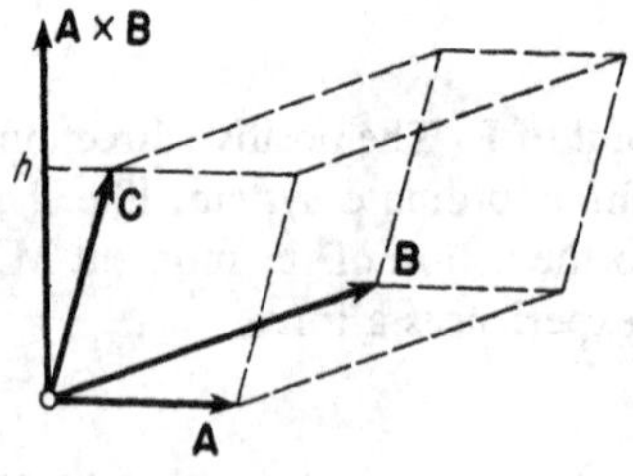

FIG. 1.16. The scalar triple product of three vectors.

Using formulas (1.21) and (1.24), we deduce the following simple representation of the scalar triple product in terms of the

[9] Here $\mathbf{E}$ and $\mathbf{H}$ are the values of the fields at the position of the charge. In general, of course, $\mathbf{E}$ and $\mathbf{H}$ vary from point to point.

components of the vectors **A**, **B** and **C** in a rectangular coordinate system:

$$(\mathbf{A}\times\mathbf{B})\cdot\mathbf{C}=\begin{vmatrix}\mathbf{i}_1 & \mathbf{i}_2 & \mathbf{i}_3\\ A_1 & A_2 & A_3\\ B_1 & B_2 & B_3\end{vmatrix}\cdot(C_1\mathbf{i}_1+C_2\mathbf{i}_2+C_3\mathbf{i}_3)=\begin{vmatrix}C_1 & C_2 & C_3\\ A_1 & A_2 & A_3\\ B_1 & B_2 & B_3\end{vmatrix}. \qquad (1.26)$$

It follows from (1.26) and a familiar property of determinants that the scalar triple product is invariant under cyclic permutations of the vectors **A**, **B** and **C**:

$$(\mathbf{A}\times\mathbf{B})\cdot\mathbf{C}=(\mathbf{B}\times\mathbf{C})\cdot\mathbf{A}=(\mathbf{C}\times\mathbf{A})\cdot\mathbf{B}. \qquad (1.27)$$

Moreover, the scalar triple product vanishes if two of the vectors **A**, **B** and **C** are identical (or parallel):

$$(\mathbf{A}\times\mathbf{B})\cdot\mathbf{A}=(\mathbf{A}\times\mathbf{B})\cdot\mathbf{B}=(\mathbf{A}\times\mathbf{A})\cdot\mathbf{B}=0.$$

Three vectors **A**, **B** and **C** are coplanar (linearly dependent) if and only if they span a parallelepiped of zero volume, i.e., if and only if their scalar triple product vanishes:

$$(\mathbf{A}\times\mathbf{B})\cdot\mathbf{C}=\begin{vmatrix}A_1 & A_2 & A_3\\ B_1 & B_2 & B_3\\ C_1 & C_2 & C_3\end{vmatrix}=0. \qquad (1.28)$$

It follows that three vectors **A**, **B** and **C** form a basis if and only if $(\mathbf{A}\times\mathbf{B})\cdot\mathbf{C}\neq 0$. The basis is said to be *right-handed* if $(\mathbf{A}\times\mathbf{B})\cdot\mathbf{C}>0$ and *left-handed* if $(\mathbf{A}\times\mathbf{B})\cdot\mathbf{C}<0$. A typical right-handed basis is shown in Fig. 1.14(a) and a typical left-handed basis in Fig. 1.14(b).

1.5.2. The vector triple product. By the *vector triple product* of three vectors **A**, **B** and **C**, we mean the vector $\mathbf{A}\times(\mathbf{B}\times\mathbf{C})$. Clearly $\mathbf{A}\times(\mathbf{B}\times\mathbf{C})$ is perpendicular to **A** and lies in the plane of **B** and **C** (see Fig. 1.17).

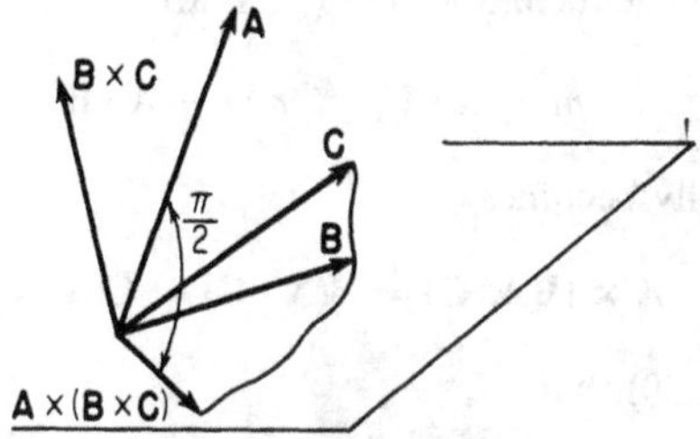

FIG. 1.17. The vector triple product of three vectors.

Suppose the vectors **A**, **B** and **C** are noncollinear [otherwise $\mathbf{A} \times (\mathbf{B} \times \mathbf{C})$ vanishes trivially]. Then $\mathbf{A} \times (\mathbf{B} \times \mathbf{C})$ has a unique expansion of the form

$$\mathbf{A} \times (\mathbf{B} \times \mathbf{C}) = m\mathbf{B} + n\mathbf{C} \tag{1.29}$$

(recall Sec. 1.4.2). To determine the scalars m and n, we introduce the vectors

$$\mathbf{D} = \mathbf{B} \times \mathbf{C}, \qquad \mathbf{E} = \mathbf{A} \times \mathbf{D} = \mathbf{A} \times (\mathbf{B} \times \mathbf{C}),$$

with components D_1, D_2, D_3 and E_1, E_2, E_3 in the same rectangular coordinate system used to define the components of **A**, **B** and **C**. Then, according to (1.23),

$$E_1 = A_2D_3 - A_3D_2,$$

and moreover

$$D_2 = (\mathbf{B} \times \mathbf{C})_2 = B_3C_1 - B_1C_3,$$

$$D_3 = (\mathbf{B} \times \mathbf{C})_3 = B_1C_2 - B_2C_1.$$

It follows that

$$\begin{aligned} E_1 &= A_2(B_1C_2 - B_2C_1) - A_3(B_3C_1 - B_1C_3) \\ &= B_1(A_2C_2 + A_3C_3) - C_1(A_2B_2 + A_3B_3). \end{aligned}$$

Adding and subtracting $A_1B_1C_1$ from the right-hand side and using (1.21), we find that

$$\begin{aligned} E_1 &= B_1(A_1C_1 + A_2C_2 + A_3C_3) - C_1(A_1B_1 + A_2B_2 + A_3B_3) \\ &= B_1(\mathbf{A} \cdot \mathbf{C}) - C_1(\mathbf{A} \cdot \mathbf{B}). \end{aligned}$$

In just the same way, it turns out that

$$E_2 = B_2(\mathbf{A} \cdot \mathbf{C}) - C_2(\mathbf{A} \cdot \mathbf{B}),$$

$$E_3 = B_3(\mathbf{A} \cdot \mathbf{C}) - C_3(\mathbf{A} \cdot \mathbf{B}).$$

Therefore the coefficients m and n in (1.28) are

$$m = \mathbf{A} \cdot \mathbf{C}, \qquad n = -\mathbf{A} \cdot \mathbf{B},$$

so that (1.29) finally becomes

$$\mathbf{A} \times (\mathbf{B} \times \mathbf{C}) = \mathbf{B}(\mathbf{A} \cdot \mathbf{C}) - \mathbf{C}(\mathbf{A} \cdot \mathbf{B}). \tag{1.30}$$

It follows from (1.30) that

$$(\mathbf{A} \times \mathbf{B}) \times \mathbf{C} = \mathbf{B}(\mathbf{A} \cdot \mathbf{C}) - \mathbf{A}(\mathbf{B} \cdot \mathbf{C}).$$

1.5.3. "Division" of vectors. The solution of equations usually leads to the operation of division, an operation which in the case of vectors is not unique. This difficulty appears even in the case of the scalar product (the simplest of the products introduced above). In fact, thinking of division as the inverse of multiplication, let

$$\mathbf{a} \cdot \mathbf{x} = m \qquad (\mathbf{a} \neq 0),$$

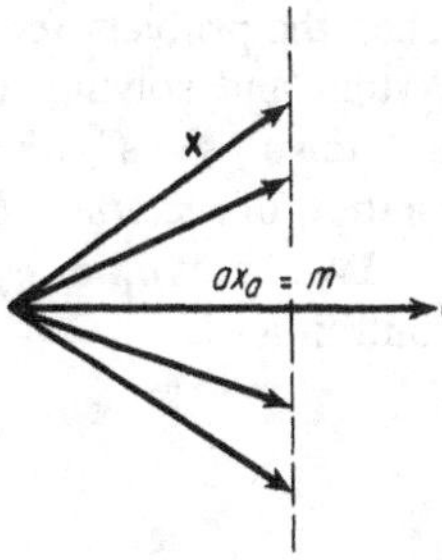

FIG. 1.18. Illustrating the equation $\mathbf{a} \cdot \mathbf{x} = ax_a = m$.

where $\mathbf{x}$ is an unknown vector. This equation has infinitely many solutions, since it merely determines the *projection* of $\mathbf{x}$ onto the direction of the given vector $\mathbf{a}$ (see Fig. 1.18). Hence the operation of division is best avoided altogether in vector algebra.

1.6. Reciprocal Bases and Related Topics

1.6.1. Reciprocal bases. Given any vector $\mathbf{A}$, let $\mathbf{i}_1$, $\mathbf{i}_2$, $\mathbf{i}_3$ be three orthonormal vectors. Then

$$\mathbf{A} = (\mathbf{A} \cdot \mathbf{i}_1)\mathbf{i}_1 + (\mathbf{A} \cdot \mathbf{i}_2)\mathbf{i}_2 + (\mathbf{A} \cdot \mathbf{i}_3)\mathbf{i}_3 \tag{1.31}$$

[cf. (1.20)]. The generalization of (1.31) to the case of three orthogonal vectors $\mathbf{e}_1$, $\mathbf{e}_2$, $\mathbf{e}_3$ (not necessarily of unit length) is immediate. We need merely note that the vectors

$$\mathbf{i}_1 = \frac{\mathbf{e}_1}{e_1}, \quad \mathbf{i}_2 = \frac{\mathbf{e}_2}{e_2}, \quad \mathbf{i}_3 = \frac{\mathbf{e}_3}{e_3} \tag{1.32}$$

($e_i = |\mathbf{e}_i|$, $i = 1, 2, 3$) are orthonormal, and then substitute (1.32) into (1.31), obtaining

$$\mathbf{A} = \frac{\mathbf{A} \cdot \mathbf{e}_1}{e_1^2}\mathbf{e}_1 + \frac{\mathbf{A} \cdot \mathbf{e}_2}{e_2^2}\mathbf{e}_2 + \frac{\mathbf{A} \cdot \mathbf{e}_3}{e_3^2}\mathbf{e}_3. \tag{1.33}$$

We now consider the even more general problem of expanding an arbitrary vector $\mathbf{A}$ with respect to three noncoplanar vectors $\mathbf{e}_1$, $\mathbf{e}_2$, $\mathbf{e}_3$, which are in general neither orthogonal nor of unit length. Let the expansion coefficients be A^1, A^2, A^3, so that[10]

$$\mathbf{A} = A^1\mathbf{e}_1 + A^2\mathbf{e}_2 + A^3\mathbf{e}_3.$$

---

[10] Again we point out that a superscript on a coordinate, a component or a vector has nothing to do with the exponent of a quantity raised to a power (cf. footnote 5, p. 10). The significance of this notation will be apparent in Sec. 1.6.3.

Then the problem reduces to projecting **A** onto the axes of some coordinate system and solving the resulting system of three scalar equations for the unknowns A^1, A^2, A^3. This important problem can be solved directly by the method of *reciprocal bases*.

Two bases $\mathbf{e}_1, \mathbf{e}_2, \mathbf{e}_3$ and $\mathbf{e}^1, \mathbf{e}^2, \mathbf{e}^3$ are said to be *reciprocal* if they satisfy the condition

$$\mathbf{e}_i \cdot \mathbf{e}^k = \begin{cases} 0 & \text{if } i \neq k, \\ 1 & \text{if } i = k. \end{cases} \tag{1.34}$$

In other words, each vector of one basis is perpendicular to two vectors of the other basis, i.e., the two vectors whose indices have different values.[11] Moreover (1.34) implies

$$|\mathbf{e}_i|\,|\mathbf{e}^i| \cos(\mathbf{e}_i, \mathbf{e}^i) = 1 > 0$$

and hence $\cos(\mathbf{e}_i, \mathbf{e}^i) > 0$. Hence each vector of one basis makes an acute angle (possibly a right angle) with the vector of the other basis whose index has the same value.

If we construct the parallelepipeds spanned by the two bases [of volumes $|V| = |\mathbf{e}_1 \cdot (\mathbf{e}_2 \times \mathbf{e}_3)|$ and $|V'| = |\mathbf{e}^1 \cdot (\mathbf{e}^2 \times \mathbf{e}^3)|$], then the faces of each parallelepiped are perpendicular to the edges of the other. Since (1.34) implies

$$|\mathbf{e}^i| = \frac{1}{|\mathbf{e}_i| \cos(\mathbf{e}_i, \mathbf{e}^i)},$$

the magnitude of each vector of one basis equals the reciprocal of the corresponding parallel altitude of the parallelepiped spanned by the reciprocal basis (see Fig. 1.19).

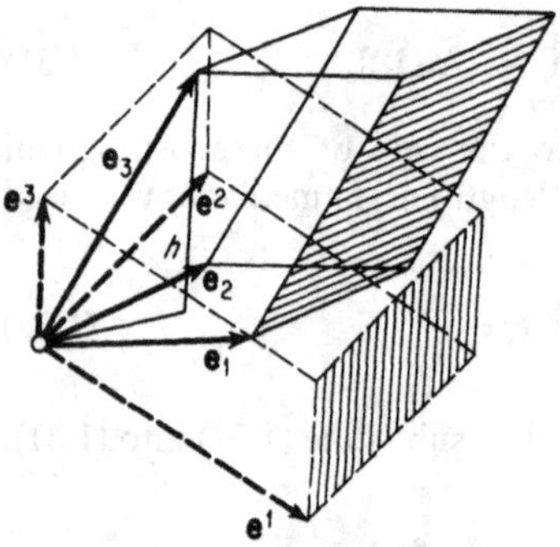

FIG. 1.19. Reciprocal bases and their parallelepipeds. The magnitude of $\mathbf{e}^3$ equals the reciprocal of the parallel altitude h.

To explicitly construct the reciprocal basis $\mathbf{e}^1, \mathbf{e}^2, \mathbf{e}^3$ corresponding to a given basis $\mathbf{e}_1, \mathbf{e}_2, \mathbf{e}_3$, we proceed as follows: The vector $\mathbf{e}^1$ must be perpendicular to the vectors $\mathbf{e}_2$ and $\mathbf{e}_3$. Therefore

$$\mathbf{e}^1 = m(\mathbf{e}_2 \times \mathbf{e}_3),$$

where the scalar m can be determined from the condition

$$\mathbf{e}_1 \cdot \mathbf{e}^1 = 1,$$

i.e.,

$$m\mathbf{e}_1 \cdot (\mathbf{e}_2 \times \mathbf{e}_3) = 1.$$

[11] The term *index* will be used to denote either a subscript or a superscript.

Since $\mathbf{e}_1 \cdot (\mathbf{e}_2 \times \mathbf{e}_3) \neq 0$ (the vectors $\mathbf{e}_1$, $\mathbf{e}_2$, $\mathbf{e}_3$ are noncoplanar, being basis vectors), we have

$$\mathbf{e}^1 = \frac{\mathbf{e}_2 \times \mathbf{e}_3}{\mathbf{e}_1 \cdot (\mathbf{e}_2 \times \mathbf{e}_3)} = \frac{\mathbf{e}_2 \times \mathbf{e}_3}{V}, \tag{1.35}$$

where $|V|$ is the volume of the parallelepiped spanned by the basis $\mathbf{e}_1$, $\mathbf{e}_2$, $\mathbf{e}_3$. Similarly, we find that

$$\mathbf{e}^2 = \frac{\mathbf{e}_3 \times \mathbf{e}_1}{V}, \qquad \mathbf{e}^3 = \frac{\mathbf{e}_1 \times \mathbf{e}_2}{V},$$

Together with (1.35), this gives

$$\mathbf{e}^i = \frac{\mathbf{e}_j \times \mathbf{e}_k}{V}, \tag{1.36}$$

where i, j, k is a cyclic permutation of 1, 2, 3.

The analogous formulas for the vectors $\mathbf{e}_1$, $\mathbf{e}_2$, $\mathbf{e}_3$ in terms of the vectors $\mathbf{e}^1$, $\mathbf{e}^2$, $\mathbf{e}^3$ are found in the same way and take the form

$$\mathbf{e}_1 = \frac{\mathbf{e}^2 \times \mathbf{e}^3}{\mathbf{e}^1 \cdot (\mathbf{e}^2 \times \mathbf{e}^3)} = \frac{\mathbf{e}^2 \times \mathbf{e}^3}{V'},$$

$$\mathbf{e}_2 = \frac{\mathbf{e}^3 \times \mathbf{e}^1}{V'},$$

$$\mathbf{e}_3 = \frac{\mathbf{e}^1 \times \mathbf{e}^2}{V'},$$

where $|V'|$ is the volume of the parallelepiped spanned by the basis $\mathbf{e}^1$, $\mathbf{e}^2$, $\mathbf{e}^3$. More concisely, we have

$$\mathbf{e}_i = \frac{\mathbf{e}^j \times \mathbf{e}^k}{V'},$$

where i, j, k is again a cyclic permutation of 1, 2, 3.

The following two properties of reciprocal bases should be noted:

1) The reciprocal basis of an orthonormal basis $\mathbf{e}_1 = \mathbf{i}_1$, $\mathbf{e}_2 = \mathbf{i}_2$, $\mathbf{e}_3 = \mathbf{i}_3$ is itself an orthonormal basis, consisting of the same vectors. In fact,

$$\mathbf{e}^1 = \frac{\mathbf{e}_2 \times \mathbf{e}_3}{\mathbf{e}_1 \cdot (\mathbf{e}_2 \times \mathbf{e}_3)} = \frac{\mathbf{i}_2 \times \mathbf{i}_3}{\mathbf{i}_1 \cdot (\mathbf{i}_2 \times \mathbf{i}_3)} = \frac{\mathbf{i}_1}{\mathbf{i}_1 \cdot \mathbf{i}_1} = \frac{\mathbf{i}_1}{1} = \mathbf{i}_1$$

[recall (1.17) and (1.22)], and similarly

$$\mathbf{e}^2 = \mathbf{i}_2, \qquad \mathbf{e}^3 = \mathbf{i}_3.$$

2) Two reciprocal bases are either both right-handed or both left-handed. This follows from the formula $VV' = 1$ (whose proof is left as an exercise).

We now return to the problem of expanding a vector **A** with respect to three noncoplanar vectors $\mathbf{e}_1$, $\mathbf{e}_2$, $\mathbf{e}_3$, i.e., of finding the coefficients A^1, A^2, A^3 in the formula

$$\mathbf{A} = A^1\mathbf{e}_1 + A^2\mathbf{e}_2 + A^3\mathbf{e}_3. \tag{1.37}$$

Let $\mathbf{e}^1$, $\mathbf{e}^2$, $\mathbf{e}^3$ be the reciprocal basis of $\mathbf{e}_1$, $\mathbf{e}_2$, $\mathbf{e}_3$. Then it follows from (1.34) that

$$\mathbf{A} \cdot \mathbf{e}^i = \sum_{k=1}^{3} A^k\mathbf{e}_k \cdot \mathbf{e}^i = A^i\mathbf{e}_i \cdot \mathbf{e}^i = A^i \qquad (i = 1, 2, 3).$$

For example,

$$A^1 = \mathbf{A} \cdot \mathbf{e}^1 = \frac{\mathbf{A} \cdot (\mathbf{e}_2 \times \mathbf{e}_3)}{\mathbf{e}_1 \cdot (\mathbf{e}_2 \times \mathbf{e}_3)}.$$

Thus (1.37) becomes

$$\mathbf{A} = (\mathbf{A} \cdot \mathbf{e}^1)\mathbf{e}_1 + (\mathbf{A} \cdot \mathbf{e}^2)\mathbf{e}_2 + (\mathbf{A} \cdot \mathbf{e}^3)\mathbf{e}_3,$$

which is the desired generalization of (1.31) and (1.32).

Using the reciprocal basis, we can easily find the vector **A** satisfying the system of equations

$$\mathbf{A} \cdot \mathbf{e}_1 = m_1, \qquad \mathbf{A} \cdot \mathbf{e}_2 = m_2, \qquad \mathbf{A} \cdot \mathbf{e}_3 = m_3. \tag{1.38}$$

In fact, it follows from (1.34) that

$$\mathbf{A} = m_1\mathbf{e}^1 + m_2\mathbf{e}^2 + m_3\mathbf{e}^3$$

is a solution of (1.38). Moreover, **A** is unique since if **A**′ were another solution, i.e., if **A**′ satisfied the system

$$\mathbf{A}' \cdot \mathbf{e}_1 = m_1, \quad \mathbf{A}' \cdot \mathbf{e}_2 = m_2, \quad \mathbf{A}' \cdot \mathbf{e}_3 = m_3, \tag{1.39}$$

then subtracting every equation of (1.39) from the corresponding equation of (1.38) would give

$$(\mathbf{A} - \mathbf{A}') \cdot \mathbf{e}_1 = (\mathbf{A} - \mathbf{A}') \cdot \mathbf{e}_2 = (\mathbf{A} - \mathbf{A}') \cdot \mathbf{e}_3 = 0.$$

But then $\mathbf{A} - \mathbf{A}' = 0$ and hence $\mathbf{A} = \mathbf{A}'$, since a vector perpendicular to every vector of the basis $\mathbf{e}_1$, $\mathbf{e}_2$, $\mathbf{e}_3$ must vanish (why?).

1.6.2. The summation convention. From now on, we will make free use of the following convention, universally encountered in the contemporary physical and mathematical literature:

1) Every letter index appearing once in an expression can take the values 1, 2 and 3. Thus A_i denotes the set of 3 quantities

$$A_1, A_2, A_3,$$

A_{ik} the set of $3^2 = 9$ quantities

$$A_{11}, A_{12}, A_{13}, A_{21}, A_{22}, A_{23}, A_{31}, A_{32}, A_{33},$$

A^{ik} the set of $3^2 = 9$ quantities

$$A^{11}, A^{12}, A^{13}, A^{21}, A^{22}, A^{23}, A^{31}, A^{32}, A^{33},$$

and so on.

2) Every letter index appearing twice in one term is regarded as being summed from 1 to 3. Thus

$$\begin{aligned} A_{ii} &= \sum_{i=1}^{3} A_{ii} = A_{11} + A_{22} + A_{33}, \\ A_iB^i &= \sum_{i=1}^{3} A_iB^i = A_1B^1 + A_2B^2 + A_3B^3, \\ A_iB^kC^i &= B^k \sum_{i=1}^{3} A_iC^i = B^k(A_1C^1 + A_2C^2 + A_3C^3). \end{aligned} \tag{1.40}$$

With this convention, we can drop the summation signs in (1.11) and (1.12), writing simply

$$\mathbf{e}'_i = \alpha_{i'}^{k}\mathbf{e}_k,$$
$$\mathbf{e}_j = \alpha_j^{k'}\mathbf{e}'_k.$$

Note that in the second of these equations, the summation is over k and *not* over k'. In other words, the prime in $\alpha_j^{k'}$ "attaches itself" to any value of k. Similarly, we can drop both summation signs in (1.13) and (1.14), obtaining

$$\mathbf{e}'_i = \alpha_{i'}^{l}\alpha_l^{k'}\mathbf{e}'_k,$$
$$\mathbf{e}_i = \alpha_i^{l'}\alpha_{l'}^{k}\mathbf{e}_k.$$

The value of a sum obviously does not depend on the particular letter chosen for an index of summation, e.g.,

$$A_{ii} = A_{kk} = A_{11} + A_{22} + A_{33}.$$

For this reason, indices of summation are often called "dummy" indices.

There is another convention that will prevail throughout this book: Let K and K' be two coordinate systems, and let P be an arbitrary point. Then the coordinates of P in K will be denoted by x_i (or by x^i if K is not a rectangular system), and the coordinates of the *same point* P in K' will be denoted by x'_i (or by x'^i if K' is not a rectangular system). The same rule also applies to components of vectors (and, more generally, of tensors). Thus, for example, A_1 and A'_1 are the components of the *same vector* $\mathbf{A}$ with respect to the "first" axis of the systems K and K', respectively, and not the components of two different vectors in the same system.

1.6.3. Covariant and contravariant components of a vector. In studying reciprocal bases, we found that the same vector $\mathbf{A}$ can be expanded as

$$\mathbf{A} = A^1\mathbf{e}_1 + A^2\mathbf{e}_2 + A^3\mathbf{e}_3 = \sum_{i=1}^{3} A^i\mathbf{e}_i \equiv A^i\mathbf{e}_i \qquad (A^i = \mathbf{A}\cdot\mathbf{e}^i) \tag{1.41}$$

with respect to the vectors of one basis $\mathbf{e}_1$, $\mathbf{e}_2$, $\mathbf{e}_3$, and as

$$\mathbf{A} = A_1\mathbf{e}^1 + A_2\mathbf{e}^2 + A_3\mathbf{e}^3 = \sum_{i=1}^{3} A_i\mathbf{e}^i \equiv A_i\mathbf{e}^i \qquad (A_i = \mathbf{A} \cdot \mathbf{e}_i) \quad (1.42)$$

with respect to the vectors of the reciprocal basis $\mathbf{e}^1$, $\mathbf{e}^2$, $\mathbf{e}^3$. The numbers A^i are called the *contravariant* components of $\mathbf{A}$, while the numbers A_i are called the *covariant* components of $\mathbf{A}$. The situation is illustrated by Fig. 1.20, which shows the covariant and contravariant components of a vector $\mathbf{A}$ lying in the plane of the vectors $\mathbf{e}_1$ and $\mathbf{e}_2$.

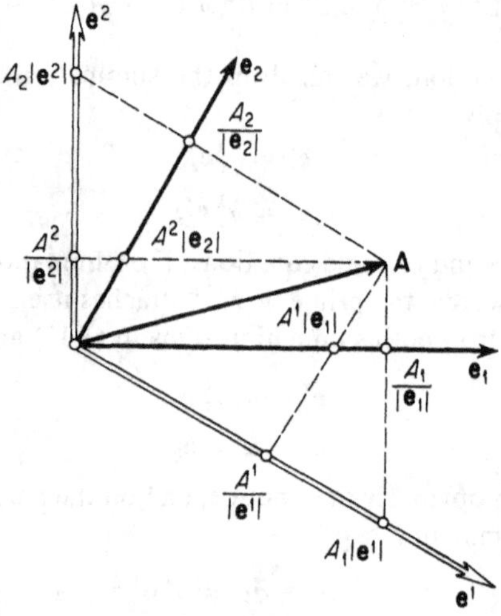

FIG. 1.20. Covariant and contravariant components of a vector in the plane:

The covariant components A_1, A_2 can be found either from the components $A_1|\mathbf{e}^1|$, $A_2|\mathbf{e}^2|$ of the vector $\mathbf{A}$ with respect to the directions of the reciprocal basis or from the projections $A_1/|\mathbf{e}_1|$, $A_2/|\mathbf{e}_2|$ of $\mathbf{A}$ onto the axes of the original basis. The contravariant components A^1, A^2 can be found either from the components $A^1|\mathbf{e}_1|$, $A^2|\mathbf{e}_2|$ of $\mathbf{A}$ with respect to the directions of the original basis or from the projections $A^1/|\mathbf{e}^1|$, $A^2/|\mathbf{e}^2|$ of $\mathbf{A}$ onto the axes of the reciprocal basis.

These designations of the components of a vector stem from the fact that the *direct* transformation of the *covariant* components involves the coefficients $\alpha_{i'}^{k}$ of the *direct* transformation, i.e.,

$$A_i' = \alpha_{i'}^{k} A_k, \qquad (1.43)$$

while the *direct* transformation of the *contravariant* components involve the coefficients $\alpha_k^{i'}$ of the *inverse* transformation:

$$A'^i = \alpha_k^{i'} A^k. \tag{1.44}$$

To see this, let **A** have covariant components A_i and contravariant components A^i in the coordinate system defined by the basis $\mathbf{e}_1, \mathbf{e}_2, \mathbf{e}_3$, and let **A** have covariant components A'_i and contravariant components A'^i in another coordinate system defined by the basis $\mathbf{e}'_1, \mathbf{e}'_2, \mathbf{e}'_3$. First we observe that formulas (1.11) and (1.12), p. 13 imply

$$\alpha_{i'}^k = \mathbf{e}'_i \cdot \mathbf{e}^k, \qquad \alpha_i^{k'} = \mathbf{e}_i \cdot \mathbf{e}'^k. \tag{1.45}$$

Then, forming the scalar product of the vector $\mathbf{e}'_i$ and both sides of the expansion $\mathbf{A} = A_k \mathbf{e}^k$, we obtain

$$\mathbf{A} \cdot \mathbf{e}'_i = A_k \mathbf{e}'_i \cdot \mathbf{e}^k,$$

which reduces to (1.43) because of (1.42) and (1.45). Similarly, forming the scalar product of the vector $\mathbf{e}'^i$ and both sides of the expansion $\mathbf{A} = A^k \mathbf{e}_k$, we obtain

$$\mathbf{A} \cdot \mathbf{e}'^i = A^k \mathbf{e}_k \cdot \mathbf{e}'^i,$$

which reduces to (1.44) because of (1.41) and (1.45). We leave it as an exercise for the reader to verify that the inverses of the formula (1.43) and (1.44) are

$$A_i = \alpha_i^{k'} A'_k, \qquad A^i = \alpha_{k'}^i A'^k.$$

In connection with the concept of contravariant components, it should be noted that the coordinates of a point in an oblique coordinate system should be written with superscripts: x^1, x^2, x^3. This is immediately clear if we recall from Fig. 1.7, p. 10 that the coordinates are the contravariant components of the radius vector of the point:

$$\mathbf{r} = x^1 \mathbf{e}_1 + x^2 \mathbf{e}_2 + x^3 \mathbf{e}_3 \equiv x^k \mathbf{e}_k.$$

1.6.4. Physical components of a vector. It will be recalled that operations on vectors are defined purely geometrically, with vectors thought of as directed line segments of length proportional to the magnitude of the quantities they represent and without regard for the physical dimensions of these quantities.[12] Clearly the physical dimensions of the components A^i and A_i of the same vector $\mathbf{A} = A^i \mathbf{e}_i = A_i \mathbf{e}^i$ are different, in a way determined by the dimensions of the basis vectors and the relations $\mathbf{e}_i \cdot \mathbf{e}^i = 1$. However, it is possible to introduce "physical" components of vectors, whose dimensions

[12] Except for the obvious requirement that all vectors appearing as terms in the same equation have the same physical dimensions.

coincide with those of the vectors themselves. This is done by defining a new "unit basis"

$$\mathbf{e}_i^* = \frac{\mathbf{e}_i}{|\mathbf{e}_i|}$$

(the vectors $\mathbf{e}_i^*$ are all of unit length) and its reciprocal basis

$$\mathbf{e}^{*i} = \mathbf{e}^i \, |\mathbf{e}_i| \, .$$

Then

$$\mathbf{A} = A_i^* \mathbf{e}^{*i} = A^{*i} \mathbf{e}_i^*,$$

where, by definition, A_i^* and A^{*i} are the *physical components* of $\mathbf{A}$.

The relation between the physical components of a vector and its covariant and contravariant components is easily found to be

$$A_i^* = \frac{A_i}{|\mathbf{e}_i|}, \qquad A^{*i} = A^i \, |\mathbf{e}_i| \qquad \text{(no summation over } i\text{)}.$$

As can be seen from Fig. 1.21, the A^{*i} are the *components* (parallel projections)

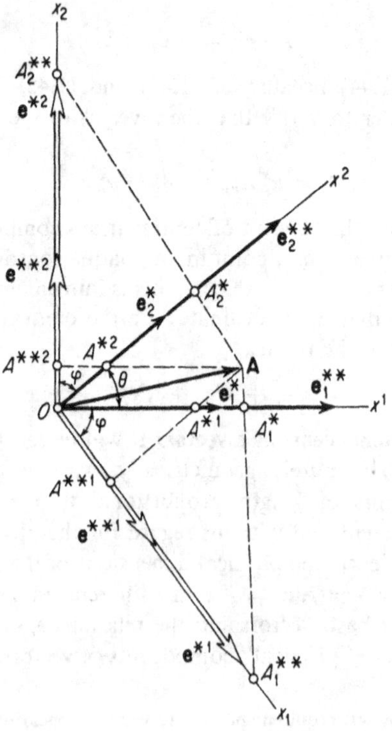

FIG. 1.21. The "physical" components of a vector.

of **A** onto the directions of the vectors $\mathbf{e}_i^*$, while the A_i^* are the (orthogonal) *projections* of **A** onto the same directions.

We can just as well start from the reciprocal basis, defining the unit basis as

$$\mathbf{e}^{**i} = \frac{\mathbf{e}^i}{|\mathbf{e}^i|}$$

and its reciprocal as

$$\mathbf{e}_i^{**} = \mathbf{e}_i\,|\mathbf{e}^i|\,.$$

We then have

$$\mathbf{e}^{**i} = \mathbf{e}^{*i}\cos(\mathbf{e}^i, \mathbf{e}_i), \qquad \mathbf{e}_i^{**} = \frac{\mathbf{e}_i^*}{\cos(\mathbf{e}^i, \mathbf{e}_i)}$$

and

$$A_i^{**} = \frac{A_i^*}{\cos(\mathbf{e}^i, \mathbf{e}_i)}, \qquad A^{**i} = A^{*i}\cos(\mathbf{e}^i, \mathbf{e}_i),$$

i.e., both definitions are essentially equivalent, with everything reducing to the choice of units of measurement.

It should be emphasized that calculations are almost always done with ordinary covariant and contravariant components, with the transcription to physical components being made only at the end (if necessary).

1.6.5. Relation between covariant and contravariant components. Taking the scalar product of (1.41) with $\mathbf{e}_i$ and the scalar product of (1.42) with $\mathbf{e}^i$, we obtain

$$\mathbf{A}\cdot\mathbf{e}_i = A^k(\mathbf{e}_i\cdot\mathbf{e}_k), \qquad (1.46)$$
$$\mathbf{A}\cdot\mathbf{e}^i = A_k(\mathbf{e}^i\cdot\mathbf{e}^k).$$

Then introducing the notation

$$\begin{aligned} \mathbf{e}_i\cdot\mathbf{e}_k &= g_{ik} = g_{ki}, \\ \mathbf{e}^i\cdot\mathbf{e}^k &= g^{ik} = g^{ki}, \\ \mathbf{e}_i\cdot\mathbf{e}^k &= g_i^{\cdot k} = \begin{cases} 0 & \text{if } i \neq k, \\ 1 & \text{if } i = k, \end{cases} \end{aligned} \qquad (1.47)$$

we can write (1.46) in the form

$$A_i = g_{ik}A^k, \qquad (1.48)$$
$$A^i = g^{ik}A_k. \qquad (1.49)$$

These formulas express the covariant components of the vector **A** in terms of its contravariant components, and vice versa. As will be shown in Sec. 2.9.2, the nine quantities g_{ik} form a second-order tensor, and so do g^{ik} and $g_i^{\cdot k}$.

The quantities g_{ik} (or g^{ik}) describe the fundamental geometric characteristics of a space "arithmetized" by introducing the coordinate system with basis $\mathbf{e}_1$, $\mathbf{e}_2$, $\mathbf{e}_3$ and corresponding coordinates x^1, x^2, x^3. To see this, let ds be the arc length between two infinitely close points x^i and $x^i + dx^i$, and let the vector $d\mathbf{r}$ joining the two points have covariant components dx_i and contravariant components dx^i. Then

$$(ds)^2 = |d\mathbf{r}|^2 = d\mathbf{r} \cdot d\mathbf{r} = \mathbf{e}_i\, dx^i \cdot \mathbf{e}_k\, dx^k = \mathbf{e}_i\, dx^i \cdot \mathbf{e}^k\, dx_k = \mathbf{e}^i\, dx_i \cdot \mathbf{e}^k\, dx_k,$$

or

$$\begin{aligned} (ds)^2 &= g_{ik}\, dx^i\, dx^k, \\ (ds)^2 &= g^{ik}\, dx_i\, dx_k, \\ (ds)^2 &= dx_i\, dx^i \end{aligned} \tag{1.50}$$

with the notation (1.47). The formulas (1.50) express the square of the element of arc length in the given coordinate system in terms of g_{ik} (or g^{ik}), i.e., the quantities g_{ik} (or g^{ik}) "determine the metric" of the given space and hence are known as the *metric tensor*.

To find the relation between the quantities g_{ik} and g^{ik}, we regard (1.48) as a system of three linear equations in the unknowns A^1, A^2, A^3, with solution

$$A^i = \frac{\sum_{k=1}^{3} G^{ik} A_k}{G} = \frac{G^{ik} A_k}{G}. \tag{1.51}$$

Here

$$G = \det g_{ik} = \begin{vmatrix} g_{11} & g_{12} & g_{13} \\ g_{21} & g_{22} & g_{23} \\ g_{31} & g_{32} & g_{33} \end{vmatrix}$$

and G^{ik} is the cofactor of g_{ik} in the determinant G. This quantity can be written in the form

$$G^{ik} = \begin{vmatrix} g_{ps} & g_{pt} \\ g_{rs} & g_{rt} \end{vmatrix},$$

where i, p, r and k, s, t are both cyclic permutations of 1, 2, 3. Thus, for example,

$$G^{11} = \begin{vmatrix} g_{22} & g_{23} \\ g_{32} & g_{33} \end{vmatrix}, \qquad G^{12} = \begin{vmatrix} g_{23} & g_{21} \\ g_{33} & g_{31} \end{vmatrix}, \qquad G^{13} = \begin{vmatrix} g_{21} & g_{22} \\ g_{31} & g_{32} \end{vmatrix}.$$

A comparison of (1.51) with (1.49) now gives the formula

$$g^{ik} = \frac{G^{ik}}{G}, \tag{1.52}$$

expressing g^{ik} in terms of the g_{ik}. In just the same way, it can be shown that

$$g_{ik} = \frac{G_{ik}}{G'},$$

where

$$G' = \det g^{ik}, \qquad G_{ik} = \begin{vmatrix} g^{ps} & g^{pt} \\ g^{rs} & g^{rt} \end{vmatrix}.$$

On the other hand, by direct calculation based on (1.47) and (1.36), we obtain

$$\begin{aligned} g^{ik} = \mathbf{e}^i \cdot \mathbf{e}^k &= \frac{1}{V^2}(\mathbf{e}_p \times \mathbf{e}_r) \cdot (\mathbf{e}_s \times \mathbf{e}_t) = \frac{1}{V^2}[(\mathbf{e}_p \times \mathbf{e}_r) \times \mathbf{e}_s] \cdot \mathbf{e}_t \\ &= \frac{1}{V^2}[\mathbf{e}_r(\mathbf{e}_p \cdot \mathbf{e}_s) - \mathbf{e}_p(\mathbf{e}_r \cdot \mathbf{e}_s)] \cdot \mathbf{e}_t \\ &= \frac{1}{V^2}[(\mathbf{e}_p \cdot \mathbf{e}_s)(\mathbf{e}_r \cdot \mathbf{e}_t) - (\mathbf{e}_p \cdot \mathbf{e}_t)(\mathbf{e}_r \cdot \mathbf{e}_s)] \\ &= \frac{1}{V^2}\begin{vmatrix} \mathbf{e}_p \cdot \mathbf{e}_s & \mathbf{e}_p \cdot \mathbf{e}_t \\ \mathbf{e}_r \cdot \mathbf{e}_s & \mathbf{e}_r \cdot \mathbf{e}_t \end{vmatrix} = \frac{1}{V^2}\begin{vmatrix} g_{ps} & g_{pt} \\ g_{rs} & g_{rt} \end{vmatrix}, \end{aligned} \tag{1.53}$$

where we have also used (1.27) and (1.30). Comparing (1.53) and (1.52), we find that

$$G = V^2, \qquad V = \pm\sqrt{G}, \tag{1.54}$$

where, according to Sec. 1.5.1, the plus sign is chosen if the given basis $\mathbf{e}_1, \mathbf{e}_2, \mathbf{e}_3$ is right-handed. Similarly, it can be shown that

$$G' = V'^2, \qquad V' = \pm\sqrt{G'}.$$

In particular,

$$GG' = 1$$

since $VV' = 1$. Thus the volume of the parallelepiped spanned by the basis $\mathbf{e}_1, \mathbf{e}_2, \mathbf{e}_3$ (assumed to be right-handed) equals $\sqrt{G}$, while that of the parallelepiped spanned by the reciprocal basis $\mathbf{e}^1, \mathbf{e}^2, \mathbf{e}^3$ equals $\sqrt{G'}$.

1.6.6. The case of orthogonal bases. Orthogonal bases are particularly important since the coordinate systems most commonly used in physics and applied mathematics are orthogonal. In this case, the original basis $\mathbf{e}_1, \mathbf{e}_2, \mathbf{e}_3$

and its reciprocal $\mathbf{e}^1, \mathbf{e}^2, \mathbf{e}^3$ are both orthogonal,[13] and it follows from (1.47) that

$$g_{ik} = g^{ki} = 0 \qquad \text{if} \quad i \neq k.$$

As a result, (1.48) and (1.49) become

$$A_1 = g_{11}A^1, \quad A_2 = g_{22}A^2, \quad A_3 = g_{33}A^3,$$

$$A^1 = g^{11}A_1, \quad A^2 = g^{22}A_2, \quad A^3 = g^{33}A_3,$$

and hence

$$g_{11} = \frac{1}{g^{11}}, \quad g_{22} = \frac{1}{g^{22}}, \quad g_{33} = \frac{1}{g^{33}}.$$

Moreover

$$(ds)^2 = \sum_{i=1}^{3} (h_i\, dx^i)^2,$$

where the quantities

$$h_1 = \sqrt{g_{11}}, \quad h_2 = \sqrt{g_{22}}, \quad h_3 = \sqrt{g_{33}}$$

are called the *metric coefficients.*

Remark 1. The physical components A_i^* and A^{*i}, the covariant components A_i and the contravariant components A^i all coincide in a rectangular coordinate system with orthonormal basis vectors. Moreover, the physical components A_i^* and A^{*i} coincide in any orthogonal system (why?).

Remark 2. The notation introduced in this section is governed by the following rule whose mnemonic character helps to keep things straight in writing formulas: *Summation can only take place over "dummy" indices in different positions,*[14] where two indices are said to be in different positions if one is a subscript and the other a superscript. For example, the expressions A_iB^i, $g^{ik}A_k$ represent sums, but not the expressions A^kB^k, $g^{ik}A^k$. In this regard, we sometimes describe the equation

$$A^i = g^{ik}A_k$$

as the operation of "raising" an index and the equation

$$A_i = g_{ik}A^k$$

as the operation of "lowering" an index (the "operator" is the set of nine coefficients g_{ik} or g^{ik}).

---

[13] In fact, the two bases coincide if $\mathbf{e}_1, \mathbf{e}_2, \mathbf{e}_3$ is orthonormal.

[14] "Dummy" in the sense that any other letter will do just as well, e.g. $A^i = g^{ik}A_k = g^{il}A_l = g^{im}A_m$.

The above rule will be particularly useful later in dealing with algebraic operations on tensors written in curvilinear coordinates. Of course, in rectangular coordinates, the expressions A_{ii}, A_iB_i, etc. can be thought of as sums, as in the first of the equations (1.40).

1.7. Variable Vectors

1.7.1. Vector functions of a scalar argument. A vector, just like a scalar, can vary with respect to both spatial position $\mathbf{r}$ and time t, giving rise to a *vector function* $\mathbf{A} = \mathbf{A}(\mathbf{r}, t)$. Here we confine ourselves to the case of a *vector function of a scalar argument*, i.e., a rule assigning a unique value of a vector $\mathbf{A}$ to each admissible value of a scalar t (usually, but not necessarily, the time). If $\mathbf{A}$ is a function of t, then so are all its components, as well as its magnitude and direction.

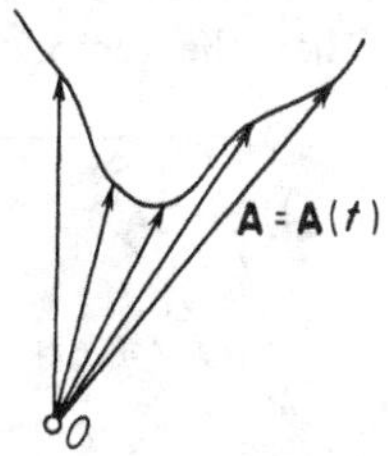

FIG. 1.22. The hodograph of $\mathbf{A} = \mathbf{A}(t)$.

Suppose the vector $\mathbf{A} = \mathbf{A}(t)$ is drawn from a fixed point O. Then as t varies, the end point of $\mathbf{A}$ traces out a curve called the *hodograph* of $\mathbf{A}$ (see Fig. 1.22). If $\mathbf{A}$ varies only in magnitude, the hodograph reduces to a straight line as in Fig. 1.23(a), while if $\mathbf{A}$ varies only in direction, the hodograph reduces to a curve lying on the surface of a sphere as in Fig. 1.23(b). If the hodograph is a straight line, then in general both the magnitude and the direction of $\mathbf{A}$ can vary, as shown in

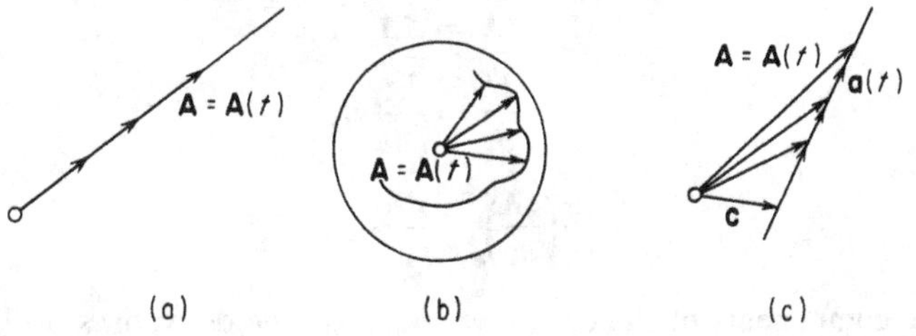

FIG. 1.23. (a) The hodograph of a vector varying only in magnitude:

$$\frac{\mathbf{A}}{|\mathbf{A}|} = \textbf{const};$$

(b) The hodograph of a vector varying only in direction:

$$|\mathbf{A}| = \text{const};$$

(c) Representation of a vector whose hodograph is a straight line.

Fig. 1.23(c), but **A** can always be represented as the sum of a constant vector and a vector varying only in magnitude, i.e.,

$$\mathbf{A} = \mathbf{A}(t) = \mathbf{c} + \mathbf{a}(t),$$

where[15]

$$\mathbf{c} = \mathbf{const}, \qquad \frac{\mathbf{a}}{|\mathbf{a}|} = \mathbf{const}.$$

1.7.2. The derivative of a vector function. Given a vector function $\mathbf{A}(t)$, suppose there exists a constant vector $\mathbf{A}_0$ such that

$$\lim_{t \to t_0} |\mathbf{A}(t) - \mathbf{A}_0| = 0.$$

Then the vector $\mathbf{A}_0$ is called the *limit* of $\mathbf{A}(t)$ as t approaches t_0. By the derivative $d\mathbf{A}/dt$ of a vector function $\mathbf{A}(t)$ we mean the limit

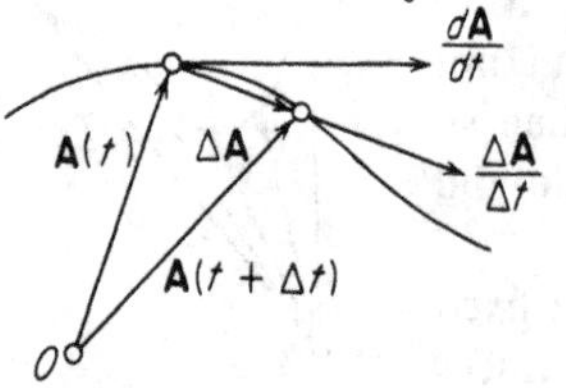

FIG. 1.24. If $\mathbf{A}(t)$ is a vector function of a scalar argument, then the derivative $d\mathbf{A}/dt$ is a vector directed along the tangent to the hodograph of $\mathbf{A}(t)$.

$$\lim_{\Delta t \to 0} \frac{\mathbf{A}(t + \Delta t) - \mathbf{A}(t)}{\Delta t} = \lim_{\Delta t \to 0} \frac{\Delta \mathbf{A}}{\Delta t}, \quad (1.55)$$

provided it exists. Since the vector $\Delta\mathbf{A}/\Delta t$ is directed along the secant to the hodograph, its limit as $\Delta t \to 0$, namely the vector $d\mathbf{A}/dt$, is directed along the tangent to the hodograph of **A** (see Fig. 1.24).

Let $A_k = A_k(t)$ be the components of $\mathbf{A}(t)$ with respect to a fixed rectangular coordinate system which is independent of the argument t and has orthonormal basis vectors $\mathbf{i}_1, \mathbf{i}_2, \mathbf{i}_3$. Then

$$\mathbf{A} = A_k \mathbf{i}_k,$$

$$\frac{d\mathbf{A}}{dt} = \frac{dA_k}{dt}\mathbf{i}_k,$$

$$\left(\frac{d\mathbf{A}}{dt}\right)_k = \frac{dA_k}{dt},$$

i.e., the components of the derivative $d\mathbf{A}/dt$ are the derivatives of the components of the vector function $\mathbf{A}(t)$, provided the coordinate system is independent of t. Moreover, the derivative $d\mathbf{A}/dt$ has magnitude

$$\left|\frac{d\mathbf{A}}{dt}\right| = \sqrt{\left(\frac{dA_1}{dt}\right)^2 + \left(\frac{dA_2}{dt}\right)^2 + \left(\frac{dA_3}{dt}\right)^2}.$$

[15] By $\alpha = \text{const}$, $\mathbf{c} = \mathbf{const}$, we mean that α is a constant scalar and $\mathbf{c}$ a constant vector.

Let $\mathbf{r}$ be the radius vector giving the position of a moving particle, and let t be the time. Then the motion of the particle is characterized by a vector function $\mathbf{r} = \mathbf{r}(t)$. The velocity $\mathbf{v}$ and acceleration $\mathbf{a}$ of the particle at time t are given by the first and second derivatives of $r(t)$:

$$\mathbf{v}(t) = \frac{d\mathbf{r}}{dt},$$

$$\mathbf{a}(t) = \frac{d\mathbf{v}}{dt} = \frac{d^2\mathbf{r}}{dt^2}.$$

The following rules for differentiating vector functions are an immediate consequence of the definition (1.55):

(1) $$\frac{d}{dt}(\mathbf{A} \pm \mathbf{B}) = \frac{d\mathbf{A}}{dt} \pm \frac{d\mathbf{B}}{dt},$$

(2) $$\frac{d}{dt}(c\mathbf{A}) = \frac{dc}{dt}\mathbf{A} + c\frac{d\mathbf{A}}{dt} \qquad (c \text{ a scalar})$$

(3) $$\frac{d}{dt}(\mathbf{A} \cdot \mathbf{B}) = \frac{d\mathbf{A}}{dt} \cdot \mathbf{B} + \mathbf{A} \cdot \frac{d\mathbf{B}}{dt},$$

(4) $$\frac{d}{dt}(\mathbf{A} \times \mathbf{B}) = \frac{d\mathbf{A}}{dt} \times \mathbf{B} + \mathbf{A} \times \frac{d\mathbf{B}}{dt}.$$

For example, rule 4 is proved by observing that

$$\begin{aligned}
\frac{d}{dt}(\mathbf{A} \times \mathbf{B}) &= \lim_{\Delta t \to 0} \frac{(\mathbf{A} + \Delta\mathbf{A}) \times (\mathbf{B} + \Delta\mathbf{B}) - \mathbf{A} \times \mathbf{B}}{\Delta t} \\
&= \lim_{\Delta t \to 0} \frac{(\Delta\mathbf{A} \times \mathbf{B}) + (\mathbf{A} \times \Delta\mathbf{B}) + (\Delta\mathbf{A} \times \Delta\mathbf{B})}{\Delta t} \\
&= \lim_{\Delta t \to 0} \frac{\Delta\mathbf{A}}{\Delta t} \times \mathbf{B} + \lim_{\Delta t \to 0} \mathbf{A} \times \frac{\Delta\mathbf{B}}{\Delta t} + \lim_{\Delta t \to 0} \frac{\Delta\mathbf{A}}{\Delta t} \times \Delta\mathbf{B} \\
&= \frac{d\mathbf{A}}{dt} \times \mathbf{B} + \mathbf{A} \times \frac{d\mathbf{B}}{dt}.
\end{aligned}$$

In using this rule, we must always bear in mind that the vector product is noncommutative.

1.7.3. The integral of a vector function. By the (*indefinite*) *integral* of the vector function $\mathbf{A}(t)$, we mean the vector function $\mathbf{B}(t) \equiv \int \mathbf{A}(t)\,dt$ whose derivative equals $\mathbf{A}(t)$, i.e., such that

$$\frac{d\mathbf{B}}{dt} = \mathbf{A}(t).$$

Thus

$$\mathbf{B}(t) = \int \mathbf{A}(t)\,dt + \mathbf{C},$$

where **C** is a constant vector. In a fixed coordinate system which does not depend on t, the components of the integral $\mathbf{B}(t)$ are completely determined by the integrals of the components of the vector function $\mathbf{A}(t)$, i.e.,

$$B_i = \int A_i(t)\, dt + C_i.$$

Similarly, we can introduce the concept of the definite integral of a vector function.

SOLVED PROBLEMS

Problem 1. Find the formulas describing the transformation from one rectangular coordinate system to another.

Solution. Suppose we introduce two rectangular coordinate systems K and K', with orthonormal bases $\mathbf{i}_1, \mathbf{i}_2, \mathbf{i}_3$ and $\mathbf{i}_1', \mathbf{i}_2', \mathbf{i}_3'$ (see Fig. 1.25). Then the

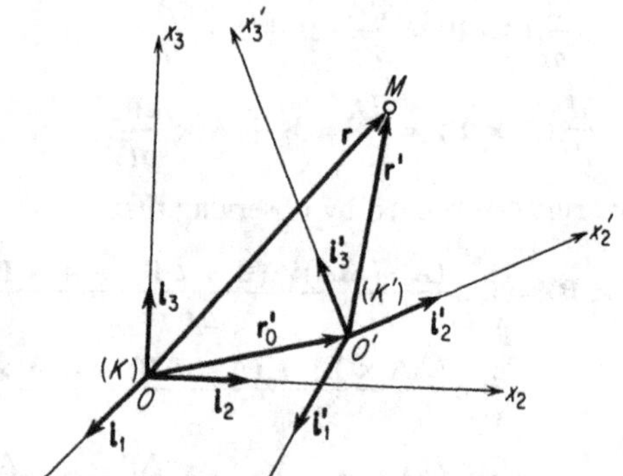

FIG. 1.25. Transformation of rectangular coordinates.

problem consists of expressing the coordinates x_1, x_2, x_3 of an arbitrary point M in the system K in terms of its coordinates x_1', x_2', x_3' in the system K', and vice versa.

Let $\mathbf{r}$ and $\mathbf{r}'$ be the radius vectors of the point M in the systems K and K'. Moreover, let the origin O' of the system K' have radius vector $\mathbf{r}_0'$ and coordinates x_{0k}' in the system K, while the origin O of the system K has radius vector $\mathbf{r}_0 = -\mathbf{r}_0'$ and coordinates x_{0k} in the system K'. Finally, let $\alpha_{j'k}$ be the cosine of the angle between the jth axis of the system K' and the kth axis of the system K, so that

$$\alpha_{j'k} = \cos(x_j', x_k) = \mathbf{i}_{j'} \cdot \mathbf{i}_k. \tag{1.56}$$

Then

$$\mathbf{r} = \mathbf{r}' + \mathbf{r}_0',$$

$$\mathbf{r}' = \mathbf{r} + \mathbf{r}_0$$

and hence

$$x_k\mathbf{i}_k = x_k'\mathbf{i}_k' + x_{0k}'\mathbf{i}_k, \tag{1.57}$$

$$x_k'\mathbf{i}_k' = x_k\mathbf{i}_k + x_{0k}\mathbf{i}_k' \tag{1.58}$$

(with summation over the index k). Taking the scalar product of (1.57) with $\mathbf{i}_l$ and the scalar product of (1.58) with $\mathbf{i}_l'$, and using (1.17) and (1.56), we obtain the formulas

$$\begin{aligned} x_l &= (\mathbf{i}_k' \cdot \mathbf{i}_l)x_k' + x_{0l}' = \alpha_{k'l}x_k' + x_{0l}', \\ x_l' &= (\mathbf{i}_l' \cdot \mathbf{i}_k)x_k + x_{0l} = \alpha_{l'k}x_k + x_{0l}, \end{aligned} \tag{1.59}$$

expressing the transformation from the system K to the system K'.

The coefficients $\alpha_{k'l}$, $\alpha_{l'k}$ satisfy certain *orthogonality conditions*, which can be obtained by using formula (1.20) to expand the basis vectors $\mathbf{i}_k$ of the system K with respect to the basis vectors $\mathbf{i}_k'$ of the system K', and vice versa. In fact, setting $\mathbf{A} = \mathbf{i}_k'$ in (1.20) gives

$$\mathbf{i}_k' = \alpha_{k'l}\mathbf{i}_l \tag{1.60}$$

and similarly

$$\mathbf{i}_k = \alpha_{l'k}\mathbf{i}_{l'}. \tag{1.61}$$

Taking the scalar product of (1.61) with $\mathbf{i}_m$ and the scalar product of (1.60) with $\mathbf{i}_m'$, we obtain

$$\begin{aligned} \mathbf{i}_k \cdot \mathbf{i}_m &= \alpha_{l'k}\alpha_{l'm}, \\ \mathbf{i}_k' \cdot \mathbf{i}_m' &= \alpha_{k'l}\alpha_{m'l}. \end{aligned} \tag{1.61}$$

Finally, using (1.17) and introducing the *Kronecker delta*

$$\delta_{km} = \mathbf{i}_k \cdot \mathbf{i}_m = \begin{cases} 0 & \text{if } k \neq m, \\ 1 & \text{if } k = m, \end{cases}$$

$$\delta_{km}' = \mathbf{i}_k' \cdot \mathbf{i}_m' = \begin{cases} 0 & \text{if } k \neq m, \\ 1 & \text{if } k = m, \end{cases}$$

we can write these conditions in the form

$$\begin{aligned} \alpha_{l'k}\alpha_{l'm} &= \delta_{km}, \\ \alpha_{k'l}\alpha_{m'l} &= \delta_{km}'. \end{aligned} \tag{1.62}$$

Problem 2. Use vectors to derive the law of cosines.

Solution. If ABC is a triangle with sides $\overrightarrow{AB} = \mathbf{c}$, $\overrightarrow{AC} = \mathbf{b}$, $\overrightarrow{CB} = \mathbf{a}$, then

$$\mathbf{a} + \mathbf{b} = \mathbf{c}. \tag{1.63}$$

Squaring (1.63), we obtain

$$a^2 + b^2 + 2\mathbf{a} \cdot \mathbf{b} = c^2.$$

But

$$\mathbf{a} \cdot \mathbf{b} = ab \cos(\mathbf{a}, \mathbf{b}) = ab \cos(\pi - \alpha) = -ab \cos \alpha$$

where $\alpha = \angle ACB$, and hence

$$c^2 = a^2 + b^2 - 2ab \cos \alpha.$$

Problem 3. Prove that

$$\cos(\alpha - \beta) = \cos \alpha \cos \beta + \sin \alpha \sin \beta.$$

Hint. Apply the formula

$$\cos(\mathbf{a}, \mathbf{b}) = \frac{\mathbf{a} \cdot \mathbf{b}}{|\mathbf{a}|\,|\mathbf{b}|}$$

to two unit vectors **a** and **b** lying in the xy-plane and making angles α and β with the x-axis.

Problem 4. Express the scalar product of two vectors in terms of their covariant and contravariant components.

Solution. By definition

$$\mathbf{A} \cdot \mathbf{B} = A^i\mathbf{e}_i \cdot B^k\mathbf{e}_k = A_i\mathbf{e}^i \cdot B_k\mathbf{e}^k = A_i\mathbf{e}^i \cdot B^k\mathbf{e}_k = A^i\mathbf{e}_i \cdot B_k\mathbf{e}^k,$$

and hence

$$\mathbf{A} \cdot \mathbf{B} = g_{ik}A^iB^k = g^{ik}A_iB_k = A_iB^i = A^iB_i$$

because of (1.47). Since

$$|\mathbf{A}| = A = \sqrt{\mathbf{A} \cdot \mathbf{A}} = \sqrt{g_{ik}A^iA^k} = \sqrt{g^{ik}A_iA_k} = \sqrt{A_iA^i}$$

and similarly for **B**, the angle between **A** and **B** is given by

$$\cos(\mathbf{A}, \mathbf{B}) = \frac{g_{ik}A^iB^k}{\sqrt{g_{ik}A^iA^k}\sqrt{g_{ik}B^jB^k}} = \frac{g^{ik}A_iB_k}{\sqrt{g^{ik}A_iA_k}\sqrt{g^{ik}B_iB_k}}$$

$$= \frac{A_iB^i}{\sqrt{A_iA^i}\sqrt{B_iB^i}} = \frac{A^iB_i}{\sqrt{A_iA^i}\sqrt{B_iB^i}}.$$

Problem 5. Find the vector product of two vectors in an oblique coordinate system.

Solution. By definition,

$$\mathbf{C} = \mathbf{A} \times \mathbf{B} = A^i\mathbf{e}_i \times B^k\mathbf{e}_k = (A^1\mathbf{e}_1 + A^2\mathbf{e}_2 + A^3\mathbf{e}_3) \times (B^1\mathbf{e}_1 + B^2\mathbf{e}_2 \times B^3\mathbf{e}_3)$$

$$= A^1B^1(\mathbf{e}_1 \times \mathbf{e}_1) + A^1B^2(\mathbf{e}_1 \times \mathbf{e}_2) + A^1B^3(\mathbf{e}_1 \times \mathbf{e}_3) + A^2B^1(\mathbf{e}_2 \times \mathbf{e}_1) + \cdots$$

$$+ A^3B^3(\mathbf{e}_3 \times \mathbf{e}_3) = (A^jB^k - A^kB^j)(\mathbf{e}_j \times \mathbf{e}_k).$$

But according to (1.36) and (1.54),

$$\mathbf{e}_j \times \mathbf{e}_k = V\mathbf{e}^i, \qquad V = \sqrt{G},$$

and hence

$$\mathbf{C} = \mathbf{A} \times \mathbf{B} = C_i \mathbf{e}^i$$

where

$$C_i = \sqrt{G}(A^j B^k - A^k B^j).$$

In the same way, it can be shown that

$$C^i = \frac{1}{\sqrt{G}}(A_j B_k - A_k B_j).$$

Problem 6. Use vectors to derive the basic formulas of spherical trigonometry.

Solution. Let ABC be a triangle on the sphere of unit radius, and let $OABC$ be the trihedral angle subtended by ABC (see Fig. 1.26). Let α, β, γ be the angles of this triangle, and let a, b, c be the lengths of its sides. Since the sphere has unit radius, a, b and c equal the plane angles BOC, AOC and AOB.

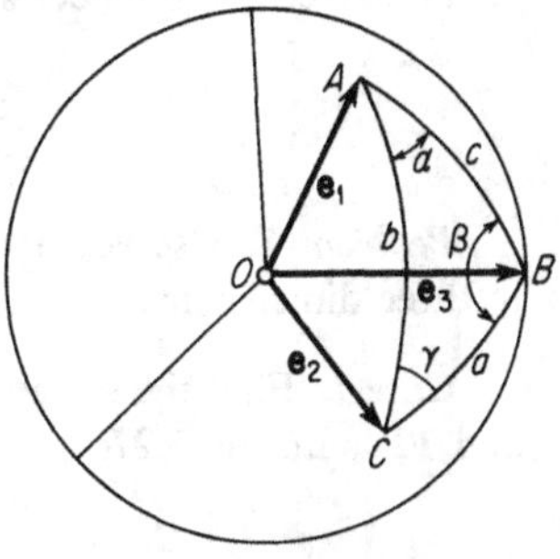

FIG. 1.26. Illustrating the formulas of spherical trigonometry.

To find the relations between the angles α, β, γ of the spherical triangle and its sides a, b, c (the face angles of the trihedral angle $OABC$), we introduce unit vectors $\mathbf{e}_1$, $\mathbf{e}_2$, $\mathbf{e}_3$ drawn from the center of the sphere to the vertices of the spherical triangle, as shown in the figure. The angle α between the planes OAC and OAB equals the angle between the normals to the planes, and hence

$$\cos\alpha = \frac{(\mathbf{e}_1 \times \mathbf{e}_2) \cdot (\mathbf{e}_1 \times \mathbf{e}_3)}{|\mathbf{e}_1 \times \mathbf{e}_2|\,|\mathbf{e}_1 \times \mathbf{e}_3|}.$$

Since

$$|\mathbf{e}_1 \times \mathbf{e}_2| = \sin b, \qquad |\mathbf{e}_1 \times \mathbf{e}_3| = \sin c,$$

it follows from (1.27) and (1.30) that

$$\begin{aligned}\cos\alpha &= \frac{\mathbf{e}_1 \cdot [\mathbf{e}_2 \times (\mathbf{e}_1 \times \mathbf{e}_3)]}{\sin b \sin c}\\ &= \frac{\mathbf{e}_1 \cdot [\mathbf{e}_1(\mathbf{e}_2 \cdot \mathbf{e}_3) - \mathbf{e}_3(\mathbf{e}_2 \cdot \mathbf{e}_1)]}{\sin b \sin c}\\ &= \frac{\cos a - \cos c \cos b}{\sin b \sin c},\end{aligned}$$

and hence

$$\cos a = \cos b \cos c + \sin b \sin c \cos \alpha.$$

In just the same way, we find that

$$\cos b = \cos c \cos a + \sin c \sin a \cos \beta,$$

$$\cos c = \cos a \cos b + \sin a \sin b \cos \gamma.$$

Another group of formulas is obtained by calculating the sines of the angles α, β and γ, e.g.,

$$\sin \alpha = \frac{|(\mathbf{e}_1 \times \mathbf{e}_2) \times (\mathbf{e}_1 \times \mathbf{e}_3)|}{|\mathbf{e}_1 \times \mathbf{e}_2|\,|\mathbf{e}_1 \times \mathbf{e}_3|}.$$

As an exercise, the reader should prove that

$$\frac{\sin \alpha}{\sin a} = \frac{\sin \beta}{\sin b} = \frac{\sin \gamma}{\sin c}.$$

***Problem** 7.* Use vector notation to write the equation of a straight line in three dimensions.

Case 1. Find the equation of a straight line going through two points A and B, as in Fig. 1.27(a).

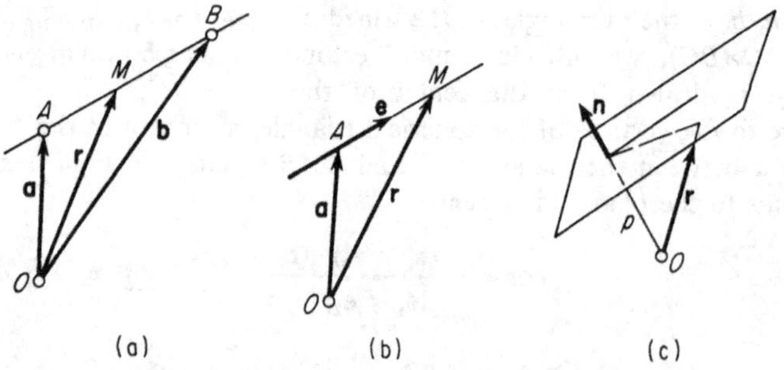

FIG. 1.27. (a) A line passing through two given points; (b) A line specified by a point and a direction; (c) Illustrating the normal equation of a plane.

Solution. Let $\mathbf{a}$ and $\mathbf{b}$ be the radius vectors of A and B with respect to some origin. Then the condition for an arbitrary point M, with radius vector $\mathbf{r}$, to lie on the line going through A and B is that the vectors $\mathbf{r} - \mathbf{a}$ and $\mathbf{b} - \mathbf{a}$ be parallel, i.e., that

$$\mathbf{r} - \mathbf{a} = \lambda(\mathbf{b} - \mathbf{a}).$$

Thus, if λ is regarded as a parameter, the equation of the line takes the form

$$\mathbf{r} = \mathbf{a} + \lambda(\mathbf{b} - \mathbf{a}). \tag{1.64}$$

The parameter λ can be eliminated by taking the vector product of (1.64) with $\mathbf{b} - \mathbf{a}$. This gives

$$(\mathbf{r} - \mathbf{a}) \times (\mathbf{b} - \mathbf{a}) = 0$$

or

$$\mathbf{r} \times (\mathbf{b} - \mathbf{a}) = \mathbf{a} \times \mathbf{b}.$$

Case 2. Find the equation of a straight line going through a given point A parallel to a given vector $\mathbf{e}$, as in Fig. 1.27(b).

Solution. Let $\mathbf{a}$ be the position vector of A. Since $\mathbf{r} - \mathbf{a}$ and $\mathbf{e}$ must be parallel, we have

$$\mathbf{r} = \mathbf{a} + \lambda\mathbf{e}, \tag{1.65}$$

where λ is a parameter. To eliminate λ, we take the vector product of (1.65) with $\mathbf{e}$, obtaining

$$\mathbf{r} \times \mathbf{e} = \mathbf{a} \times \mathbf{e}.$$

Case 3. Find the equation of a straight line going through a given point A perpendicular to two given vectors $\mathbf{e}_1$ and $\mathbf{e}_2$.

Answer. If $\mathbf{a}$ is the position vector of A, then

$$\mathbf{r} = \mathbf{a} + \lambda(\mathbf{e}_1 \times \mathbf{e}_2),$$

$$(\mathbf{r} - \mathbf{a}) \times (\mathbf{e}_1 \times \mathbf{e}_2) = 0.$$

Problem 8. Write a condition for the points A, B and C to be collinear.

Solution. Let $\mathbf{a}$, $\mathbf{b}$ and $\mathbf{c}$ be the position vectors of A, B, C with respect to some origin O. From the preceding problem we find that

$$\frac{c_1 - a_1}{b_1 - a_1} = \frac{c_2 - a_2}{b_2 - a_2} = \frac{c_3 - a_3}{b_3 - a_3},$$

where a_i, b_i, c_i are the coordinates of A, B, C with respect to a rectangular coordinate system with origin O.

As an exercise, show that this condition can be written in the form

$$(\mathbf{a} \times \mathbf{b}) + (\mathbf{b} \times \mathbf{c}) + (\mathbf{c} \times \mathbf{a}) = 0.$$

What is the geometric meaning of this condition?

Problem 9. Write the equation of a plane in vector form.

Case 1. Find the equation of the plane going through three given points A, B and C, with position vectors $\mathbf{a}$, $\mathbf{b}$ and $\mathbf{c}$.

Solution. Since the vectors $\mathbf{r} - \mathbf{a}$, $\mathbf{b} - \mathbf{a}$ and $\mathbf{c} - \mathbf{a}$ are coplanar, we have

$$\mathbf{r} - \mathbf{a} = \lambda(\mathbf{b} - \mathbf{a}) + \mu(\mathbf{c} - \mathbf{a}), \tag{1.66}$$

where λ and μ are parameters. To eliminate λ and μ, we first take the vector product of (1.66) with $\mathbf{c} - \mathbf{a}$ and then the scalar product with $\mathbf{b} - \mathbf{a}$, obtaining

$$[(\mathbf{r} - \mathbf{a}) \times (\mathbf{c} - \mathbf{a})] \cdot (\mathbf{b} - \mathbf{a}) = 0.$$

This condition could have been written down at once by starting from the coplanarity condition (1.28).

Case 2. Find the equation of the plane going through two points A and B, with position vectors $\mathbf{a}$ and $\mathbf{b}$, parallel to a given vector $\mathbf{e}$.

Answer. $$[(\mathbf{r} - \mathbf{a}) \times (\mathbf{b} - \mathbf{a})] \cdot \mathbf{e} = 0.$$

Case 3. Find the equation of the plane going through the point A, with position vector $\mathbf{a}$, parallel to two vectors $\mathbf{e}_1$ and $\mathbf{e}_2$.

Answer. $$[(\mathbf{r} - \mathbf{a}) \times \mathbf{e}_1] \cdot \mathbf{e}_2 = 0.$$

Case 4. Find the equation of the plane with unit normal $\mathbf{n}$ whose distance from the origin O equals p, as in Fig. 1.27(c).

Solution. If $\mathbf{n}$ points in the direction away from O, then

$$\mathbf{r} \cdot \mathbf{n} = r \cos(\mathbf{r}, \mathbf{n}) = p.$$

This is called the normal form of the equation of a plane.

Problem 10. Prove that infinitesimal rotations are vectors.

Solution. As in Fig. 1.3(a), p. 5, suppose the first rotation $\boldsymbol{\alpha}_1$ of the sphere carries the point A_1 into the point A_2, and the second rotation $\boldsymbol{\alpha}_2$ carries A_2 into A_3. If the angle α_1 is small, we can write

$$\overrightarrow{OA_2} = \overrightarrow{OA_1} + \overrightarrow{A_1A_2} = \overrightarrow{OA_1} + (\boldsymbol{\alpha}_1 \times \overrightarrow{OA_1}), \tag{1.67}$$

since

$$|\boldsymbol{\alpha}_1 \times OA_1| = \alpha_1 \, |OA_1| = \widehat{A_1A_2} = |\overrightarrow{A_1A_2}|$$

and moreover $\boldsymbol{\alpha}_1 \times \overrightarrow{OA_1}$ has the same direction as the vector $\overrightarrow{A_1A_2}$. Similarly

$$\overrightarrow{OA_3} = \overrightarrow{OA_2} + \overrightarrow{A_2A_3} = \overrightarrow{OA_2} + (\boldsymbol{\alpha}_2 \times \overrightarrow{OA_2}). \tag{1.68}$$

Substituting (1.68) into (1.67), we obtain

$$\begin{aligned}\overrightarrow{OA_3} &= \overrightarrow{OA_1} + (\boldsymbol{\alpha}_1 \times \overrightarrow{OA_1}) + \boldsymbol{\alpha}_2 \times [\overrightarrow{OA_1} + (\boldsymbol{\alpha}_1 \times \overrightarrow{OA_1})] \\ &= \overrightarrow{OA_1} + (\boldsymbol{\alpha}_1 + \boldsymbol{\alpha}_2) \times \overrightarrow{OA_1} + \boldsymbol{\alpha}_2 \times (\boldsymbol{\alpha}_1 \times \overrightarrow{OA_1}) \qquad (1.69) \\ &= \overrightarrow{OA_1} + (\boldsymbol{\alpha}_2 + \boldsymbol{\alpha}_1) \times \overrightarrow{OA_1} + \boldsymbol{\alpha}_2 \times (\boldsymbol{\alpha}_1 \times \overrightarrow{OA_1}).\end{aligned}$$

If (and only if) α_1 and α_2 are infinitesimal, we can drop the second-order terms in (1.69), obtaining

$$\overrightarrow{OA_3} = \overrightarrow{OA_1} + (\boldsymbol{\alpha}_1 + \boldsymbol{\alpha}_2) \times \overrightarrow{OA_1} = \overrightarrow{OA_1} + (\boldsymbol{\alpha}_2 + \boldsymbol{\alpha}_1) \times \overrightarrow{OA_1}. \quad (1.70)$$

On the other hand, if $\boldsymbol{\alpha}_3$ is the (infinitesimal) rotation carrying A_1 into A_3 directly, then

$$\overrightarrow{OA_3} = \overrightarrow{OA_1} + \boldsymbol{\alpha}_3 \times \overrightarrow{OA_1}. \quad (1.71)$$

Comparing (1.70) and (1.71), we obtain

$$\boldsymbol{\alpha}_3 = \boldsymbol{\alpha}_1 + \boldsymbol{\alpha}_2 = \boldsymbol{\alpha}_2 + \boldsymbol{\alpha}_1.$$

Thus infinitesimal rotations are vectors, since they obey the laws of vector algebra (the resultant of two rotations is the geometric sum of the separate rotations, and the sum is independent of the order of the terms).

Problem 11. Find the velocity of an arbitrary point of a rigid body rotating about a fixed point O (see Fig. 1.28).

Solution. Suppose that during the time interval Δt, the body undergoes an infinitesimal rotation $\Delta\varphi$ causing an arbitrary point M of the body to experience a displacement $\Delta\mathbf{r}$. Then, according to Prob. 10,

$$\Delta\mathbf{r} = \Delta\boldsymbol{\varphi} \times \mathbf{r}, \quad (1.72)$$

where $\mathbf{r}$ is the radius vector of M. Dividing (1.72) by Δt and taking the limit as $\Delta t \to 0$, we obtain

$$\mathbf{v} = \boldsymbol{\omega} \times \mathbf{r},$$

where

$$\mathbf{v} = \lim_{\Delta t \to 0} \frac{\Delta\mathbf{r}}{\Delta t}$$

is the velocity of the point M, and

$$\boldsymbol{\omega} = \lim_{\Delta t \to 0} \frac{\Delta\boldsymbol{\varphi}}{\Delta t}$$

is the instantaneous *angular velocity* of the body about the point O.

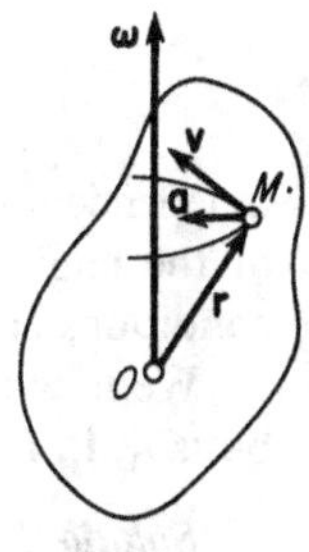

FIG. 1.28. Velocity **v** and centripetal acceleration **a** of a rigid body rotating about a fixed point O.

Problem 12 (The Eulerian angles). Let K and K' be two (right-handed) rectangular coordinate systems with the same origin O and orthonormal bases

$\mathbf{i}_1$, $\mathbf{i}_2$, $\mathbf{i}_3$ and $\mathbf{i}'_1$, $\mathbf{i}'_2$, $\mathbf{i}'_3$, respectively. Then K can be carried into K' by making three rotations in succession:

1) Through an angle ψ (called the angle of *precession*) about the x_3-axis;
2) Through an angle θ (called the angle of *nutation*) about the line ON, characterized by the unit vector $\mathbf{n}$;
3) Through an angle φ (called the angle of *pure rotation*) about the x'_3-axis

(see Fig. 1.29). The angles ψ, θ and φ are called the *Eulerian angles*. The fact that the position of K' with respect to K can be specified by just three

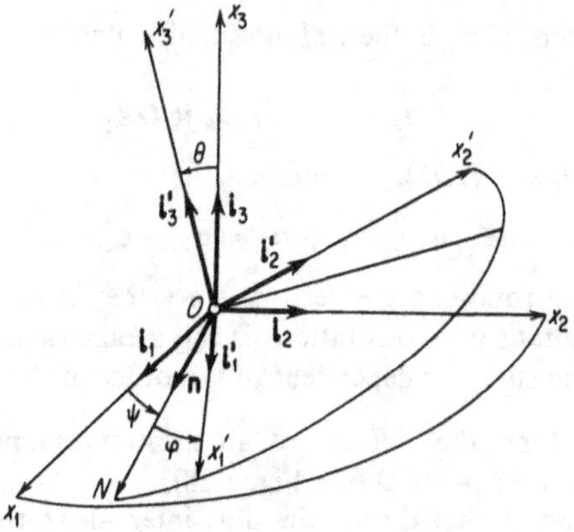

FIG. 1.29. The Eulerian angles.

independent parameters like ψ, θ, φ is hardly surprising, since the nine cosines of the angles between the axes of K and K' satisfy the six orthogonality conditions (1.62).

We now pose the problem of expressing the basis $\mathbf{i}'_1$, $\mathbf{i}'_2$, $\mathbf{i}'_3$ in terms of the basis $\mathbf{i}_1$, $\mathbf{i}_2$, $\mathbf{i}_3$ and the Eulerian angles ψ, θ, φ.

Solution. It will be recalled from Prob. 1 that

$$\mathbf{i}'_1 = \alpha_{1'1}\mathbf{i}_1 + \alpha_{1'2}\mathbf{i}_2 + \alpha_{1'3}\mathbf{i}_3,$$

$$\mathbf{i}'_2 = \alpha_{2'1}\mathbf{i}_1 + \alpha_{2'2}\mathbf{i}_2 + \alpha_{2'3}\mathbf{i}_3,$$

$$\mathbf{i}'_3 = \alpha_{3'1}\mathbf{i}_1 + \alpha_{3'2}\mathbf{i}_2 + \alpha_{3'3}\mathbf{i}_3.$$

Consider the spherical triangle formed on the unit sphere by the end points of the vectors $\mathbf{i}'_1$, $\mathbf{n}$ and $\mathbf{i}_1$. Then, according to the formulas derived in Prob. 6,

$$\begin{aligned}\alpha_{1'1} &= \cos(\mathbf{i}'_1, \mathbf{i}_1) = \cos\psi\cos\varphi + \sin\psi\sin\varphi\cos(\pi - \theta)\\ &= \cos\psi\cos\varphi - \sin\psi\sin\varphi\cos\theta.\end{aligned}$$

Similarly, an examination of the spherical triangles formed by the end points of $\mathbf{i}_1'$, $\mathbf{n}$, $\mathbf{i}_2$ and $\mathbf{i}_1'$, $\mathbf{n}$, $\mathbf{i}_3$ shows that

$$\alpha_{1'2} = \cos(\mathbf{i}_1', \mathbf{i}_2) = \cos\left(\frac{\pi}{2} - \psi\right)\cos\varphi + \sin\left(\frac{\pi}{2} - \psi\right)\sin\varphi\cos\theta$$

$$= \sin\psi\cos\varphi + \cos\psi\sin\varphi\cos\theta,$$

$$\alpha_{1'3} = \cos(\mathbf{i}_1', \mathbf{i}_3) = \cos\varphi\cos\frac{\pi}{2} + \sin\varphi\sin\frac{\pi}{2}\cos\left(\frac{\pi}{2} - \theta\right)$$

$$= \sin\varphi\sin\theta.$$

It follows that

$$\mathbf{i}_1' = \mathbf{i}_1(\cos\psi\cos\varphi - \sin\psi\sin\varphi\cos\theta) + \mathbf{i}_2(\sin\psi\cos\varphi + \cos\psi\sin\varphi\cos\theta) + \mathbf{i}_3\sin\varphi\sin\theta.$$

In the same way, it turns out that

$$\mathbf{i}_2' = \mathbf{i}_1(-\cos\psi\sin\varphi - \sin\psi\cos\varphi\cos\theta) + \mathbf{i}_2(\cos\psi\cos\varphi\cos\theta - \sin\psi\sin\varphi) + \mathbf{i}_3\sin\theta\cos\varphi,$$

$$\mathbf{i}_3' = \mathbf{i}_1\sin\psi\sin\theta - \mathbf{i}_2\cos\psi\sin\theta + \mathbf{i}_3\cos\theta.$$

The reader should also verify that the formulas expressing the basis $\mathbf{i}_1, \mathbf{i}_2, \mathbf{i}_3$ in terms of the basis $\mathbf{i}_1', \mathbf{i}_2', \mathbf{i}_3'$ and the Eulerian angles are

$$\mathbf{i}_1 = \mathbf{i}_1'(\cos\psi\cos\varphi - \sin\psi\sin\varphi\cos\theta) + \mathbf{i}_2'(-\cos\psi\sin\varphi - \sin\psi\cos\varphi\cos\theta) + \mathbf{i}_3'\sin\psi\sin\theta,$$

$$\mathbf{i}_2 = \mathbf{i}_1'(\sin\psi\cos\varphi + \cos\psi\sin\varphi\cos\theta) + \mathbf{i}_2'(\cos\psi\cos\varphi\cos\theta - \sin\psi\sin\varphi) - \mathbf{i}_3'\cos\psi\sin\theta,$$

$$\mathbf{i}_3 = \mathbf{i}_1'\sin\varphi\sin\theta + \mathbf{i}_2'\cos\varphi\sin\theta + \mathbf{i}_3'\cos\theta.$$

Problem 13. Given a system of n electric charges $e_1, e_2, \ldots, e_n$, let $\mathbf{r}_k$ be the radius vector of e_k $(k = 1, 2, \ldots, n)$ with respect to some origin O. Then the vector

$$\mathbf{p} = \sum_{k=1}^{n} e_k\mathbf{r}_k$$

is called the *dipole moment* of the system of charges. Moreover, by analogy with the center of mass, the point C with radius vector

$$\mathbf{R} = \frac{\mathbf{p}}{\sum_{k=1}^{n} e_k} = \frac{\sum_{k=1}^{n} e_k\mathbf{r}_k}{\sum_{k=1}^{n} e_k}$$

is called the *center of charge* of the system. The point C can be defined only if

$$\sum_{k=1}^{n} e_k \neq 0.$$

If, on the other hand,

$$\sum_{k=1}^{n} e_k = 0,$$

the system of charges is said to be *neutral.*

Prove that the dipole moment of a neutral system is independent of the origin O, and express this moment in terms of the centers of the systems of positive and negative charges making up the original system.

Solution. Let

$$\mathbf{p} = \sum_{k=1}^{n} e_k \mathbf{r}_k$$

be the dipole moment with respect to an origin O, while

$$\mathbf{p}' = \sum_{k=1}^{n} e_k \mathbf{r}_k'$$

is the dipole moment with respect to another origin O', where $\overrightarrow{OO'} = \mathbf{r}_0$. Then $\mathbf{r}_k'$, the radius vector of e_k with respect to O', equals $\mathbf{r}_k + \mathbf{r}_0$, and hence

$$\mathbf{p}' = \sum_{k=1}^{n} e_k \mathbf{r}_k' = \sum_{k=1}^{n} e_k(\mathbf{r}_k + \mathbf{r}_0) = \sum_{k=1}^{n} e_k \mathbf{r}_k + \mathbf{r}_0 \sum_{k=1}^{n} e_k.$$

But

$$\sum_{k=1}^{n} e_k = 0$$

since the system is neutral, and hence

$$\mathbf{p}' = \sum_{k=1}^{n} e_k \mathbf{r}_k = \mathbf{p}.$$

Now let the system consist of positive charges e_k^+ and negative charges e_k^-, so that

$$\sum_{k=1}^{n} e_k = \sum e_k^+ + \sum e_k^-,$$

and let

$$\sum e_k^+ = -\sum e_k^- = Q.$$

Then, by definition, the centers of the systems of positive and negative charges have position vectors

$$\mathbf{R}^+ = \frac{\sum e_k^+ \mathbf{r}_k^+}{\sum e_k^+}, \qquad \mathbf{R}^- = \frac{\sum e_k^- \mathbf{r}_k^-}{\sum e_k^-},$$

where $\mathbf{r}_k^+$ and $\mathbf{r}_k^-$ have the obvious meaning. Hence the dipole moment of the original neutral system equals

$$\mathbf{p} = \sum_{k=1}^{n} e_k \mathbf{r}_k = \sum e_k^+ \mathbf{r}_k^+ + \sum e_k^- \mathbf{r}_k^-$$
$$= \mathbf{R}^+ \sum e_k^+ + \mathbf{R}^- \sum e_k^- = Q(\mathbf{R}^+ - \mathbf{R}^-).$$

***Problem 14* (*Collision of particles*).** Suppose two particles of identical mass have velocities $\mathbf{v}_1$, $\mathbf{v}_2$ before colliding and velocities $\mathbf{v}_1'$, $\mathbf{v}_2'$ after colliding (see Fig. 1.30). Suppose the collision is governed by the action of central

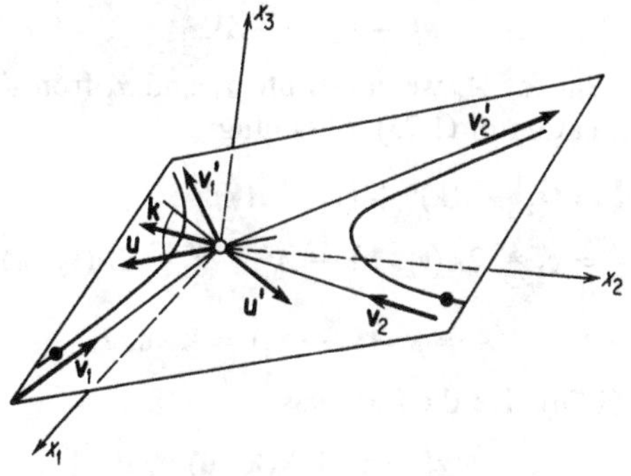

FIG. 1.30. Collision of two particles.

forces. Then, as is familiar from mechanics, the trajectories of the two particles lie in a fixed plane in a coordinate system in which the center of mass is at rest. Moreover, the collision conserves both momentum and kinetic energy (the potential energy before and after the collision equals zero). It follows that

$$\begin{aligned} \mathbf{v}_1 + \mathbf{v}_2 &= \mathbf{v}_1' + \mathbf{v}_2', \\ v_1^2 + v_2^2 &= v_1'^2 + v_2'^2. \end{aligned} \tag{1.73}$$

Express $\mathbf{v}_1'$, $\mathbf{v}_2'$ in terms of $\mathbf{v}_1$, $\mathbf{v}_2$ and show that the relative velocities

$$\begin{aligned} \mathbf{u} &= \mathbf{v}_2 - \mathbf{v}_1, \\ \mathbf{u}' &= \mathbf{v}_2' - \mathbf{v}_1' \end{aligned} \tag{1.74}$$

before and after the collision have the same magnitude.

Solutions. The system (1.73) consists of four scalar equations in the six components of the velocities $\mathbf{v}_1'$ and $\mathbf{v}_2'$. Thus we can solve for $\mathbf{v}_1'$ and $\mathbf{v}_2'$ explicitly in terms of $\mathbf{v}_1$ and $\mathbf{v}_2$, provided we introduce two extra parameters characterizing the geometry of the collision, i.e., the position of the plane of the trajectories in some system of rectangular coordinates x_1, x_2, x_3. In other words, the collision of two particles is completely characterized by specifying two geometric parameters. We introduce these parameters in the guise of the unit vector $\mathbf{k}$ ($|\mathbf{k}| = 1$) pointing in the direction of change of the velocity of the first particle, i.e.,

$$\mathbf{v}_1' - \mathbf{v}_1 = A\mathbf{k} \tag{1.75}$$

(see Fig. 1.30). Any two independent angles between $\mathbf{k}$ and the axes of the x_1, x_2, x_3 system can then be regarded as the geometric parameters of the collision.

It follows from the first of the equations (1.73) that

$$\mathbf{v}_2' - \mathbf{v}_2 = -A\mathbf{k}. \tag{1.76}$$

To get an expression for A, we substitute $\mathbf{v}_1'$ and $\mathbf{v}_2'$ from (1.75) and (1.76) into the second equation of (1.73), obtaining

$$\begin{aligned} v_1^2 + v_2^2 &= (\mathbf{v}_1 + A\mathbf{k})^2 + (\mathbf{v}_2 - A\mathbf{k})^2 \\ &= v_1^2 + 2A(\mathbf{v}_1 \cdot \mathbf{k}) + A^2 + v_2^2 - 2A(\mathbf{v}_2 \cdot \mathbf{k}) + A^2 \end{aligned}$$

which implies

$$A = \mathbf{k} \cdot (\mathbf{v}_2 - \mathbf{v}_1) = \mathbf{k} \cdot \mathbf{u}.$$

Then (1.75) and (1.76) give the formulas

$$\begin{aligned} \mathbf{v}_1' &= \mathbf{v}_1 + \mathbf{k}(\mathbf{k} \cdot \mathbf{u}), \\ \mathbf{v}_2' &= \mathbf{v}_2 - \mathbf{k}(\mathbf{k} \cdot \mathbf{u}), \end{aligned} \tag{1.77}$$

expressing the velocities $\mathbf{v}_1'$ and $\mathbf{v}_2'$ after the collision in terms of the velocities $\mathbf{v}_1$ and $\mathbf{v}_2$ before the collision and the vector $\mathbf{k}$.

Subtracting the second of the equations (1.74) from the first and using (1.77), we find that

$$\mathbf{u}' - \mathbf{u} = \mathbf{v}_2' - \mathbf{v}_1' - (\mathbf{v}_2 - \mathbf{v}_1) = -2\mathbf{k}(\mathbf{k} \cdot \mathbf{u})$$

or

$$\mathbf{u}' = \mathbf{u} - 2\mathbf{k}(\mathbf{k} \cdot \mathbf{u}). \tag{1.78}$$

Squaring (1.78), we obtain

$$u'^2 = u^2 - 4(\mathbf{k} \cdot \mathbf{u})^2 + 4k^2(\mathbf{k} \cdot \mathbf{u})^2$$

or

$$u'^2 = u^2,$$

i.e., the relative velocity of the particles has the same magnitude before and after the collision. Taking the scalar product of (1.78) with $\mathbf{k}$, we get

$$\mathbf{u}' \cdot \mathbf{k} = -\mathbf{u} \cdot \mathbf{k}.$$

It follows that the vector $\mathbf{k}$ bisects the angle between $\mathbf{u}$ and $-\mathbf{u}'$, as shown in the figure.

Problem 15. Pursuing the study of collisions between two particles, let the particles have different masses m_1 and m_2. Show that the relative velocity $\mathbf{u} = \mathbf{v}_2 - \mathbf{v}_1$ is still preserved under the collision.

Solution. The momenta of the particles before and after collision are given by

$$\mathbf{p}_1 = m_1\mathbf{v}_1, \quad \mathbf{p}_2 = m_2\mathbf{v}_2, \quad \mathbf{p}_1' = m_1\mathbf{v}_1', \quad \mathbf{p}_2' = m_2\mathbf{v}_2'$$

in terms of their masses and the velocities before and after collision. The law of conservation of momentum is now

$$\mathbf{p}_1 + \mathbf{p}_2 = \mathbf{p}_1' + \mathbf{p}_2',$$

while the law of conservation of (kinetic) energy is

$$p_1^2 + mp_2^2 = p_1'^2 + mp_2'^2 \qquad \left(m = \frac{m_1}{m_2}\right). \tag{1.79}$$

Introducing the unit vector $\mathbf{k}$ defined by

$$\mathbf{p}_1' - \mathbf{p}_1 = A\mathbf{k}, \qquad \mathbf{p}_2' - \mathbf{p}_2 = -A\mathbf{k},$$

we determine A from (1.79), obtaining

$$A = \frac{2}{1+m}\,\mathbf{k}\cdot(m\mathbf{p}_2 - \mathbf{p}_1) = \frac{2m_1}{1+m}\,(\mathbf{k}\cdot\mathbf{u}).$$

It follows that

$$\mathbf{p}_1' = \mathbf{p}_1 + \frac{2}{1+m}\,\mathbf{k}[\mathbf{k}\cdot(m\mathbf{p}_2 - \mathbf{p}_1)] = \mathbf{p}_1 + \frac{2m_1}{1+m}\,\mathbf{k}(\mathbf{k}\cdot\mathbf{u}),$$

$$\mathbf{p}_2' = \mathbf{p}_2 - \frac{2}{1+m}\,\mathbf{k}[\mathbf{k}\cdot(m\mathbf{p}_2 - \mathbf{p}_1)] = \mathbf{p}_2 - \frac{2m_1}{1+m}\,\mathbf{k}(\mathbf{k}\cdot\mathbf{u}).$$

Therefore

$$m\mathbf{p}_2' - \mathbf{p}_1' = m\mathbf{p}_2 - \mathbf{p}_1 - 2\mathbf{k}[\mathbf{k}\cdot(m\mathbf{p}_2 - \mathbf{p}_1)]$$

or

$$\mathbf{u}' = \mathbf{u} - 2\mathbf{k}(\mathbf{k}\cdot\mathbf{u}), \tag{1.80}$$

where $\mathbf{u}' = \mathbf{v}_2' - \mathbf{v}_1'$. Squaring (1.80), we obtain

$$u'^2 = u^2.$$

Problem 16. Prove that if the force acting on a moving particle is always directed along the tangent to the trajectory, then the trajectory is a straight line.

Solution. The force equals

$$\mathbf{F} = m\mathbf{a} = m\,\frac{d^2\mathbf{r}}{dt^2}.$$

If **F** is directed along the tangent, i.e., along

$$\mathbf{v} = \frac{d\mathbf{r}}{dt},$$

then

$$m\mathbf{a} = \alpha\mathbf{v}$$

and hence

$$\frac{d^2\mathbf{r}}{dt^2} + \beta\frac{d\mathbf{r}}{dt} = 0, \tag{1.81}$$

where α and β are constants. Integrating (1.81), we obtain

$$\frac{d\mathbf{r}}{dt} + \beta\mathbf{r} = \mathbf{b}, \tag{1.82}$$

where **b** is a constant vector. The solution of (1.82) can be written in the form

$$\mathbf{r} = \mathbf{c}f(t) + \frac{\mathbf{b}}{\beta}, \tag{1.83}$$

where **c** is another constant vector and f satisfies the differential equation

$$f'(t) + \beta f(t) = 0.$$

It follows from (1.83) that the trajectory of the moving point is a straight line (recall Prob. 7). The character of the motion along the straight line is determined by the function $f(t)$.

Problem 17. Prove that if the trajectory $\mathbf{r} = \mathbf{r}(t)$ of a moving particle is such that

$$\frac{d\mathbf{r}}{dt}\cdot\left(\frac{d^2\mathbf{r}}{dt^2}\times\frac{d^3\mathbf{r}}{dt^3}\right) = 0, \tag{1.84}$$

then $\mathbf{r} = \mathbf{r}(t)$ is a plane curve.

Solution. It follows from (1.84) that

$$\frac{d^3\mathbf{r}}{dt^3} = \alpha\frac{d^2\mathbf{r}}{dt^2} + \beta\frac{d\mathbf{r}}{dt}, \tag{1.85}$$

where α and β are constants. Integrating (1.85), we obtain

$$\frac{d^2\mathbf{r}}{dt^2} = \alpha\frac{d\mathbf{r}}{dt} + \beta\mathbf{r} + \mathbf{b}, \tag{1.86}$$

where **b** is a constant vector. The solution of (1.86) can be written in the form

$$\mathbf{r} = -\frac{1}{\beta}\mathbf{b} + f_1(t)\mathbf{h}_1 + f_2(t)\mathbf{h}_2, \tag{1.87}$$

where $\mathbf{h}_1$ and $\mathbf{h}_2$ are constant vectors and f_1, f_2 are two independent solutions of the differential equation

$$f''(t) = \alpha f'(t) + \beta f(t).$$

Clearly (1.87) is the equation of a plane curve (see Prob. 9).

Problem 18. Show that the trajectory of a particle moving under the influence of gravitational attraction is a conic section.

Solution. The equation of motion of the particle is of the form

$$\frac{d\mathbf{v}}{dt} = -\alpha\frac{\mathbf{r}}{r^3} = -\alpha\frac{\mathbf{r}_0}{r^2}, \tag{1.88}$$

where α is a constant, $\mathbf{v} = d\mathbf{r}/dt$ and

$$\mathbf{r}_0 = \frac{\mathbf{r}}{r}$$

is the "unit radius vector." Taking the vector product of (1.88) with $\mathbf{r}$, we get the first integral of the equation of motion [cf. Exercise 15(c)]:

$$\mathbf{r} \times \mathbf{v} = \mathbf{h} = \textbf{const.}$$

The vector $\mathbf{h}$ can be written as

$$\mathbf{h} = \mathbf{r} \times \mathbf{v} = \mathbf{r} \times \frac{d\mathbf{r}}{dt} = r\mathbf{r}_0 \times \frac{d}{dt}(r\mathbf{r}_0) = r\mathbf{r}_0 \times \left(\frac{dr}{dt}\mathbf{r}_0 + r\frac{d\mathbf{r}_0}{dt}\right)$$

But then

$$\frac{d\mathbf{v}}{dt} \times \mathbf{h} = -\frac{\alpha\mathbf{r}_0}{r^2} \times \mathbf{h} = \alpha\frac{d\mathbf{r}_0}{dt},$$

where we have used (1.30) and the fact that

$$\mathbf{r}_0 \cdot \frac{d\mathbf{r}_0}{dt} = \frac{1}{2}\frac{d}{dt}(\mathbf{r}_0 \cdot \mathbf{r}_0) = \frac{1}{2}\frac{d}{dt}r_0^2 = 0.$$

Therefore, since $\mathbf{h} = \textbf{const}$,

$$\frac{d}{dt}(\mathbf{v} \times \mathbf{h}) = \alpha\frac{d\mathbf{r}_0}{dt}. \tag{1.89}$$

Integrating (1.89), we obtain

$$\mathbf{v} \times \mathbf{h} = \alpha\mathbf{r}_0 + \mathbf{P},$$

where $\mathbf{P}$ is a constant vector. Hence

$$(\mathbf{v} \times \mathbf{h}) \cdot \mathbf{r} = \alpha r + \mathbf{r} \cdot \mathbf{P} = \alpha r + rP\cos\theta,$$

where θ is the angle between the variable vector $\mathbf{r} = \mathbf{r}(t)$ and the constant vector $\mathbf{P}$. Since

$$(\mathbf{v} \times \mathbf{h}) \cdot \mathbf{r} = \mathbf{h} \cdot (\mathbf{r} \times \mathbf{v}) = \mathbf{h} \cdot \mathbf{h} = h^2$$

[see (1.27)], we have

$$h^2 = \alpha r + rP \cos \theta$$

or

$$r = \frac{\dfrac{h^2}{\alpha}}{1 + \dfrac{P}{\alpha} \cos \theta}. \tag{1.90}$$

As is familiar from analytic geometry, (1.90) is the equation of a conic section (an ellipse, parabola or hyperbola) in polar coordinates.

EXERCISES

1. Prove that the projection of a sum of vectors onto any axis equals the sum of the projections of the vectors onto the same axis.

2. Given the vectors

$$\mathbf{A} = \mathbf{i}_1 + 2\mathbf{i}_2 + 3\mathbf{i}_3, \qquad \mathbf{B} = 4\mathbf{i}_1 + 5\mathbf{i}_2 + 6\mathbf{i}_3,$$

$$\mathbf{C} = 3\mathbf{i}_1 + 2\mathbf{i}_2 + \mathbf{i}_3, \qquad \mathbf{D} = 6\mathbf{i}_1 + 5\mathbf{i}_2 + 4\mathbf{i}_3,$$

where $\mathbf{i}_1$, $\mathbf{i}_2$, $\mathbf{i}_3$ are an orthonormal basis, find

a) The sums and differences

$$\mathbf{A} + \mathbf{B} + \mathbf{C} + \mathbf{D}, \qquad \mathbf{A} + \mathbf{B} - \mathbf{C} - \mathbf{D},$$

$$\mathbf{A} - \mathbf{B} + \mathbf{C} - \mathbf{D}, \qquad -\mathbf{A} + \mathbf{B} - \mathbf{C} + \mathbf{D};$$

b) The angles between **A**, **B**, **C**, **D** and the basis vectors;

c) The magnitudes of the vectors **A**, **B**, **C**, **D**.

3. Find the sum of three vectors of length a drawn

a) From a common vertex of a cube along three of its sides;

b) From a common vertex of a regular tetrahedron along three of its sides.

4. Given a system of n particles of masses $m_1, m_2, \ldots, m_n$, let $\mathbf{r}_k$ be the radius vector of the kth particle ($k = 1, 2, \ldots, n$) with respect to some origin O. Then the *center of mass* of the system has radius vector

$$\mathbf{R} = \frac{\sum_{k=1}^{n} m_k \mathbf{r}_k}{\sum_{k=1}^{n} m_k}.$$

Find the center of mass of each of the following systems:

a) Masses equal to 1, 2, 3 at the vertices of an equilateral triangle of side length a;

b) Masses equal to 1, 2, 3, 4 at the vertices of a square of side length a;

c) Masses equal to 1, 2, 3, 4 at the lower vertices of a cube of side length a, and masses equal to 5, 6, 7, 8 at the upper vertices.

5. A parallelogram has acute angle $\pi/3$ and side lengths $a = 3$, $b = 5$. Thinking of the corresponding sides as vectors $\mathbf{a}$ and $\mathbf{b}$, find

a) The vectors $\mathbf{a} + \mathbf{b}$ and $\mathbf{a} - \mathbf{b}$ (what is their geometric meaning?);

b) The area of the parallelogram;

c) The projection of each side onto the direction of the other.

6. Let $\mathbf{A}$, $\mathbf{B}$, $\mathbf{C}$ and $\mathbf{D}$ be the same as in Exercise 2. Find

a) $(\mathbf{A} + \mathbf{B}) \cdot (\mathbf{C} + \mathbf{D})$;

b) The angles made by $\mathbf{A}$ with $\mathbf{B}$, $\mathbf{C}$ and $\mathbf{D}$;

c) The projection of $\mathbf{A}$ onto the directions of $\mathbf{B}$, $\mathbf{C}$ and $\mathbf{D}$;

d) The vector products $\mathbf{A} \times \mathbf{B}$, $\mathbf{B} \times \mathbf{C}$, $\mathbf{C} \times \mathbf{D}$ and the angles they make with $\mathbf{D}$;

e) The areas of the parallelograms spanned by the vectors $\mathbf{A}$, $\mathbf{B}$ and by the vectors $\mathbf{C}$, $\mathbf{D}$, and also the lengths of the diagonals of these parallelograms.

7. Show that the vectors $\mathbf{A}$, $\mathbf{B}$, $\mathbf{C}$ and $\mathbf{D}$ of Exercise 2 are coplanar.

8. Let $\mathbf{i}_1$, $\mathbf{i}_2$, $\mathbf{i}_3$ be a right-handed orthonormal basis. Verify that the vectors

$$\mathbf{A} = \mathbf{i}_1 + 2\mathbf{i}_2 + 3\mathbf{i}_3, \quad B = 4\mathbf{i}_1 + 5\mathbf{i}_2, \quad C = 3\mathbf{i}_1 + 2\mathbf{i}_2 + \mathbf{i}_3$$

form a basis. Is this basis right-handed or left-handed? Find

a) The volume of the parallelepiped spanned by $\mathbf{A}$, $\mathbf{B}$ and $\mathbf{C}$;

b) The vectors forming two diagonals of the parallelepiped (drawn from the end points of $\mathbf{A}$) and the lengths of these vectors;

c) The area of the diagonal cross section of the parallelepiped going through the vector $\mathbf{A}$.

9. Suppose the midpoints of the sides of an arbitrary quadrilateral are joined (in order) by straight line segments. Show that the resulting figure is a parallelogram.

Hint. If the sides of the quadrilateral are represented by vectors $\mathbf{a}$, $\mathbf{b}$, $\mathbf{c}$ and $\mathbf{d}$, then $\mathbf{a} + \mathbf{b} + \mathbf{c} + \mathbf{d} = 0$.

10. Given four points with radius vectors $\mathbf{a}$, $\mathbf{b}$, $\mathbf{c}$ and $\mathbf{d}$, suppose

$$[(\mathbf{d} - \mathbf{a}) \times (\mathbf{c} - \mathbf{a})] \cdot (\mathbf{b} - \mathbf{a}) = 0.$$

Prove that the points are coplanar.

11. Let $\mathbf{i}_1$, $\mathbf{i}_2$, $\mathbf{i}_3$ be an orthonormal basis. Is

$$\mathbf{a}_1 = 2\mathbf{i}_1 + \mathbf{i}_2 - 3\mathbf{i}_3, \quad \mathbf{a}_2 = \mathbf{i}_1 - 4\mathbf{i}_3, \quad \mathbf{a}_3 = 4\mathbf{i}_1 + 3\mathbf{i}_2 - \mathbf{i}_3$$

a basis? How about

$$\mathbf{b}_1 = \mathbf{i}_1 - 3\mathbf{i}_2 + 2\mathbf{i}_3, \quad \mathbf{b}_2 = 2\mathbf{i}_1 - 4\mathbf{i}_2 - \mathbf{i}_3, \quad \mathbf{b}_3 = 3\mathbf{i}_1 + 2\mathbf{i}_2 - \mathbf{i}_3?$$

12. Let $\mathbf{b}_1, \mathbf{b}_2, \mathbf{b}_3$ be the same as in Exercise 11. Is

$$\mathbf{B}_1 = 2\mathbf{b}_1 - 3\mathbf{b}_2 + \mathbf{b}_3,$$
$$\mathbf{B}_2 = 3\mathbf{b}_1 - 5\mathbf{b}_2 + 2\mathbf{b}_3,$$
$$\mathbf{B}_3 = 4\mathbf{b}_1 - 5\mathbf{b}_2 + \mathbf{b}_3$$

a basis?

13. Prove formula (1.30), p. 22 without introducing a coordinate system.

14. Prove that

$$\frac{d}{dt}[\mathbf{A}\cdot(\mathbf{B}\times\mathbf{C})] = \frac{d\mathbf{A}}{dt}\cdot(\mathbf{B}\times\mathbf{C}) + \mathbf{A}\cdot\left(\frac{d\mathbf{B}}{dt}\times\mathbf{C}\right) + \mathbf{A}\cdot\left(\mathbf{B}\times\frac{d\mathbf{C}}{dt}\right).$$

15. Prove that

a) $\dfrac{d}{dt}\left[\mathbf{A}\cdot\left(\dfrac{d\mathbf{A}}{dt}\times\dfrac{d^2\mathbf{A}}{dt^2}\right)\right] = \mathbf{A}\cdot\left(\dfrac{d\mathbf{A}}{dt}\times\dfrac{d^3\mathbf{A}}{dt^3}\right);$

b) $\displaystyle\int \mathbf{A}\times\frac{d^2\mathbf{A}}{dt^2}\,dt = \mathbf{A}\times\frac{d\mathbf{A}}{dt} + \mathbf{C} \qquad (\mathbf{C} = \text{const});$

c) $\mathbf{r}\times\dfrac{d\mathbf{r}}{dt} = \mathbf{C}$ if $\dfrac{d^2\mathbf{r}}{dt^2} = \mathbf{r}f(r).$

16. Using the formula $\mathbf{v} = \boldsymbol{\omega}\times\mathbf{r}$ (see Prob. 11, p. 45), find the linear velocity $\mathbf{v}$ of the center of a rectangle of side lengths $a = 2$ cm and $b = 4$ cm rotating about one of its vertices if the instantaneous angular velocity $\boldsymbol{\omega}$ equals 5 radians per second and points along

a) The short side; b) The long side.

17. The *moment* $\mathbf{M}_0$ *of a force* $\mathbf{F}$ *with respect to a point* O is given by the expression

$$\mathbf{M}_0 = \mathbf{r}\times\mathbf{F},$$

where $\mathbf{r}$ is the radius vector of the initial point of $\mathbf{F}$ with respect to O. The projection of $\mathbf{M}_0$ onto an axis u going through O, i.e., the quantity

$$M_u = \mathbf{M}_0\cdot\mathbf{u}_0 = (\mathbf{r}\times\mathbf{F})\cdot\mathbf{u}_0$$

where $\mathbf{u}_0$ is a unit vector directed along u, is called the *moment of* $\mathbf{F}$ *with respect to the axis* u. Prove that M_u is independent of the position of O on u.

18. Find the moment of a force of 5 dynes directed along one side of a cube of side length 2 cm with respect to

a) All vertices of the cube;

b) All axes going through the given side.

19. Given a system of n particles of masses $m_1, m_2, \ldots, m_n$, let $\mathbf{r}_k$ be the radius vector and $\mathbf{v}_k$ the velocity of the kth particle ($k = 1, 2, \ldots, n$) with respect to some origin O. Then the vector

$$\mathbf{L}_0 = \sum_{k=1}^{n}\mathbf{r}_k\times m_k\mathbf{v}_k$$

is called the *angular momentum of the system with respect to O*. Given a cube of side length a cm, find the angular momentum with respect to every vertex of the cube of two particles of masses $m_1 = 1$ g and $m_2 = 2$ g moving in opposite directions with speed 3 cm/sec along two opposite sides of the cube.

20. Let P be the parallelogram spanned by the vectors $\mathbf{a}$ and $\mathbf{b}$. Then P has diagonals $\mathbf{a} + \mathbf{b}$ and $\mathbf{a} - \mathbf{b}$. Prove that

a) The sum of the squares of the diagonals of P equals the sum of the squares of the sides of P;

b) The diagonals of P are perpendicular if and only if P is a rhombus;

c) The area of the parallelogram P' spanned by the diagonals of P is twice as large as the area of P.

21. Suppose n springs with stiffnesses $C_1, C_2, \ldots, C_n$ are fastened at n points $M_1, M_2, \ldots, M_n$ and joined at a common point M (see Fig. 1.31). Find the equilibrium position of M.

Ans. If $\mathbf{r}_M$ is the radius vector of M and $\mathbf{r}_k$ that of M_k, then

$$\mathbf{R} = \frac{\sum_{k=1}^{n} C_k \mathbf{r}_k}{\sum_{k=1}^{n} C_k}.$$

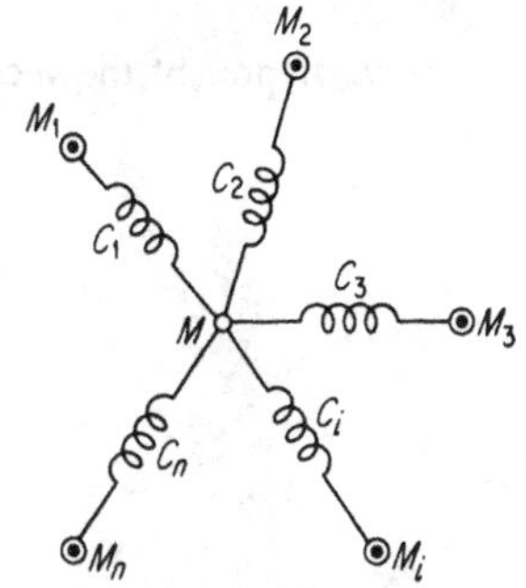

FIG. 1.31. Illustrating Exercise 21.

22. Verify the following identities:

a) $\mathbf{a} \times (\mathbf{b} \times \mathbf{c}) + \mathbf{b} \times (\mathbf{c} \times \mathbf{a}) + \mathbf{c} \times (\mathbf{a} \times \mathbf{b}) = 0$;

b) $(\mathbf{a} \times \mathbf{b}) \cdot (\mathbf{c} \times \mathbf{d}) = \begin{vmatrix} \mathbf{a} \cdot \mathbf{c} & \mathbf{a} \cdot \mathbf{d} \\ \mathbf{b} \cdot \mathbf{c} & \mathbf{b} \cdot \mathbf{d} \end{vmatrix}$;

c) $$(\mathbf{a} \times \mathbf{b}) \times (\mathbf{c} \times \mathbf{d}) = \mathbf{b}[\mathbf{a} \cdot (\mathbf{c} \times \mathbf{d})] - \mathbf{a}[\mathbf{b} \cdot (\mathbf{c} \times \mathbf{d})]$$
$$= \mathbf{c}[\mathbf{a} \cdot (\mathbf{b} \times \mathbf{d})] - \mathbf{d}[\mathbf{a} \cdot (\mathbf{b} \times \mathbf{c})];$$

d) $(\mathbf{a} \times \mathbf{b}) \cdot (\mathbf{c} \times \mathbf{d}) + (\mathbf{b} \times \mathbf{c}) \cdot (\mathbf{a} \times \mathbf{d}) + (\mathbf{c} \times \mathbf{a}) \cdot (\mathbf{b} \times \mathbf{d}) = 0$.

23. Given the basis

$$\mathbf{e}_1 = -4\mathbf{i}_1 + 2\mathbf{i}_2, \quad \mathbf{e}_2 = 3\mathbf{i}_1 + 3\mathbf{i}_2, \quad \mathbf{e}_3 = 2\mathbf{i}_3,$$

where $\mathbf{i}_1, \mathbf{i}_2, \mathbf{i}_3$ is an orthonormal basis, find the covariant and contravariant components of the vector joining the origin to the point (1, 1, 1).

24. Express the scalar triple product $(\mathbf{A} \times \mathbf{B}) \cdot \mathbf{C}$ in terms of the covariant and contravariant components of the vectors $\mathbf{A}$, $\mathbf{B}$ and $\mathbf{C}$.

25. A scalar function $f(\mathbf{A})$ of a vector argument $\mathbf{A}$ is said to be *linear* if

$$f(c\mathbf{A}) = cf(\mathbf{A}), \qquad f(\mathbf{A} + \mathbf{B}) = f(\mathbf{A}) + f(\mathbf{B}),$$

where $\mathbf{A}$ and $\mathbf{B}$ are arbitrary vectors and c is an arbitrary scalar. Prove that the most general function of this kind is of the form

$$f(\mathbf{A}) = \alpha A_1 + \beta A_2 + \gamma A_3,$$

where A_1, A_2, A_3 are the components of $\mathbf{A}$ and α, β, γ are scalars.

26. Given a tetrahedron T, let $\mathbf{S}_i$ be the vector perpendicular to the ith face of T $(i = 1, 2, 3, 4)$, of magnitude equal to the area of the face. Prove that

$$\mathbf{S}_1 + \mathbf{S}_2 + \mathbf{S}_3 + \mathbf{S}_4 = 0.$$

Hint. Represent the vectors $\mathbf{S}_i$ as vector products.

2

THE TENSOR CONCEPT

2.1. Preliminary Remarks

It will be recalled from Sec. 1.1 that a scalar is a quantity whose specification (in any coordinate system) requires just one number. On the other hand, a vector (originally defined as a directed line segment) is a quantity whose specification requires three numbers, namely its components with respect to some basis (see Sec. 1.6). Scalars and vectors are both special cases of a more general object called a *tensor of order n*, whose specification in any given coordinate system requires 3^n numbers, again called the *components* of the tensor.[1] In fact, scalars are tensors of order 0, with $3^0 = 1$ components, and vectors are tensors of order 1, with $3^1 = 3$ components.

Of course, a tensor of order n is much more than just a set of 3^n numbers. The key property of a tensor, which will emerge in the course of this chapter, is the *transformation law* of its components, i.e., the way its components in one coordinate system are related to its components in another coordinate system. The precise form of this transformation law is a consequence of the physical or geometric meaning of the tensor.

Suppose we have a law involving components $a, b, c, \ldots$ of various physical quantities with respect to some three-dimensional coordinate system K. Then it is an empirical fact that the law has the same form when written in terms of the components $a', b', c', \ldots$ of the same quantities with respect to another coordinate system K' which is shifted relative to K ("space is homogeneous") or rotated with respect to K ("space is isotropic"). In other words, properly formulated physical laws are "invariant under shifts and rotations" (see Sec. 2.7).

[1] In writing 3^n we have in mind three-dimensional tensors. More generally, an m-dimensional tensor of order n has m^n components (see Exercise 17, p. 133).

Remark. Properly formulated physical laws must also be independent of the choice of units. In dimensional analysis, for example, one often uses the fact that the ratio of two values of the same physical quantity cannot depend on the units of measurement.

2.2. Zeroth-Order Tensors (Scalars)

We begin by sharpening the definition of a scalar given in Sec. 1.1: By a scalar is meant a quantity uniquely specified in any coordinate system by a single real number (the "component" or "value" of the scalar) which is *invariant under changes of the coordinate system*, i.e., which does not change when the coordinate system is changed. Thus if φ is the value of a scalar in one coordinate system and φ' its value in another coordinate system, then $\varphi' = \varphi$.

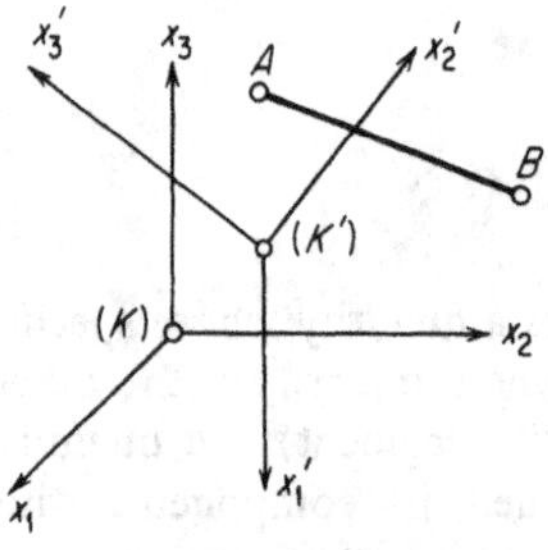

FIG. 2.1. Illustrating the invariance of the distance between two points A and B.

Example. Let A and B be two points with coordinates x_i^A, x_i^B in one rectangular coordinate system K and coordinates $x_i'^A$, $x_i'^B$ in another rectangular coordinate system K' (see Fig. 2.1), and let Δs be the distance beween A and B, i.e., the length of the line segment AB.[2] Then Δs is a scalar, i.e., its value $\Delta s'$ in the system K' equals its value Δs in the system K.

This geometrically obvious fact can also be verified by direct calculation. Let

$$\Delta x_i = x_i^B - x_i^A, \qquad \Delta x_i' = x_i'^B - x_i'^A \qquad (i = 1, 2, 3),$$

and let the transformation from K to K' be given by

$$x_i' = \alpha_{i'k} x_k + x_{0i}$$

as in Prob. 1, p. 38, where x_{i0}' are the coordinates of the old origin in the new system and $\alpha_{i'k} = \cos(x_i', x_k)$ is the cosine of the angle between the ith axis of the new system and the kth axis of the old system. Then

$$\begin{aligned}\Delta x_i' = x_i'^B - x_i'^A &= \alpha_{i'k} x_k^B + x_{0i} - \alpha_{i'k} x_k^A - x_{0i} \\ &= \alpha_{i'k}(x_k^B - x_k^A) = \alpha_{i'k}\,\Delta x_k.\end{aligned}$$

---

[2] Here and henceforth we assume that the units of measurement are the same in all coordinate systems under consideration.

By the Pythagorean theorem,

$$(\Delta s')^2 = \sum_{i=1}^{3} (\Delta x_i')^2,$$

and hence

$$(\Delta s')^2 = \sum_{i=1}^{3} \alpha_{i'k} \Delta x_k \alpha_{i'l} \Delta x_l = \alpha_{i'k}\alpha_{i'l} \Delta x_k \Delta x_l.$$

Therefore, because of the orthogonality condition

$$\alpha_{i'k}\alpha_{i'l} = \delta_{kl}$$

[recall (1.62)],

$$(\Delta s')^2 = \delta_{kl} \Delta x_k \Delta x_l = \sum_{k=1}^{3} (\Delta x_k)^2.$$

But the quantity on the right is just $(\Delta s)^2$, and hence

$$\Delta s' = \Delta s,$$

as asserted.

Remark. For the time being, we confine ourselves to the case of rectangular coordinate systems. More general coordinate systems will be considered in Secs. 2.8 and 2.9. Tensors written in rectangular coordinate systems are often called *Cartesian tensors.*

2.3. First-Order Tensors (Vectors)

As already noted in Sec. 2.1, three numbers (scalars) are required to specify a vector (like displacement, acceleration, force, etc.) rather than a single number as in the case of a scalar (like density, pressure, temperature, etc.). However, a vector is much more than just a set of $3^1 = 3$ scalars. For example, the state of an ideal gas is uniquely specified by two numbers (the density and temperature, say), but these numbers are invariant under changes of coordinate system, being scalars. On the other hand, a displacement in the plane is also determined by two numbers (the differences between the abscissas and ordinates of the initial and final points), but under changes of coordinates these numbers transform according to a definite law. More generally, the three components of a vector in space transform according to a definite law guaranteeing that the new components always determine the same vector.

To find this law, let Δx_i and $\Delta x_i'$ be the differences between the rectangular coordinates of two points A and B in two rectangular coordinate systems K and K'. Then, it will be recalled from Sec. 2.2 that

$$\Delta x_i' = \alpha_{i'k} \Delta x_k, \tag{2.1}$$

where $\alpha_{i'k}$ is the cosine of the angle between the ith axis of K' and the kth axis of K. Similarly, suppose a vector **A** has components A_i in K and components A'_i in K' (see Fig. 2.2). Then, being a directed line segment, **A** is completely determined by its initial and final points, and hence its components A_i, A'_i must transform just like the coordinate differences Δx_i, $\Delta x'_i$. This leads to the following definition: By a vector is meant a quantity uniquely

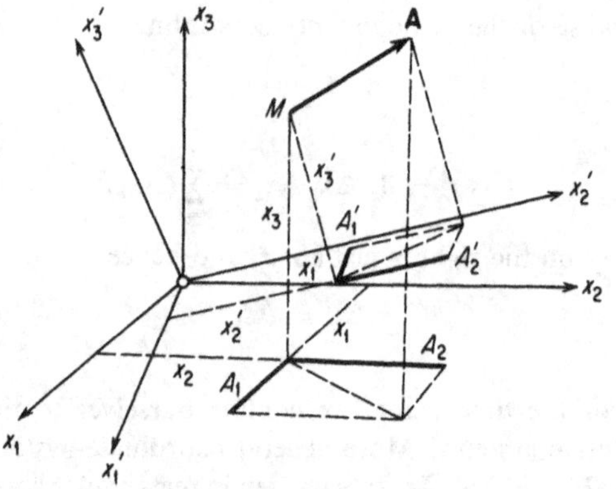

FIG. 2.2. Illustrating the change of components of a vector under changes of the coordinate system.

specified in any coordinate system by three real numbers (the *components* of the vector) which transform under changes of the coordinate system according to the law

$$A'_i = \alpha_{i'k} A_k, \tag{2.2}$$

where A_k, A'_i are the components of the vector in the old and new coordinate systems K and K', respectively, and $\alpha_{i'k}$ is the cosine of the angle between the ith axis of K' and the kth axis of K.

Remark 1. Given the components of a vector in one (rectangular) coordinate system, we can use (2.2) to determine its components in any other coordinate system. In particular, a vector vanishing in one coordinate system vanishes in any other coordinate system.[3]

Remark 2. The definition (2.2) of a vector is equivalent to the definition of a vector as a directed line segment. However, (2.2) has the advantage of being easily generalized to the case of tensors of arbitrary order.

2.3.1. Examples. Suppose the coordinates x_i of a point P in a system K are functions of time:

$$x_i = x_i(t).$$

[3] Note that a vector vanishes if and only if all its components vanish.

Then the displacement of P in time Δt is given by

$$x_i(t + \Delta t) - x_i(t). \tag{2.3}$$

The quantities (2.3) determine a vector (the displacement vector), since they become

$$x_i'(t' + \Delta t') - x_i'(t')$$

in a new coordinate system K'. But using (2.1) and the fact that $t' = t$, $\Delta t' = \Delta t$ (time is a scalar), we find that

$$x_i'(t' + \Delta t') - x_i'(t') = \alpha_{i'k}[x_k(t + \Delta t) - x_k(t)],$$

i.e., the quantities (2.3) transform like a vector. Similarly, the ratios

$$\frac{x_i(t + \Delta t) - x_i(t)}{\Delta t}$$

determine a vector (the average velocity of P during the interval from t to $t + \Delta t$). Moreover, the limits

$$v_i = \lim_{\Delta t \to 0} \frac{x_i(t + \Delta t) - x_i(t)}{\Delta t},$$

provided they exist, also determine a vector (the instantaneous velocity of P at time t). In fact,

$$v_i' = \lim_{\Delta t' \to 0} \frac{x_i'(t' + \Delta t') - x_i'(t')}{\Delta t'}$$

$$= \alpha_{i'k} \lim_{\Delta t \to 0} \frac{x_k(t + \Delta t) - x_k(t)}{\Delta t} = \alpha_{i'k} v_k$$

(the quantities $\alpha_{i'k}$ are independent of t), which is again the transformation law of a vector.

In just the same way, it is easily verified that the limits

$$a_i = \lim_{\Delta t \to 0} \frac{v_i(t + \Delta t) - v_i(t)}{\Delta t}$$

determine a vector (the instantaneous acceleration of P at time t). Therefore, since Newton's second law

$$F_i = ma_i$$

holds in all coordinate systems, the force F_i must also be a vector.

2.4. Second-Order Tensors

Second-order tensors are next in order of complexity after scalars and vectors. By a *second-order tensor* is meant a quantity uniquely specified by nine real numbers (the *components* of the tensor) which transform under changes of the coordinate system according to the law

$$A'_{ik} = \alpha_{i'l}\alpha_{k'm}A_{lm}, \tag{2.4}$$

where A_{lm}, A'_{ik} are the components of the tensor in the old and new coordinate systems K and K', respectively, and $\alpha_{i'l}$ is the cosine of the angle between the ith axis of K' and the lth axis of K (similarly for $\alpha_{k'm}$). Note the sense in which (2.4) generalizes (2.2).

Remark 1. Given the components of a second-order tensor in one (rectangular) coordinate system, we can use (2.4) to determine its components in any other coordinate system. In particular, if all the components of a tensor vanish in one coordinate system, they also vanish in any other coordinate system.

Remark 2. The components of a second-order tensor are often written as a *matrix:*

$$\|A_{ik}\| = \begin{Vmatrix} A_{11} & A_{12} & A_{13} \\ A_{21} & A_{22} & A_{23} \\ A_{31} & A_{32} & A_{33} \end{Vmatrix}.$$

2.4.1. Examples. We now give some examples illustrating the meaning of the transformation law (2.4).

Example 1. Given two vectors **A** and **B**, there are nine products of a component of **A** with a component of **B**:

$$A_iB_k \qquad (i, k = 1, 2, 3).$$

Suppose we transform to a new coordinate system K', in which **A** and **B** have components A'_i and B'_k. Then, by (2.2),

$$A'_i = \alpha_{i'l}A_l, \qquad B'_k = \alpha_{k'm}B_m,$$

and hence

$$A'_iB'_k = \alpha_{i'l}\alpha_{k'm}A_lB_m. \tag{2.5}$$

Comparing (2.5) and (2.4), we find that A_iB_k is a second-order tensor.

Example 2. The equation of a quadric surface (e.g., an ellipsoid) centered at the origin is of the form

$$A_{ik}x_ix_k = 1 \tag{2.6}$$

in the old system K and of the form

$$A'_{ik}x'_ix'_k = 1 \tag{2.7}$$

in a new system K' with the same origin as K. To find the relation between the old coefficients A_{ik} and the new coefficients A'_{ik}, we note that

$$x'_i = \alpha_{i'l}x_l, \qquad x'_k = \alpha_{k'm}x_m$$

and

$$x_l = \alpha_{i'l}x'_i, \qquad x_m = \alpha_{k'm}x'_k, \tag{2.8}$$

as in formula (1.59), p. 39. Substituting (2.8) into (2.6), with i, k replaced by l, m, we obtain

$$A_{lm}x_lx_m = A_{lm}\alpha_{i'l}x'_i\alpha_{k'm}x'_k = 1. \tag{2.9}$$

Then comparing (2.9) and (2.7), we get

$$A'_{ik} = \alpha_{i'l}\alpha_{k'm}A_{lm},$$

which is identical with (2.4). It follows that A_{ik} is a second-order tensor.

Example 3. A vector function $\mathbf{B} = f(\mathbf{A})$ of a vector argument $\mathbf{A}$ is said to be *linear* if each component B_i is linear. It follows from Exercise 25, p. 58 that the most general function of this kind is of the form

$$B_1 = a_{11}A_1 + a_{12}A_2 + a_{13}A_3,$$

$$B_2 = a_{21}A_1 + a_{22}A_2 + a_{23}A_3,$$

$$B_3 = a_{31}A_1 + a_{32}A_2 + a_{33}A_3,$$

or more concisely

$$B_i = a_{ik}A_k. \tag{2.10}$$

The coefficients a_{ik} transform like a second-order tensor. In fact, in a new coordinate system, (2.10) is replaced by

$$B'_i = a'_{ik}A'_k. \tag{2.11}$$

Multiplying (2.10) by $\alpha_{l'i}$ and summing over i, we get

$$\alpha_{l'i}B_i = a_{ik}\alpha_{l'i}A_k, \tag{2.12}$$

where, by (2.2), the quantity on the left is the lth component of the vector $\mathbf{B}$ in the system K'. Thus (2.12) becomes

$$B'_l = a_{ik}\alpha_{l'i}A_k.$$

But

$$A_k = \alpha_{m'k}A'_m$$

(why?), and hence

$$B'_l = \alpha_{l'i}\alpha_{m'k}a_{ik}A'_m$$

or equivalently

$$B'_i = \alpha_{i'l}\alpha_{k'm}a_{lm}A'_k. \tag{2.13}$$

Since the vector **B** is arbitrary, a comparison of (2.13) and (2.10) shows

$$a'_{ik} = \alpha_{i'l}\alpha_{k'm}a_{lm},$$

i.e., the coefficients a_{ik} have the same transformation law as (2.4) and hence define a second-order tensor.

2.4.2. The stress tensor. The state of stress of an elastic medium is specified once we know the force acting on an arbitrary element of area $d\sigma$

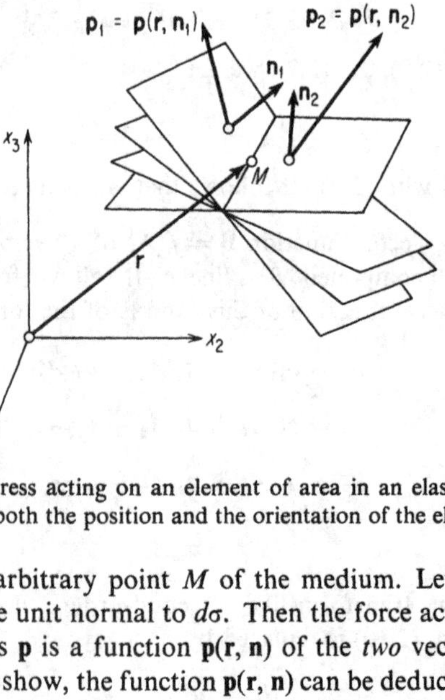

FIG. 2.3. The stress acting on an element of area in an elastic medium depends on both the position and the orientation of the element.

passing through an arbitrary point M of the medium. Let **r** be the radius vector of M and **n** the unit normal to $d\sigma$. Then the force acting on $d\sigma$ equals **p** $d\sigma$, where the stress **p** is a function **p**(**r**, **n**) of the *two* vectors **r** and **n** (see Fig. 2.3). As we now show, the function **p**(**r**, **n**) can be deduced from a certain second-order tensor called the *stress tensor*, which depends on **r** *but not on* **n**.

To this end, we construct an elementary tetrahedron about the point M with its edges directed along the axes of a rectangular coordinate system K (see Fig. 2.4). Let $d\sigma_1$, $d\sigma_2$, $d\sigma_3$ denote the areas of the faces perpendicular to the axes x_1, x_2, x_3, and let $d\sigma_n$ denote the area of the inclined face with unit exterior normal **n**. Moreover, let $\mathbf{p}_{-1}\, d\sigma_1$, $\mathbf{p}_{-2}\, d\sigma_2$, $\mathbf{p}_{-3}\, d\sigma_3$ and $\mathbf{p}_n\, d\sigma_n$ be the forces exerted by the rest of the medium on the areas $d\sigma_1$, $d\sigma_2$, $d\sigma_3$ and $d\sigma_n$, respectively. Here the minus

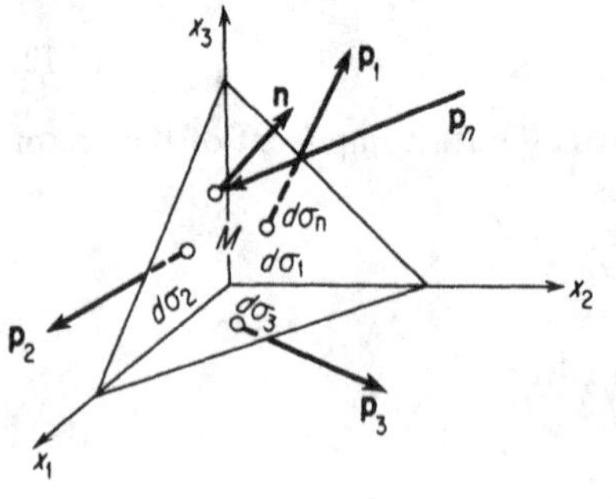

FIG. 2.4. Stresses on the faces of a tetrahedron.

signs mean that the stresses $\mathbf{p}_{-1}$, $\mathbf{p}_{-2}$ and $\mathbf{p}_{-3}$ act on the outside faces of the tetrahedron, whose exterior normals point in the directions opposite to those of the coordinate axes. By the law of action and reaction, the forces $\mathbf{p}_1\, d\sigma_1$, $\mathbf{p}_2\, d\sigma_2$, $\mathbf{p}_3\, d\sigma_3$ acting on the inside faces of the tetrahedron are equal and opposite to those acting on the outside faces, and hence

$$\mathbf{p}_1 = -\mathbf{p}_{-1}, \quad \mathbf{p}_2 = -\mathbf{p}_{-2}, \quad \mathbf{p}_3 = -\mathbf{p}_{-3}.$$

Now let $\mathbf{a}$ be the acceleration of the center of mass of the tetrahedron, and let $\mathbf{f}$ be the body force per unit mass. Then, by Newton's second law,

$$\begin{aligned}\mathbf{a}\, dm &= \mathbf{f}\, dm + \mathbf{p}_n\, d\sigma_n + \mathbf{p}_{-1}\, d\sigma_1 + \mathbf{p}_{-2}\, d\sigma_2 + \mathbf{p}_{-3}\, d\sigma_3 \\ &= \mathbf{f}\, dm + \mathbf{p}_n\, d\sigma_n - \mathbf{p}_1\, d\sigma_1 - \mathbf{p}_2\, d\sigma_2 - \mathbf{p}_3\, d\sigma_3,\end{aligned}$$

where dm is the mass of the tetrahedron.[4] In the limit as the tetrahedron shrinks to the point M, we find that

$$\mathbf{p}_n\, d\sigma_n = \mathbf{p}_1\, d\sigma_1 + \mathbf{p}_2\, d\sigma_2 + \mathbf{p}_3\, d\sigma_3 = \sum_{i=1}^{3} \mathbf{p}_i\, d\sigma_i,$$

since the terms containing dm are proportional to the volume of the tetrahedron and hence are of a higher order of smallness compared to the terms containing elements of area. Therefore, since

$$d\sigma_i = d\sigma_n \cos(\mathbf{n}, x_i) = n_i\, d\sigma_n,$$

the stress on an element of area with unit normal $\mathbf{n}$ is given by

$$\mathbf{p}_n = \sum_{i=1}^{3} \mathbf{p}_i n_i \equiv \mathbf{p}_i n_i.$$

Projecting $\mathbf{p}_n$ onto the axes of the system K, we obtain

$$p_{nk} = p_{ik} n_i, \qquad (2.14)$$

where p_{ik} $(i, k = 1, 2, 3)$ is a set of nine normal $(i = k)$ and tangential $(i \neq k)$ stresses acting on three orthogonal elements of area at the point M (see Fig. 2.5). Although themselves independent of the orientation $\mathbf{n}$ of the area on which the stress acts, these nine quantities, which depend only on the point M, allow us to determine $\mathbf{p}_n$ for arbitrary

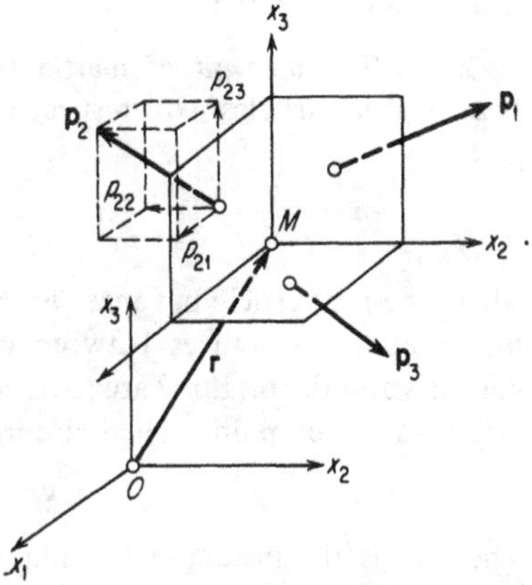

FIG. 2.5. The stress tensor as a set of three stress vectors $\mathbf{p}_1$, $\mathbf{p}_2$, $\mathbf{p}_3$ acting on three orthogonal elements of area. The projections of these vectors onto the coordinate axes give the nine components of the stress tensor.

[4] No summation over n is implied in the expression $\mathbf{p}_n\, d\sigma_n$.

n. Thus the physical quantity with components p_{ik}, called the *stress tensor*, uniquely specifies the state of stress at every point of the elastic medium.

It only remains to verify the tensor character of p_{ik}. Since the definition of p_{ik} involves no restriction on the normal **n**, we can assume without loss of generality that the ith axis of the new coordinate system K' is directed along **n**, so that

$$\mathbf{n} = \mathbf{i}'_i$$

(K and K' have orthonormal bases $\mathbf{i}_1, \mathbf{i}_2, \mathbf{i}_3$ and $\mathbf{i}'_1, \mathbf{i}'_2, \mathbf{i}'_3$, respectively). Then projecting **n** onto the lth axis of K gives

$$n_l = \mathbf{n} \cdot \mathbf{i}_l = \mathbf{i}'_i \cdot \mathbf{i}_l = \alpha_{i'l},$$

where $\alpha_{i'l}$ is the cosine of the angle between the ith axis of K' and the lth axis of K, and hence

$$\mathbf{p}_n \equiv \mathbf{p}'_i = \mathbf{p}_l n_l = \alpha_{i'l}\mathbf{p}_l = \alpha_{i'l}\mathbf{i}_m p_{lm}.$$

Finally, projecting $\mathbf{p}'_i$ onto the kth axis of K', we obtain

$$\mathbf{p}'_i \cdot \mathbf{i}'_k = \alpha_{i'l}(\mathbf{i}_m \cdot \mathbf{i}'_k)p_{lm}$$

or

$$p'_{ik} = \alpha_{i'l}\alpha_{k'm}p_{lm}. \tag{2.15}$$

Comparing (2.15), and (2.4), we find that p_{ik} transforms like a second-order tensor, as anticipated.

2.4.3. The moment of inertia tensor. The angular momentum **L** of a system of n particles with respect to the origin of a coordinate system K is given by

$$\mathbf{L} = \sum_{j=1}^{n} m_j(\mathbf{r}_j \times \mathbf{v}_j), \tag{2.16}$$

where the jth particle has mass m_j, radius vector $\mathbf{r}_j$ and velocity $\mathbf{v}_j$. Suppose that both the distances between particles and the distances between the particles and the origin O are fixed, so that the system is a rigid body with the origin as a fixed point. Then, according to Prob. 11, p. 45,

$$\mathbf{v}_j = \boldsymbol{\omega} \times \mathbf{r}_j,$$

where $\boldsymbol{\omega}$ is the instantaneous angular velocity of the system. Substituting this expression for $\mathbf{v}_j$ into (2.16), we obtain

$$\mathbf{L} = \sum_{j=1}^{n} m_j[\mathbf{r}_j \times (\boldsymbol{\omega} \times \mathbf{r}_j)] = \sum_{j=1}^{n} m_j[\boldsymbol{\omega}(\mathbf{r}_j \cdot \mathbf{r}_j) - \mathbf{r}_j(\boldsymbol{\omega} \cdot \mathbf{r}_j)],$$

where we have used formula (1.30). Projecting **L** onto the axes of K, we obtain

$$L_i = \sum_{j=1}^{n} m_j(\omega_i x_l^{(j)} x_l^{(j)} - x_i^{(j)} \omega_k x_k^{(j)}) \qquad \text{(summation over } k \text{ and } l),$$

where the jth particle has coordinates $x_i^{(j)}$. Writing $\omega_i = \delta_{ik}\omega_k$ (δ_{ik} is the Kronecker delta defined on p. 39), we have

$$L_i = \omega_k \sum_{j=1}^{n} m_j(\delta_{ik} x_l^{(j)} x_l^{(j)} - x_i^{(j)} x_k^{(j)}) = \omega_k I_{ik},$$

where

$$I_{ik} = \sum_{j=1}^{n} m_j(\delta_{ik} x_l^{(j)} x_l^{(j)} - x_i^{(j)} x_k^{(j)}). \tag{2.17}$$

Suppose the system has moments of inertia $I_{x_1}, I_{x_2}, I_{x_3}$ about the coordinate axes and products of inertia $I_{x_1x_2}, I_{x_1x_3}, I_{x_2x_3}$. Then these moments are related to the nine quantities I_{ik} ($i, k = 1, 2, 3$) as follows:

$$I_{11} = \sum_{j=1}^{n} m_j[(x_2^{(j)})^2 + (x_3^{(j)})^2] = I_{x_1},$$

$$I_{22} = \sum_{j=1}^{n} m_j[(x_1^{(j)})^2 + (x_3^{(j)})^2] = I_{x_2},$$

$$I_{33} = \sum_{j=1}^{n} m_j[(x_1^{(j)})^2 + (x_2^{(j)})^2] = I_{x_3},$$

$$I_{12} = I_{21} = -\sum_{j=1}^{n} m_j x_1^{(j)} x_2^{(j)} = -I_{x_1x_2},$$

$$I_{13} = I_{31} = -\sum_{j=1}^{n} m_j x_1^{(j)} x_3^{(j)} = -I_{x_1x_3},$$

$$I_{23} = I_{32} = -\sum_{j=1}^{n} m_j x_2^{(j)} x_3^{(j)} = -I_{x_2x_3}.$$

The quantities I_{ik} form a second-order tensor, called the *moment of inertia tensor* (about the origin O). To see this, we note that (2.17) becomes

$$I'_{ik} = \sum_{j=1}^{n} m_j(\delta'_{ik} x_l'^{(j)} x_l'^{(j)} - x_i'^{(j)} x_k'^{(j)})$$

in another rectangular coordinate system K' with the same origin. But

$$x_l'^{(j)} x_l'^{(j)} = x_l^{(j)} x_l^{(j)}$$

(invariance of the scalar product), while

$$x_i'^{(j)} x_k'^{(j)} = \alpha_{i'r}\alpha_{k's} x_r^{(j)} x_s^{(j)}.$$

Moreover

$$\alpha_{l'i}\alpha_{l'k} = \delta_{ik},$$
$$\alpha_{i'r}\alpha_{k'r} = \delta'_{ik}$$

in terms of the Kronecker delta [recall (1.62)], and hence

$$\delta'_{ik} = \alpha_{i'r}\alpha_{k's}\delta_{rs} \tag{2.18}$$

since

$$\alpha_{k'r} = \alpha_{k's}\delta_{rs}.$$

Therefore

$$
\begin{aligned}
I'_{ik} &= \sum_{j=1}^{n} m_j(\alpha_{i'r}\alpha_{k's}\delta_{rs}x_l^{(j)}x_l^{(j)} - \alpha_{i'r}\alpha_{k's}x_r^{(j)}x_s^{(j)}) \\
&= \alpha_{i'r}\alpha_{k's}\sum_{j=1}^{n} m_j(\delta_{rs}x_l^{(j)}x_l^{(j)} - x_r^{(j)}x_s^{(j)}) \\
&= \alpha_{i'r}\alpha_{k's}I_{rs},
\end{aligned}
$$

which is the same transformation law as (2.4).

It follows from (2.18) that the Kronecker delta δ_{ik} is also a second-order tensor. The tensor δ_{ik} is often called the *unit tensor*, since its matrix is of the form

$$
\begin{Vmatrix} 1 & 0 & 0 \\ 0 & 1 & 0 \\ 0 & 0 & 1 \end{Vmatrix}
$$

in every rectangular coordinate system.

2.4.4. The deformation tensor. Given any two neighboring points A and B of an elastic body, suppose a deformation carries A and B into new positions A' and B'. Let A and B have radius vectors $\mathbf{r}$ and $\mathbf{r} + \Delta\mathbf{r}$, while A' and

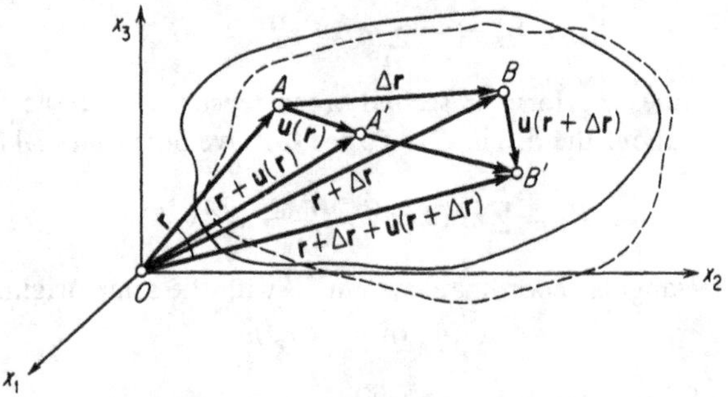

FIG. 2.6. Deformation of an elastic body.

B' have radius vectors $\mathbf{r} + \mathbf{u}(\mathbf{r})$ and $\mathbf{r} + \Delta\mathbf{r} + \mathbf{u}(\mathbf{r} + \Delta\mathbf{r})$, as shown in Fig. 2.6, where the vectors $\mathbf{u}(\mathbf{r})$ and $\mathbf{u}(\mathbf{r} + \Delta\mathbf{r})$ describe the displacement of the points A and B as a result of the deformation.

As shown in the figure, the relative position of the points is given by $\Delta\mathbf{r}$ before the deformation and by

$$
\Delta\mathbf{r}' = A'B' = \Delta\mathbf{r} + \mathbf{u}(\mathbf{r} + \Delta\mathbf{r}) - \mathbf{u}(\mathbf{r})
$$

after the deformation. The change in magnitude of $\Delta\mathbf{r}$ by can be found calculating the quantity $(\Delta r')^2 - (\Delta r)^2$. Suppose $\mathbf{u}$ is a sufficiently smooth function of position, with components $u_i = u_i(x_1, x_2, x_3)$. Then

$$\Delta x_i' = \Delta x_i + u_i(x_1 + \Delta x_1, x_2 + \Delta x_2, x_3 + \Delta x_3) - u_i(x_1, x_2, x_3),$$

or

$$\Delta x_i' = \Delta x_i + \frac{\partial u_i}{\partial x_k} \Delta x_k \tag{2.19}$$

after using Taylor's theorem and neglecting terms of the second order of smallness. Noting that

$$\Delta x_i' \Delta x_i' = (\Delta r')^2, \qquad \Delta x_i \Delta x_i = (\Delta r)^2,$$

we square (2.19), obtaining

$$\begin{aligned}(\Delta r')^2 - (\Delta r)^2 &= 2\frac{\partial u_i}{\partial x_k}\Delta x_i \Delta x_k + \frac{\partial u_i}{\partial x_k}\frac{\partial u_i}{\partial x_l}\Delta x_k \Delta x_l \\ &= \left(\frac{\partial u_i}{\partial x_k} + \frac{\partial u_k}{\partial x_i} + \frac{\partial u_l}{\partial x_i}\frac{\partial u_l}{\partial x_k}\right)\Delta x_i \Delta x_k \\ &= 2u_{ik}\Delta x_i \Delta x_k,\end{aligned}$$

where

$$u_{ik} = \frac{1}{2}\left(\frac{\partial u_i}{\partial x_k} + \frac{\partial u_k}{\partial x_i} + \frac{\partial u_l}{\partial x_i}\frac{\partial u_l}{\partial x_k}\right). \tag{2.20}$$

Thus the change in the distance between any two points of the elastic body is uniquely determined by the quantity u_{ik}, called the *deformation tensor.*

To verify the tensor character of u_{ik}, we transform to a new coordinate coordinate system K', obtaining

$$u_{ik}' = \frac{1}{2}\left(\frac{\partial u_i'}{\partial x_k'} + \frac{\partial u_k'}{\partial x_i'} + \frac{\partial u_l'}{\partial x_i'}\frac{\partial u_l'}{\partial x_k'}\right).$$

It follows from the formula

$$x_i = \alpha_{k'i}x_k' + x_{0i}'$$

describing the transformation from K' back to the old coordinate system K (x_{0i}' are the coordinates of the new origin in the old system) that

$$\frac{\partial x_i}{\partial x_k'} = \alpha_{k'i}. \tag{2.21}$$

Repeatedly using (2.2), (2.21) and the chain rule for partial differentiation, we find that

$$u'_{ik} = \frac{1}{2}\left[\frac{\partial}{\partial x_n}(\alpha_{i'm}u_m)\frac{\partial x_n}{\partial x'_k} + \frac{\partial}{\partial x_m}(\alpha_{k'n}u_n)\frac{\partial x_m}{\partial x'_i}\right.$$

$$\left. + \frac{\partial}{\partial x_m}(\alpha_{l'r}u_r)\frac{\partial x_m}{\partial x'_i}\frac{\partial}{\partial x_n}(\alpha_{l's}u_s)\frac{\partial x_n}{\partial x'_k}\right]$$

$$= \frac{1}{2}\left(\alpha_{i'm}\frac{\partial u_m}{\partial x_n}\alpha_{k'n} + \alpha_{k'n}\frac{\partial u_n}{\partial x_m}\alpha_{i'm} + \alpha_{l'r}\frac{\partial u_r}{\partial x_m}\alpha_{i'm}\alpha_{l's}\frac{\partial u_s}{\partial x_n}\alpha_{k'n}\right)$$

$$= \alpha_{i'm}\alpha_{k'n}\frac{1}{2}\left(\frac{\partial u_m}{\partial x_n} + \frac{\partial u_n}{\partial x_m} + \delta_{rs}\frac{\partial u_r}{\partial x_m}\frac{\partial u_s}{\partial x_n}\right)$$

$$= \alpha_{i'm}\alpha_{k'n}\frac{1}{2}\left(\frac{\partial u_m}{\partial x_n} + \frac{\partial u_n}{\partial x_m} + \frac{\partial u_r}{\partial x_m}\frac{\partial u_r}{\partial x_n}\right),$$

i.e.,

$$u'_{ik} = \alpha_{i'm}\alpha_{k'n}u_{mn}.$$

It follows that u_{ik} is a second-order tensor [recall (2.4)]. In the linear theory of elasticity, the term $(\partial u_l/\partial x_i)(\partial u_l/\partial x_k)$ is dropped in (2.20), leaving just

$$u_{ik} = \frac{1}{2}\left(\frac{\partial u_i}{\partial x_k} + \frac{\partial u_k}{\partial x_i}\right).$$

2.4.5. The rate of deformation tensor. Suppose the velocity at the point M of a moving fluid (liquid or gas) is $\mathbf{v} = \mathbf{v}(M)$. Then it can be shown that the motion of any element of the fluid is the sum of a "quasi-rigid" motion in which the element acts like part of a rigid body and a deformational motion.[5] The latter is determined by the *rate of deformation tensor*

$$v_{ik} = \frac{1}{2}\left(\frac{\partial v_i}{\partial x_k} + \frac{\partial v_k}{\partial x_i}\right), \tag{2.22}$$

in the sense that the part of the velocity of the element of the fluid at M relative to the point O which is due entirely to the element's ability to undergo deformation is equal to

$$v_i^{\text{def}}(M) = v_{ik}(O)\,\Delta x_k, \tag{2.23}$$

where

$$\Delta r = \sqrt{(\Delta x_1)^2 + (\Delta x_2)^2 + (\Delta x_3)^2}$$

[5] See e.g., R. Aris, *Vectors, Tensors and the Basic Equations of Fluid Mechanics*, Prentice-Hall, Inc., Englewood Cliffs, N.J. (1962), p. 89.

is the distance between the points M and O, $\mathbf{v}^{\mathrm{def}}(M)$ is the deformational velocity of the element at the point M, and $v_{ik}(O)$ is the value of the tensor (2.22) at the point O. The tensor character of v_{ik} is verified in just the same way as that of the deformation tensor u_{ik} in Sec. 2.4.4.

We now investigate the physical meaning of the components of the tensor (2.22). Consider two points of a fluid element which have positions A, B before the deformation and positions A', B' after the deformation (see Fig. 2.7). Moreover, let

$$\overrightarrow{OA} = \Delta\mathbf{r}, \quad \overrightarrow{OA'} = \Delta\mathbf{r}',$$

$$\overrightarrow{OB} = \Delta\mathbf{R}, \quad \overrightarrow{OB'} = \Delta\mathbf{R}'.$$

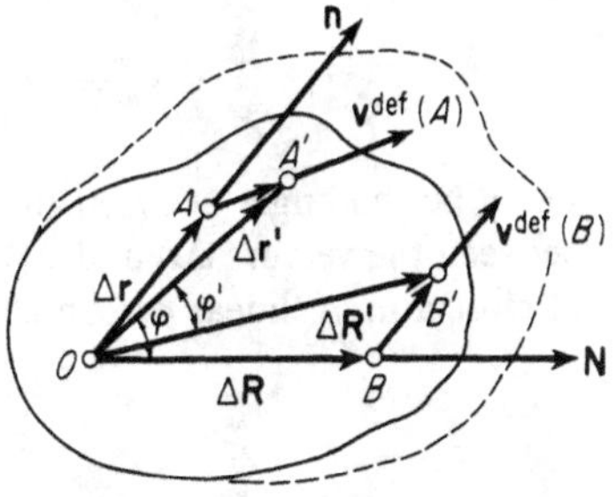

FIG. 2.7. Deformation of a fluid element.

Then the deformation of the given fluid element is characterized by the change in length of the vectors $\Delta\mathbf{r}$, $\Delta\mathbf{R}$ and the change in angle between them.

Due to the deformation of the fluid, the displacements of the points A and B in time Δt are given by

$$\overrightarrow{AA'} = \mathbf{v}^{\mathrm{def}}(A)\,\Delta t,$$

$$\overrightarrow{BB'} = \mathbf{v}^{\mathrm{def}}(B)\,\Delta t.$$

Hence, consulting the figure, we find that

$$\Delta\mathbf{r}' = \Delta\mathbf{r} + \overrightarrow{AA'} = \Delta\mathbf{r} + \mathbf{v}^{\mathrm{def}}(A)\,\Delta t,$$

$$\Delta\mathbf{R}' = \Delta\mathbf{R} + \overrightarrow{BB'} = \Delta\mathbf{R} + \mathbf{v}^{\mathrm{def}}(B)\,\Delta t.$$

Taking components and using (2.23) to express the deformation velocities in terms of the rate of deformation tensor, we get

$$\Delta x_i' = \Delta x_i + v_{ik}\,\Delta x_k\,\Delta t,$$

$$\Delta X_i' = \Delta X_i + v_{ik}\,\Delta X_k\,\Delta t,$$

where Δx_i, $\Delta x_i'$, ΔX_i, $\Delta X_i'$ are the ith components of the vectors $\Delta\mathbf{r}$, $\Delta\mathbf{r}'$, $\Delta\mathbf{R}$, $\Delta\mathbf{R}'$, respectively, and the derivatives in v_{ik} [recall (2.22)] are evaluated at the point O. Then, using the symmetry of v_{ik} ($v_{ik} = v_{ki}$) and neglecting terms of order higher than one in Δt, we form the scalar product

$$\Delta\mathbf{r}' \cdot \Delta\mathbf{R}' = \Delta x_i'\,\Delta X_i' = \Delta x_i\,\Delta X_i + 2v_{ik}\,\Delta x_i\,\Delta X_k\,\Delta t. \tag{2.24}$$

Now let $\mathbf{n}$ and $\mathbf{N}$ be the unit vectors corresponding to $\Delta\mathbf{r}$ and $\Delta\mathbf{R}$, so that

$$\mathbf{n} = \frac{\Delta\mathbf{r}}{|\Delta\mathbf{r}|} = \frac{\Delta\mathbf{r}}{\Delta r}, \qquad \mathbf{N} = \frac{\Delta\mathbf{R}}{|\Delta\mathbf{R}|} = \frac{\Delta\mathbf{R}}{\Delta R}.$$

By definition, the relative elongation of the fluid element during the time Δt in the direction $\mathbf{n}$ is

$$\varepsilon_n = \frac{\Delta r' - \Delta r}{\Delta r} = \frac{\Delta r'}{\Delta r} - 1.$$

Hence the rate of relative elongation in the same direction is

$$\dot{\varepsilon}_n = \lim_{\Delta t \to 0} \frac{\varepsilon_n}{\Delta t} = \lim_{\Delta t \to 0} \frac{\Delta r' - \Delta r}{\Delta r \, \Delta t}.$$

Let φ be the angle between the vectors $\Delta\mathbf{r}$ and $\Delta\mathbf{R}$, and let φ' be the angle between the vectors $\Delta\mathbf{r}'$ and $\Delta\mathbf{R}'$. Then the quantity $\dot{\varepsilon}_n$ characterizes the rate of elongation ("linear deformation") of the fluid element, while the quantity,

$$\dot{\gamma} = \lim_{\Delta t \to 0} \frac{\varphi' - \varphi}{\Delta t}$$

characterizes its rate of "angular deformation." Dividing (2.24) by $\Delta r \, \Delta R$, we find that

$$(1 + \varepsilon_n)(1 + \varepsilon_N) \cos \varphi' = \cos \varphi + 2v_{ik} n_i N_k \, \Delta t, \tag{2.25}$$

since

$$\frac{\Delta x_i' \, \Delta X_i'}{\Delta r \, \Delta R} = \frac{\Delta r' \, \Delta R' \cos \varphi'}{\Delta r \, \Delta R},$$

$$n_i = \frac{\Delta x_i}{\Delta r}, \qquad N_k = \frac{\Delta X_k}{\Delta R}.$$

Dropping the term of order two in ε_n and ε_N, we can write (2.25) as

$$(1 + \varepsilon_n + \varepsilon_N) \cos \varphi' = \cos \varphi + 2v_{ik} n_i N_k \, \Delta t. \tag{2.26}$$

To interpret (2.26), we examine two special cases:

1) Suppose the points A and B coincide and lie on the x_1-axis before the deformation. Then

$$\varphi = 0, \qquad \varphi' \approx 0,$$

$$\Delta\mathbf{r} = \Delta\mathbf{R} = \Delta x_1 \mathbf{i}_1,$$

$$\mathbf{n} = \mathbf{N} = \mathbf{i}_1, \qquad (n_1 = N_1 = 1, \qquad n_2 = n_3 = N_2 = N_3 = 0),$$

so that (2.26) becomes

$$1 + \varepsilon_1 + \varepsilon_2 = 1 + 2v_{11} \, \Delta t,$$

which implies

$$\dot{\varepsilon}_1 = v_{11} = \frac{\partial v_1}{\partial x_1}.$$

Similarly, choosing the points A and B before the deformation on the x_2-axis, and then on the x_3-axis, we obtain

$$\dot{\varepsilon}_2 = v_{22} = \frac{\partial v_2}{\partial x_2},$$

$$\dot{\varepsilon}_3 = v_{33} = \frac{\partial v_3}{\partial x_3}.$$

Thus the diagonal components v_{11}, v_{22}, v_{33} of the rate of deformation tensor are the rates of relative elongation of a fluid element along the three coordinate axes.

2) Suppose that before the deformation, the point A lies on the x_1-axis, while the point B lies on the x_2-axis. Then

$$\varphi = \frac{\pi}{2}, \quad \mathbf{n} = \mathbf{i}_1, \quad \mathbf{N} = \mathbf{i}_2, \quad \Delta\mathbf{r} \perp \Delta\mathbf{R},$$

so that (2.26) becomes

$$(1 + \varepsilon_1 + \varepsilon_2) \cos \varphi' = 2v_{12}\,\Delta t. \tag{2.27}$$

Let γ_{ij} be the decrease in angle (in time Δt) between two line segments "embedded" in the fluid, directed along the x_i and x_j-axes before the deformation. Then if $i = 1, j = 2$,

$$\gamma_{12} = \frac{\pi}{2} - \varphi' \approx \sin\left(\frac{\pi}{2} - \varphi'\right) = \cos \varphi'. \tag{2.28}$$

Substituting (2.28) into (2.27) and dropping small terms of the second order, we obtain

$$\gamma_{12} = 2v_{12}\,\Delta t.$$

It follows that

$$\dot{\gamma}_{12} = \lim_{\Delta t \to 0} \frac{\dfrac{\pi}{2} - \varphi'}{\Delta t} = 2v_{12} = \frac{\partial v_1}{\partial x_2} + \frac{\partial v_2}{\partial x_1},$$

and similarly

$$\dot{\gamma}_{13} = \frac{\partial v_1}{\partial x_3} + \frac{\partial v_3}{\partial x_1} = 2v_{13},$$

$$\dot{\gamma}_{23} = \frac{\partial v_2}{\partial x_3} + \frac{\partial v_3}{\partial x_2} = 2v_{23}.$$

Thus the nondiagonal components $v_{12} = v_{21}$, $v_{13} = v_{31}$, $v_{23} = v_{32}$ of the rate of deformation tensor equal half the rates of angular deformation of a fluid element, i.e., half the rates of change of the angles between line segments directed along appropriate axes before the deformation.

2.5. Higher-Order Tensors

In Secs. 2.2-2.4 we found that tensors of orders 0, 1, 2 have the transformation laws

$$\varphi' = \varphi, \quad A'_i = \alpha_{i'l}A_l, \quad A'_{ik} = \alpha_{i'l}\alpha_{k'm}A_{lm}, \tag{2.29}$$

respectively. The first formula does not involve the coefficients $\alpha_{i'k}$ at all, the right-hand side of the second formula is a homogeneous linear form in the $\alpha_{i'k}$, while the right-hand side of the third formula is a homogeneous quadratic form in the $\alpha_{i'k}$.[6] The natural generalization of (2.29) is the following: By a *tensor of order n* is meant a quantity uniquely specified by 3^n real numbers (the *components* of the tensor) which transform under changes of the coordinate system according to the law

$$A'_{i_1 i_2 \ldots i_n} = \alpha_{i_1'k_1}\alpha_{i_2'k_2} \cdots \alpha_{i_n'k_n}A_{k_1 k_2 \ldots k_n}, \tag{2.30}$$

where $A_{k_1 k_2 \ldots k_n}$, $A'_{i_1 i_2 \ldots i_n}$ are the components of the tensor in the old and new coordinate systems K and K', respectively, and $\alpha_{i_1'k_1}$ is the cosine of the angle between the i_1st axis of K' and the k_1st axis of K (similarly for $\alpha_{i_2'k_2}, \ldots, \alpha_{i_n'k_n}$). The right-hand side of (2.30) is a homogeneous form of degree n in the quantities $\alpha_{i_1'k_1}, \alpha_{i_2'k_2}, \ldots, \alpha_{i_n'k_n}$.

Remark. Given the components of a tensor of order n in one (rectangular) coordinate system, we can use (2.30) to determine its components in any other coordinate system. In particular, if all the components of a tensor vanish in one coordinate system, they also vanish in any other coordinate system.

Example 1. If **A**, **B** and **C** are three vectors, the $3^3 = 27$ quantities

$$D_{ikl} = A_i B_k C_l$$

form a tensor of order 3 (why?).

---

[6] Given n variables $x_1, x_2, \ldots, x_n$, the expression

$$c_i x_i = \sum_{i=1}^{n} c_i x_i$$

is called a *homogeneous linear form* (in the x_i), the expression

$$c_{ik}x_i x_k = \sum_{i=1}^{n}\sum_{k=1}^{n} c_{ik}x_i x_k$$

is called a *homogeneous quadratic form*, the expression

$$c_{ikl}x_i x_k x_l = \sum_{i=1}^{n}\sum_{k=1}^{n}\sum_{l=1}^{n} c_{ikl}x_i x_k x_l$$

is called a *homogeneous form of degree* 3, and so on.

Example 2. Suppose one second-order tensor A_{ik} is a linear function of another second-order tensor B_{ik},[7] so that

$$A_{ik} = \lambda_{iklm} B_{lm},$$

where λ_{iklm} is a set of $3^4 = 81$ coefficients. Just as in Example 3, p. 65, it can be shown that λ_{iklm} is a tensor of order 4, i.e.,

$$\lambda'_{iklm} = \alpha_{i'n}\alpha_{k'p}\alpha_{l'r}\alpha_{m's}\lambda_{nprs}$$

(the details are left as an exercise).

2.6. Transformation of Tensors under Rotations about a Coordinate Axis

One is often interested in coordinate transformations of a special kind, i.e., rotations about one of the coordinate axes which for simplicity we take to be the z-axis (here we write x, y, z instead of x_1, x_2, x_3). Let φ be the angle between the new x'-axis and the old x-axis (see Fig. 2.8). Then the general formula

$$x'_i = \alpha_{i'k}x_k + x_{0i}$$

[recall (1.59)] reduces to

$$\begin{aligned} x' &= x\cos\varphi + y\sin\varphi, \\ y' &= -x\sin\varphi + y\cos\varphi, \qquad (2.31) \\ z' &= z, \end{aligned}$$

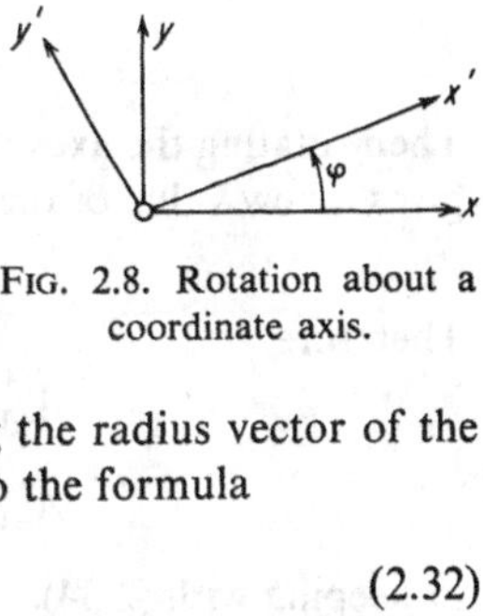

FIG. 2.8. Rotation about a coordinate axis.

Therefore the complex number $x + iy$ determining the radius vector of the point (x, y) in the xy-plane transforms according to the formula

$$x' + iy' = (x + iy)e^{-i\varphi}. \qquad (2.32)$$

In fact, writing out (2.32) in full, we have

$$\begin{aligned} x' + iy' &= (x + iy)(\cos\varphi - i\sin\varphi) \\ &= (x\cos\varphi + y\sin\varphi) + i(-x\sin\varphi + y\cos\varphi), \end{aligned}$$

and then taking real and imaginary parts, we get the first two of the formulas (2.31). On the other hand, the complex conjugate $x - iy$ transforms as follows:

$$x' - iy' = (x - iy)e^{i\varphi}.$$

---

[7] In the sense of Exercise 25, p. 58 and Example 3, p. 65.

Similarly, given a vector **A** with components A_x, A_y, A_z, the quantities $A_x + iA_y$, $A_x - iA_y$, A_z transform under the rotation (2.31) according to the formulas[8]

$$A'_x + iA'_y = (A_x + iA_y)e^{-i\varphi},$$

$$A'_x - iA'_y = (A_x - iA_y)e^{i\varphi}, \tag{2.33}$$

$$A'_z = A_z.$$

Introducing the notation

$$A_{+1} = A_x + iA_y,$$

$$A_{-1} = A_x - iA_y,$$

$$A_0 = A_z,$$

we can write (2.33) as

$$A'_\alpha = A_\alpha e^{-i\alpha\varphi} \qquad \text{(no summation over } \alpha), \tag{2.34}$$

where α takes the values $-1, 0, +1$.

The formulas (2.34) can be derived in another way by introducing the modulus and argument of the complex number A_{+1}:

$$A_{+1} = Ae^{i\gamma}, \qquad A_{-1} = Ae^{-i\gamma}.$$

Then rotating the axes through the angle φ in the counterclockwise direction gives a new value of the argument equal to

$$\gamma' = \gamma - \varphi.$$

Therefore

$$A'_{+1} = Ae^{i\gamma'} = Ae^{i\gamma}e^{-i\varphi} = A_{+1}e^{-i\varphi},$$

$$A'_{-1} = Ae^{-i\gamma'} = Ae^{-i\gamma}e^{i\varphi} = A_{-1}e^{i\varphi},$$

in keeping with (2.34).

Given two vectors **A** and **B** with components A_x, A_y, A_z and B_x, B_y, B_z respectively, we have

$$A'_\alpha = A_\alpha e^{-i\alpha}, \qquad B'_\alpha = B_\alpha e^{-i\alpha} \qquad \text{(no summation over } \alpha).$$

Moreover,

$$A'_\alpha B'_\beta = A_\alpha B_\beta e^{-i(\alpha+\beta)} \qquad \text{(no summation over } \alpha \text{ and } \beta), \tag{2.35}$$

where α and β separately take the values $-1, 0, +1$. The relation between the

---

[8] If $A_z = 0$, the vector **A** lies entirely in the xy-plane and its components transform according to the first two of the equations (2.33).

quantities $A_\alpha B_\beta$ and the components of the vectors **A** and **B** is given by

$$\begin{aligned}
A_0B_0 &= A_zB_z,\\
A_{+1}B_{+1} &= (A_x + iA_y)(B_x + iB_y) = A_xB_x - A_yB_y + i(A_xB_y + A_yB_x),\\
A_{-1}B_{-1} &= (A_x - iA_y)(B_x - iB_y) = A_xB_x - A_yB_y - i(A_xB_y + A_yB_x),\\
A_0B_{+1} &= A_z(B_x + iB_y) = A_zB_x + iA_zB_y,\\
A_0B_{-1} &= A_z(B_x - iB_y) = A_zB_x - iA_zB_y,\\
A_{+1}B_0 &= (A_x + iA_y)B_z = A_xB_z + iA_yB_z,\\
A_{-1}B_0 &= (A_x - iA_y)B_z = A_xB_z - iA_yB_z,\\
A_{+1}B_{-1} &= (A_x + iA_y)(B_x - iB_y) = A_xB_x + A_yB_y - i(A_xB_y - A_yB_x),\\
A_{-1}B_{+1} &= (A_x - iA_y)(B_x + iB_y) = A_xB_x + A_yB_y + i(A_xB_y - A_yB_x).
\end{aligned} \tag{2.36}$$

Formula (2.35) tells how certain combinations of the components of **A** and **B** transform under rotations about the z-axis. For example, setting $\alpha = +1$, $\beta = -1$, we obtain

$$A'_{+1}B'_{-1} = A_{+1}B_{-1}$$

or

$$A'_xB'_x + A'_yB'_y - i(A'_xB'_y - A'_yB'_x) = A_xB_x + A_yB_y - i(A_xB_y - A_yB_x).$$

Taking real and imaginary parts then gives

$$\begin{aligned}
A'_xB'_x + A'_yB'_y &= A_xB_x + A_yB_y,\\
A'_xB'_y - A'_yB'_x &= A_xB_y - A_yB_x.
\end{aligned}$$

The first formula together with

$$A'_zB'_z = A_zB_z,$$

obtained from (2.35) by setting $\alpha = \beta = 0$, expresses the fact that the scalar product $\mathbf{A} \cdot \mathbf{B}$ is *invariant* (i.e., does not change) under rotations of the given type (or, for that matter, under any coordinate transformation). The second formula expresses the invariance of the z-component of the vector product $\mathbf{A} \times \mathbf{B}$. This could have been predicted from the fact that the z-axis is fixed under a rotation about the z-axis. The same conclusions are obtained if we consider the case $\alpha = -1$, $\beta = +1$ instead.

Next we turn our attention to the behavior under rotations about the z-axis of a symmetric second-order tensor with components[9]

$$p_{xx},\quad p_{yy},\quad p_{zz},\quad p_{xy} = p_{yx},\quad p_{xz} = p_{zx},\quad p_{yz} = p_{zy}.$$

[9] If $p_{xz} = p_{yz} = p_{zz} = 0$, the tensor is said to be *two-dimensional*. For example, the state of stress of an elastic body is completely determined by a two-dimensional tensor if the stresses are independent of z and have no z-components.

Guided by (2.36), we form the combinations

$$\begin{aligned} p_{0,0} &= p_{zz}, \\ p_{+1,+1} &= p_{xx} - p_{yy} + 2ip_{xy}, \\ p_{-1,-1} &= p_{xx} - p_{yy} - 2ip_{xy}, \\ p_{0,+1} = p_{+1,0} &= p_{xz} + ip_{yz}, \\ p_{0,-1} = p_{-1,0} &= p_{xz} - ip_{yz}, \\ p_{+1,-1} = p_{-1,+1} &= p_{xx} + p_{yy}. \end{aligned} \tag{2.37}$$

Then the natural generalization of (2.35) is

$$p'_{\alpha\beta} = p_{\alpha\beta}e^{-i(\alpha+\beta)} \qquad \text{(no summation over } \alpha \text{ and } \beta\text{)}, \tag{2.38}$$

where α and β separately take the values $-1, 0, +1$. It follows from (2.38) that

$$\begin{aligned} p'_{0,0} &= p_{0,0}, \\ p'_{+1,+1} &= p_{+1,+1}e^{-2i\varphi}, \\ p'_{-1,-1} &= p_{-1,-1}e^{2i\varphi}, \\ p'_{0,+1} &= p_{0,+1}e^{-i\varphi}, \\ p'_{0,-1} &= p_{0,-1}e^{i\varphi}, \\ p'_{+1,-1} &= p_{+1,-1}, \end{aligned}$$

or, in terms of the components $p_{xx}, p_{yy}, p_{zz}, p_{xy}, p_{xz}, p_{yz}$,

$$\begin{aligned} p'_{zz} &= p_{zz}, \\ p'_{xx} - p'_{yy} + 2ip'_{xy} &= (p_{xx} - p_{yy} + 2ip_{xy})e^{-2i\varphi}, \\ p'_{xx} - p'_{yy} - 2ip'_{xy} &= (p_{xx} - p_{yy} - 2ip_{xy})e^{2i\varphi}, \\ p'_{xz} + ip'_{yz} &= (p_{xz} + ip_{yz})e^{-i\varphi}, \\ p'_{xz} - ip'_{yz} &= (p_{xz} - ip_{yz})e^{i\varphi}, \\ p'_{xx} + p'_{yy} &= p_{xx} + p_{yy}. \end{aligned} \tag{2.39}$$

The second, third and last equations (which do not involve the subscript z) are important in two-dimensional elasticity theory. Substituting

$$e^{-2i\varphi} = \cos 2\varphi - i \sin 2\varphi$$

into the second equation and then taking real and imaginary parts, we find that

$$\begin{aligned} p'_{xx} - p'_{yy} &= (p_{xx} - p_{yy}) \cos 2\varphi + 2p_{xy} \sin 2\varphi, \\ 2p'_{xy} &= 2p_{xy} \cos 2\varphi - (p_{xx} - p_{yy}) \sin 2\varphi. \end{aligned} \tag{2.40}$$

It follows from (2.40) and the last of the equations (2.39) that

$$p'_{xx} = \frac{p_{xx} + p_{yy}}{2} + \frac{p_{xx} - p_{yy}}{2} \cos 2\varphi + p_{xy} \sin 2\varphi,$$

$$p'_{yy} = \frac{p_{xx} + p_{yy}}{2} - \frac{p_{xx} - p_{yy}}{2} \cos 2\varphi - p_{xy} \sin 2\varphi,$$

$$p'_{xy} = -\frac{p_{xx} + p_{yy}}{2} \sin 2\varphi + p_{xy} \cos 2\varphi.$$

2.7. Invariance of Tensor Equations

Let

$$F(\varphi, \psi, \ldots, a_i, b_i, \ldots, c_{ik}, d_{ik}, \ldots) = 0 \qquad (2.41)$$

be an equation involving scalars $\varphi, \psi, \ldots$, vectors $a_i, b_i, \ldots$, second-order tensors $c_{ik}, d_{ik}, \ldots$, etc., written in a rectangular coordinate system K. Suppose we shift and rotate K, thereby obtaining a new rectangular coordinate system K'. Then (2.41) is replaced by

$$G(\varphi', \psi', a'_i, b'_i, \ldots, c'_{ik}, d'_{ik}, \ldots) = 0, \qquad (2.42)$$

where all components of scalars, vectors and tensors are now written in K'. In general, (2.41) and (2.42) are not of the same form, i.e., $F \not\equiv G$. However, supppose (2.42) has the same form as (2.41), so that (2.42) becomes

$$F(\varphi', \psi', a'_i, b'_i, \ldots, c'_{ik}, d'_{ik}, \ldots) = 0.$$

Then the equation (2.41) is said to be *invariant* under the transformation from K to K'. All properly formulated physical laws must be invariant under shifts and rotations, since real space is homogeneous and isotropic (see p. 59). In particular, all tensors appearing as (additive) terms in an equation expressing a physical law must be of the same order. Another requirement satisfied by properly formulated physical laws has already been noted in the remark on p. 60.

Example 1. According to Prob. 7, p. 42, the equation of a straight line is

$$x_k - a_k - \lambda e_k = 0 \qquad (2.43)$$

in a rectangular coordinate system K. Multiplying by $\alpha_{i'k}$ and summing over k, we obtain

$$\alpha_{i'k}x_k - \alpha_{i'k}a_k - \lambda\alpha_{i'k}e_k = 0.$$

But x_k, a_k, e_k are components of vectors, and hence

$$x'_i - a'_i - \lambda e'_i = 0, \qquad (2.44)$$

where x'_i, a'_i, e'_i are the components of the same vectors in the new system K'. Comparison of (2.43) and (2.44) shows that (2.43) is invariant under the transformation from K to K'.

Example 2. Newton's second law has the same form in any two rectangular coordinate systems, or for that matter, in any two inertial systems (moving with respect to each other with constant translational velocity). To see this, we first write Newton's law

$$F_k = \frac{d}{dt}(mv_k)$$

in one system K. We then multiply by $\alpha_{i'k}$ and sum over k, obtaining

$$\alpha_{i'k}F_k = \frac{d}{dt}(m\alpha_{i'k}v_k) = \frac{d}{dt}m(\alpha_{i'k}v_k + v_{0i}),$$

where the v_{0i} are components of a constant vector. But then

$$F'_i = \frac{d}{dt'}(m'v'_i) = \frac{d}{dt}(mv'_i),$$

since $m' = m$, $t' = t$ and

$$F'_i = \alpha_{i'k}F_k, \quad v'_i = \alpha_{i'k}v_k + v_{0i}$$

(the term v_{0i} describes the constant translational velocity).

2.8. Curvilinear Coordinates

Any three numbers q^1, q^2, q^3 uniquely specifying the position of a point M in space are called (*generalized*) *coordinates* of M.[10]

Example 1. In a rectangular coordinate system with origin O, $q^1 = x_1$, $q^2 = x_2$, $q^3 = x_3$ are the (signed) distances between M and three perpendicular planes going through O.

Example 2. Given an underlying system of rectangular coordinates x_1, x_2, x_3 with origin O, let $q^1 = R$ be the distance between M and the x_3-axis, let $q^2 = \varphi$ be the angle between the half-plane determined by the x_3-axis and the positive x_1-axis and the half-plane determined by the x_3-axis and the point M, and let $q^3 = z$ be the distance between M and the x_1x_2-plane [see Fig. 2.9(a)]. Then R, φ and z are called the *cylindrical coordinates* of M. They are related to the rectangular coordinates x_1, x_2, x_3 by the formulas

$$q^1 = R = \sqrt{x_1^2 + x_2^2}, \quad \tan q^2 = \tan\varphi = \frac{x_2}{x_1}, \quad q^3 = z = x_3 \tag{2.45}$$

$$x_1 = R\cos\varphi, \quad x_2 = R\sin\varphi, \quad x_3 = z.$$

[10] The numbers 1, 2, 3 appearing in q^1, q^2, q^3 are superscripts, not exponents.

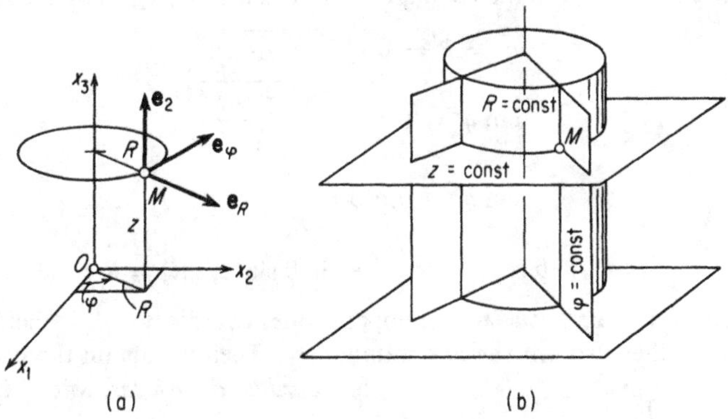

FIG. 2.9. Illustrating cylindrical coordinates.

Example 3. Given an underlying system of rectangular coordinates x_1, x_2, x_3 with origin O, let $q^1 = R$ be the distance between M and O, let $q^2 = \theta$ be the angle between the positive x_3-axis and the vector OM, and let $q^3 = \varphi$ be the angle between the half-plane determined by the x_3-axis and the positive x_1-axis and the half-plane determined by the x_3-axis and the point M [see Fig. 2.10 (a)]. Then R, φ and θ are called the *spherical coordinates* of

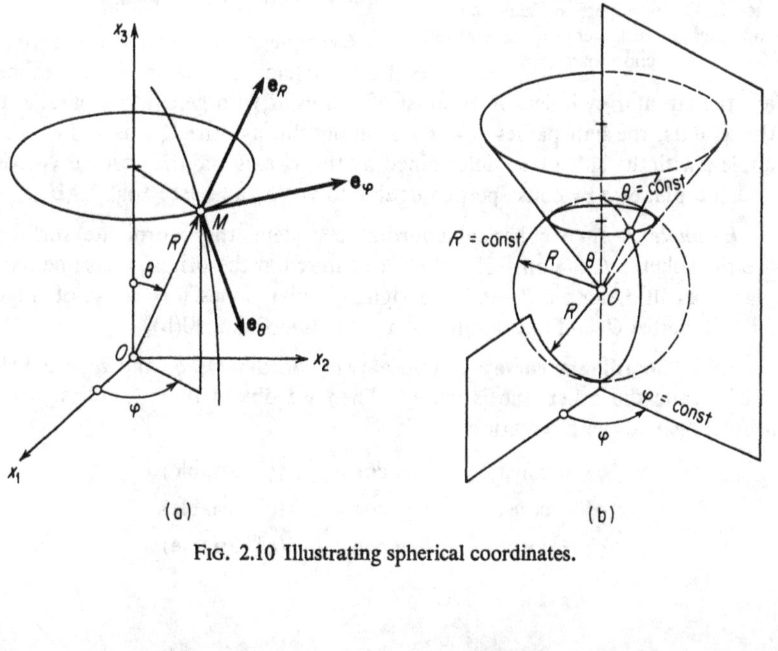

FIG. 2.10 Illustrating spherical coordinates.

M. They are related to the rectangular coordinates x_1, x_2, x_3 by the formulas

$$\begin{aligned} q^1 &= R = \sqrt{x_1^2 + x_2^2 + x_3^2}, \\ \tan q^2 &= \tan\theta = \frac{\sqrt{x_1^2 + x_2^2}}{x_3}, \\ \tan q^3 &= \tan\varphi = \frac{x_2}{x_1}, \end{aligned} \tag{2.46}$$

$$x_1 = R\sin\theta\cos\varphi, \quad x_2 = R\sin\theta\sin\varphi, \quad x_3 = R\cos\theta.$$

2.8.1. Coordinate surfaces. Suppose one coordinate q^i is held fixed, while the other two are varied continuously. Then we obtain three families of *coordinate surfaces*, with equations

$$\begin{aligned} q^1 &= \text{const} \quad (q^2, q^3 \text{ variable}), \\ q^2 &= \text{const} \quad (q^3, q^1 \text{ variable}), \\ q^3 &= \text{const} \quad (q^1, q^2 \text{ variable}). \end{aligned}$$

It will always be assumed that a unique surface of each family goes through any given point M (see Fig. 2.11), in keeping with the hypothesis that M is uniquely determined by its coordinates q^1, q^2, q^3.

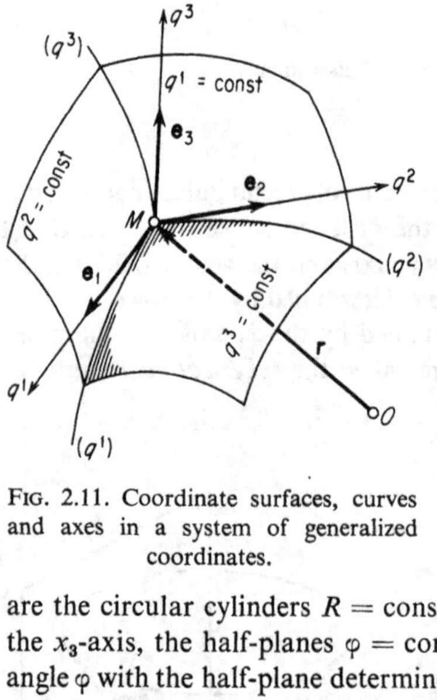

FIG. 2.11. Coordinate surfaces, curves and axes in a system of generalized coordinates.

Example 1. In a rectangular coordinate system, the coordinate surfaces are three perpendicular planes.

Example 2. In a cylindrical coordinate system, the coordinate surfaces are the circular cylinders $R = \text{const}$ of radius R with generators parallel to the x_3-axis, the half-planes $\varphi = \text{const}$ going through the x_3-axis and making angle φ with the half-plane determined by the x_3-axis and the positive x_1-axis, and the planes $z = \text{const}$ perpendicular to the x_3-axis [see Fig. 2.9(b)].

Example 3. In a spherical coordinate system, the coordinate surfaces are the spheres $R = \text{const}$ of radius R centered at the origin, the same half-planes as in Example 2, and the right circular cones $\theta = \text{const}$ of angle 2θ with vertex O and axis along the x_3-axis [see Fig. 2.10(b)].

2.8.2. Coordinate curves. Suppose two coordinates q^i and q^j are held fixed, while the other one is varied. Then we obtain three families of *coordinate curves*, with equations

$$\begin{aligned} q^2 &= \text{const}, \quad q^3 = \text{const} \quad (q^1 \text{ variable}), \\ q^3 &= \text{const}, \quad q^1 = \text{const} \quad (q^2 \text{ variable}), \\ q^1 &= \text{const}, \quad q^2 = \text{const} \quad (q^3 \text{ variable}). \end{aligned}$$

These are the curves denoted by (q^1), (q^2), (q^3), respectively, in Fig. 2.11. The coordinate curve (q^i) is clearly the intersection of two coordinate surfaces $q^j = \text{const}$ and $q^k = \text{const}$, where j and k are the values of 1, 2, 3 other than i. By the *positive direction* along the coordinate curve (q^i), we mean the direction in which a variable point of the curve moves as q^i is increased.

Example 1. In a rectangular coordinate system, the coordinate curves are perpendicular straight lines.

Example 2. In a cylindrical coordinate system, the coordinate curves are the straight lines

$$R = \text{const}, \qquad \varphi = \text{const},$$

the straight lines

$$\varphi = \text{const}, \qquad z = \text{const},$$

and the circles

$$R = \text{const}, \qquad z = \text{const}.$$

Example 3. In a spherical coordinate system, the coordinate curves are the circles

$$R = \text{const}, \qquad \varphi = \text{const},$$

the circles

$$R = \text{const}, \qquad \theta = \text{const},$$

and the straight lines

$$\varphi = \text{const}, \qquad \theta = \text{const}.$$

2.8.3. Bases and coordinate axes. By a *basis* of a system of generalized coordinates q^1, q^2, q^3, we mean any set of vectors $\mathbf{e}_1$, $\mathbf{e}_2$, $\mathbf{e}_3$ of fixed length pointing in the positive directions of the coordinate curves. The vectors $\mathbf{e}_1$, $\mathbf{e}_2$, $\mathbf{e}_3$ themselves are called *basis vectors*. Thus $\mathbf{e}_1$ is tangent to the coordinate curve (q^1) and points in the direction of increasing q^1. The basis $\mathbf{e}_1$, $\mathbf{e}_2$, $\mathbf{e}_3$ is said to be *local*, since in general it varies from point to point, as shown in Fig. 2.12(a). It should be noted that in general the basis vectors are neither perpendicular nor of unit length.

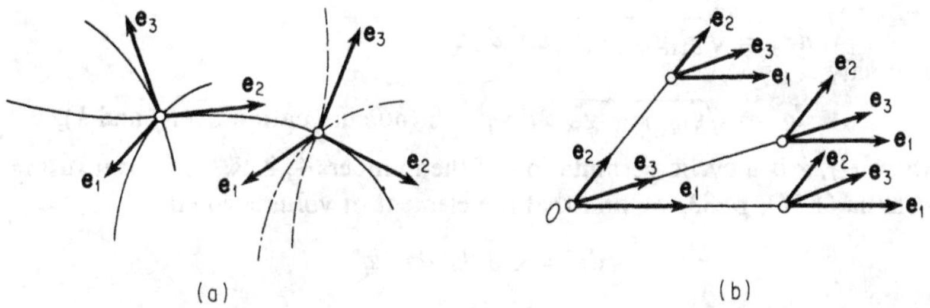

FIG. 2.12. Local bases. The local basis varies from point to point except in the case of rectangular or oblique coordinates.

The tangent to the coordinate curve (q^i) is called the q^i*-axis* $(i = 1, 2, 3)$, and the positive direction along the q^i-axis is the direction of the basis vector $\mathbf{e}_i$.

In rectangular and oblique coordinates, and only in such coordinate systems, the basis vectors do *not* vary from point to point [see Fig. 2.12(b)]. As on p. 11, a coordinate system is said to be *curvilinear* (as opposed to rectangular or oblique) if its coordinate curves are not straight lines. Thus the basis of a curvilinear coordinate system is "local" in the full sense of the word, since it does vary from point to point.

Coordinate systems whose basis vectors intersect at right angles are called *orthogonal* systems, just as on p. 11. Thus rectangular, cylindrical and spherical coordinates are all orthogonal systems. As already noted, the coordinates most widely used in the applications are orthogonal.

2.8.4. Arc length. Metric coefficients. The fundamental geometric characteristics of a space "arithmetized" by introducing the generalized coordinates q^1, q^2, q^3 are given by its metric, i.e., by the expression for the square of the element of arc length

$$(ds)^2 = g_{ik}\, dq^i\, dq^k$$

(recall Sec. 1.6.5). Here $g_{ik} = \mathbf{e}_i \cdot \mathbf{e}_k$ is the metric tensor (see Sec. 2.9.2). The element of arc length along the coordinate curve (q^i) is

$$ds_i = |\mathbf{e}_i|\, dq^i = \sqrt{g_{ii}}\, dq^i \qquad \text{(no summation over } i\text{)},$$

while the element of area in the coordinate surface $q^1 = \text{const}$ is

$$\begin{aligned} d\sigma_1 &= |\mathbf{e}_2 \times \mathbf{e}_3|\, dq^2\, dq^3 \\ &= \sqrt{(\mathbf{e}_2 \times \mathbf{e}_3) \cdot (\mathbf{e}_2 \times \mathbf{e}_3)}\, dq^2\, dq^3 \\ &= \sqrt{(\mathbf{e}_2 \cdot \mathbf{e}_2)(\mathbf{e}_3 \cdot \mathbf{e}_3) - (\mathbf{e}_2 \cdot \mathbf{e}_3)(\mathbf{e}_2 \cdot \mathbf{e}_3)}\, dq^2\, dq^3 \\ &= \sqrt{g_{22}g_{33} - g_{23}^2}\, dq^2\, dq^3. \end{aligned}$$

Similarly, we have

$$\begin{aligned} d\sigma_2 &= \sqrt{g_{33}g_{11} - g_{13}^2}\, dq^3\, dq^1, \\ d\sigma_3 &= \sqrt{g_{11}g_{22} - g_{12}^2}\, dq^1\, dq^2, \end{aligned}$$

or briefly

$$d\sigma_i = \sqrt{g_{jj}g_{kk} - g_{jk}^2}\, dq^j\, dq^k \qquad \text{(no summation over } j \text{ and } k\text{)},$$

where i, j, k is a cyclic permutation of the numbers 1, 2, 3. Moreover, using formula (1.54), p. 33, we find that the element of volume equals

$$dV = \sqrt{G}\, dq^1\, dq^2\, dq^3,$$

where

$$G = \det \|g_{ik}\|.$$

The basic quantities describing an *orthogonal* coordinate system are the metric coefficients h_1, h_2, h_3. These are defined as

$$h_1 = \sqrt{g_{11}}, \quad h_2 = \sqrt{g_{22}}, \quad h_3 = \sqrt{g_{33}},$$

and satisfy the formula

$$(ds)^2 = (h_1\, dq^1)^2 + (h_2\, dq^2)^2 + (h_3\, dq^3)^2$$

(recall Sec. 1.6.6). In this case, we have

$$ds_i = h_i\, dq^i \qquad \text{(no summation over } i\text{)},$$

$$d\sigma_i = h_j h_k\, dq^j\, dq^k \qquad \text{(no summation over } j \text{ and } k\text{)},$$

$$dV = h_1 h_2 h_3\, dq^1\, dq^2\, dq^3.$$

Example 1. In rectangular coordinates,

$$(ds)^2 = (dx_1)^2 + (dx_2)^2 + (dx_3)^2$$

and hence

$$h_1 = 1, \quad h_2 = 1, \quad h_3 = 1.$$

Example 2. In cylindrical coordinates,

$$(ds)^2 = (dR)^2 + (R\, d\varphi)^2 + (dz)^2$$

and hence

$$h_1 = 1, \quad h_2 = R, \quad h_3 = 1. \tag{2.47}$$

Example 3. In spherical coordinates,

$$(ds)^2 = (dR)^2 + (R\, d\theta)^2 + (R \sin\theta\, d\varphi)^2$$

and hence

$$h_1 = 1, \quad h_2 = R, \quad h_3 = R\sin\theta. \tag{2.48}$$

Suppose the relation between a system of generalized coordinates q^1, q^2, q^3 and an underlying system of rectangular coordinates x_1, x_2, x_3 is given by the formulas

$$q^1 = q^1(x_1, x_2, x_3), \quad q^2 = q^2(x_1, x_2, x_3), \quad q^3 = q^3(x_1, x_2, x_3) \tag{2.49}$$

and

$$x_1 = x_1(q^1, q^2, q^3), \quad x_2 = x_2(q^1, q^2, q^3), \quad x_3 = x_3(q^1, q^2, q^3), \tag{2.50}$$

where the Jacobians

$$J = \det\left\|\frac{\partial q^i}{\partial x_k}\right\|, \qquad J^{-1} = \det\left\|\frac{\partial x_i}{\partial q^k}\right\|$$

are neither zero nor infinite. For example, (2.49) and (2.50) take the form (2.45) with $J = 1/R$ in the case of cylindrical coordinates, and the form (2.46) with $J = 1/R \sin\theta$ in the case of spherical coordinates. Writing (2.50) more concisely as $\mathbf{r} = \mathbf{r}(q^1, q^2, q^3)$, where $\mathbf{r} = x_1\mathbf{i}_1 + x_2\mathbf{i}_2 + x_3\mathbf{i}_3$ is the radius vector of an arbitrary point M, we find that

$$d\mathbf{r} = \frac{\partial \mathbf{r}}{\partial q^1} dq^1 + \frac{\partial \mathbf{r}}{\partial q^2} dq^2 + \frac{\partial \mathbf{r}}{\partial q^3} dq^3 = \frac{\partial \mathbf{r}}{\partial q^i} dq^i,$$

where in using the summation convention in the last expression, we agree to regard a superscript in the denominator as a subscript in the numerator and vice versa. It follows that

$$(ds)^2 = d\mathbf{r} \cdot d\mathbf{r} = \frac{\partial \mathbf{r}}{\partial q^i} \cdot \frac{\partial \mathbf{r}}{\partial q^k} dq^i \, dq^k.$$

Therefore the vectors of the local basis are

$$\mathbf{e}_i = \frac{\partial \mathbf{r}}{\partial q^i},$$

and the metric tensor is

$$g_{ik} = \frac{\partial \mathbf{r}}{\partial q^i} \cdot \frac{\partial \mathbf{r}}{\partial q^k} = \frac{\partial x_l}{\partial q^i} \frac{\partial x_l}{\partial q^k}.$$

In the case of orthogonal coordinates, this implies the following expression for the metric coefficients:

$$h_i = \sqrt{\left(\frac{\partial x_1}{\partial q^i}\right)^2 + \left(\frac{\partial x_2}{\partial q^i}\right)^2 + \left(\frac{\partial x_3}{\partial q^i}\right)^2}. \tag{2.51}$$

As an exercise, the reader should deduce (2.47) and (2.48) from (2.45), (2.46) and (2.51).

2.9. Tensors in Generalized Coordinate Systems

2.9.1. Covariant, contravariant and mixed components of a tensor. In a generalized coordinate system, a first-order tensor (vector) **A** is uniquely determined either by its three covariant components A_i or by its three contravariant components A^i (recall Sec. 1.6.3). Under changes of basis the quantities A_i transform differently than the quantities A^i, i.e.,

$$A'_i = \alpha_{i'}^{k} A_k,$$
$$A'^i = \alpha_k^{i'} A^k$$

[see (1.43) and (1.44)]. Nevertheless, the covariant and contravariant components are not independent, and are in fact related by the formula

$$A_i = g_{ik} A^k,$$
$$A^i = g^{ik} A_k$$

[see (1.48) and (1.49)], where the coefficients g_{ik} (g^{ik}) are determined by the basis of the coordinate system in which the components of **A** are taken.

In just the same way, tensors of order two or higher can have various kinds of components in a generalized coordinate system. These components differ in the way they transform under changes of basis. Moreover, just as in the case of vectors, there exist formulas relating the various kinds of components.

Thus we now reexamine the concept of a second-order tensor, this time relaxing the requirement that the coordinate system be rectangular: By a second-order tensor is meant a quantity uniquely specified by nine numbers (the components of the tensor). These components can be *covariant* A_{ik}, *contravariant* A^{ik} or *mixed* $A_i^{\cdot k}$, $A^i_{\cdot k}$, and transform according to the formulas

$$\begin{aligned} A'_{ik} &= \alpha^l_{i'}\alpha^m_{k'}A_{lm}, \\ A'^{ik} &= \alpha^{i'}_l\alpha^{k'}_m A^{lm}, \\ A_i'^{\cdot k} &= \alpha^l_{i'}\alpha^{k'}_m A_l^{\cdot m}, \\ A'^i_{\cdot k} &= \alpha^{i'}_l\alpha^m_{k'}A^l_{\cdot m}, \end{aligned} \tag{2.52}$$

where $\alpha^k_{i'}$ and $\alpha^{k'}_i$ $(i, k = 1, 2, 3)$ are the coefficients of the direct and the inverse transformations [see (1.11) and (1.12)]. The relation between the various components of a tensor, considered in a coordinate system with metric $(ds)^2 = g_{ik}\,dx^i\,dx^k$, are given by the formulas

$$\begin{aligned} A_{ik} &= g_{il}g_{km}A^{lm} = g_{kl}A_i^{\cdot l} = g_{il}A^l_{\cdot k}, \\ A^{ik} &= g^{il}g^{km}A_{lm} = g^{il}A_l^{\cdot k} = g^{kl}A^i_{\cdot l}, \\ A_i^{\cdot k} &= g^{kl}A_{il} = g_{il}A^{lk}, \\ A^i_{\cdot k} &= g^{il}A_{lk} = g_{kl}A^{il}. \end{aligned} \tag{2.53}$$

The dot in the mixed components emphasizes the order of occurrence of the indices. Thus in $A_i^{\cdot k}$ the first index is "covariant" and the second "contravariant," while in $A^i_{\cdot k}$ the first index is contravariant and the second covariant.

2.9.2. The tensor character of g_{ik}, g^{ik} and $g_i^{\cdot k}$. We now show that the quantities g_{ik}, g^{ik}, $g_i^{\cdot k}$ defined in Sec. 1.6.5 are actually the components of a second-order tensor, called the *metric tensor*. First we observe that formulas (1.11), (1.12) and (1.47) imply the following transformation laws for g_{ik}, g^{ik}, $g_i^{\cdot k}$ under changes of basis:

$$\begin{aligned} g'_{ik} &= \mathbf{e}'_i \cdot \mathbf{e}'_k = \alpha^l_{i'}\mathbf{e}_l \cdot \alpha^m_{k'}\mathbf{e}_m = \alpha^l_{i'}\alpha^m_{k'}\mathbf{e}_l \cdot \mathbf{e}_m = \alpha^l_{i'}\alpha^m_{k'}g_{lm}, \\ g'^{ik} &= \mathbf{e}'^i \cdot \mathbf{e}'^k = \alpha^{i'}_l\alpha^{k'}_m\mathbf{e}^l \cdot \mathbf{e}^m = \alpha^{i'}_l\alpha^{k'}_m g^{lm}, \\ g_i'^{\cdot k} &= \mathbf{e}'_i \cdot \mathbf{e}'^k = \alpha^l_{i'}\alpha^{k'}_m\mathbf{e}_l \cdot \mathbf{e}^m = \alpha^l_{i'}\alpha^{k'}_m g_l^{\cdot m}. \end{aligned}$$

Comparing these formulas with (2.52), we see that the g_{ik} are the covariant components of some tensor, the g^{ik} are contravariant components of some tensor, and the $g_i^{\cdot k}$ are mixed components of some tensor. To verify that all these quantities are components of the *same* tensor, we need only show that they are connected by relations of the form (2.53). But it follows from the definition of g_{ik} and the properties of the basis $\mathbf{e}_1, \mathbf{e}_2, \mathbf{e}_3$ and its reciprocal $\mathbf{e}^1, \mathbf{e}^2, \mathbf{e}^3$ that

$$\mathbf{e}_i = g_{il}\mathbf{e}^l$$

[recall (1.42)]. Therefore

$$g_{ik} = \mathbf{e}_i \cdot \mathbf{e}_k = g_{il}\mathbf{e}^l \cdot g_{km}\mathbf{e}^m = g_{il}g_{km}\mathbf{e}^l \cdot \mathbf{e}^m = g_{il}g_{km}g^{lm},$$

$$g_{ik} = g_{il}\mathbf{e}^l \cdot \mathbf{e}_k = g_{il}g^l_{\cdot k},$$

etc., in keeping with (2.53).

Remark. The components of $g_i^{\cdot k}$ are the same as those of the Kronecker delta defined on p. 39, i.e.,

$$g_i^{\cdot k} = \begin{cases} 0 & \text{if } i \neq k, \\ 1 & \text{if } i = k \end{cases}$$

[recall (1.47)].

2.9.3. Higher-order tensors in generalized coordinates. In a generalized coordinate system, a tensor of order n has 3^n components as in Sec. 2.5, but now it can have various kinds of components, i.e., covariant, contravariant and mixed components of various kinds. For example, a third-order tensor has $3^3 = 27$ components, with mixed components $A_{ik}^{\cdot\cdot l}, A^i_{\cdot kl}, \ldots$ which transform according to the formulas

$$A'^{\cdot\cdot l}_{ik} = \alpha^m_{i'}\alpha^n_{k'}\alpha^{l'}_r A^{\cdot\cdot r}_{mn},$$

$$A'^{i\cdot l}_{\cdot k} = \alpha^{i'}_m\alpha^n_{k'}\alpha^{l'}_r A^{m\cdot r}_{\cdot n},$$

etc. Here we say that $A_{ik}^{\cdot\cdot l}$ is a mixed tensor with two covariant indices and one contravariant index (and similarly for $A^{i\cdot l}_{\cdot k}$).

The requirement for invariance of tensor equations given in Sec. 2.7, i.e., that all tensors appearing as (additive) terms in an equation expressing a physical law be of the same order, must now be strengthened by the requirement that all terms have the same "covariance." In other words, covariant components cannot be added to contravariant components, and mixed tensors can be added only if they have the same structure (like $A_i^{\cdot kl}$ and $B_i^{\cdot kl}$).

2.9.4. Physical components of a tensor. The case of orthogonal bases. The concept of "physical" components of a vector has a natural generalization to the case of tensors of order two or higher. In the general case, this is a consequence of the fact that the components of a tensor of order n can be written as a sum of products of components of n three-dimensional vectors (see Prob. 5, p. 97), In particular, the physical components A^*_{ik} and A^{*ik} of a second-order tensor are given by the formulas

$$A^*_{ik} = \frac{A_{ik}}{|\mathbf{e}_i|\,|\mathbf{e}_k|} = \frac{A_{ik}}{\sqrt{g_{ii}g_{kk}}} \qquad \text{(no summation over } i \text{ and } k),$$
$$A^{*ik} = A^{ik}\,|\mathbf{e}_i|\,|\mathbf{e}_k| = A^{ik}\sqrt{g_{ii}g_{kk}} \qquad \text{(no summation over } i \text{ and } k), \tag{2.54}$$

generalizing the expressions for A^*_i and A^{*i} on p. 30.

In the case of orthogonal bases, it follows from (2.53) and (1.47) that

$$A_{ik} = A^{ik}g_{ii}g_{kk} = A^{ik}h_i^2h_k^2 \qquad \text{(no summation over } i \text{ and } k),$$

$$A^{ik} = A_{ik}g^{ii}g^{kk} = \frac{A_{ik}}{h_i^2h_k^2} \qquad \text{(no summation over } i \text{ and } k)$$

(recall from Sec. 1.6.6 that $g_{ii}g^{ii} = 1$). In this case, we see from (2.54) that

$$A^*_{ik} = A^{*ik} = \frac{A_{ik}}{h_ih_k} = A^{ik}h_ih_k \qquad \text{(no summation over } i \text{ and } k).$$

Remark. Again, as in Remark 2, p. 34, summation can only take place over "dummy" indices in different positions, where two indices are said to be in different positions if one is a subscript and the other a superscript. For example,

$$A_{ikl} = g_{im}A^m_{.kl} = g_{im}g_{kn}A^{mn}_{..l} = g_{im}g_{kn}g_{lr}A^{mnr}.$$

The operations of "raising" and "lowering" indices have the same meaning as on p. 34. Moreover, an equation like

$$A_{ik} = g_i^{\cdot l}A_{lk}$$

is sometimes described as "renaming" an index.

2.9.5. Covariant, contravariant and mixed tensors as such. In describing a tensor arising in a physical or geometric problem, it is sometimes most natural to start from a particular set of components, say covariant components. The tensor itself is then said to be "covariant," but formulas like (2.53) can always be used to deduce all the contravariant and mixed components of the same tensor, which is to be thought of as a single object with no more than 3^n independent components (if it is of order n).

Example 1. Let $f(x^1, x^2, x^3)$ be a scalar function of the generalized coordinates x^1, x^2, x^3 (not to be confused with the rectangular coordinates x_1, x_2, x_3). Then, by the chain rule for partial differentiation,

$$\frac{\partial f}{\partial x'^i} = \frac{\partial f}{\partial x^k}\frac{\partial x^k}{\partial x'^i} = \alpha^k_{i'}\frac{\partial f}{\partial x^k},$$

where we use the fact that

$$x^k = \alpha^k_{i'} x'^i$$

(see p. 29). Therefore the three quantities $\partial f/\partial x^i$ transform like the covariant components of a vector [see (1.43)], and in this sense form a "covariant vector." However, it would be more accurate to describe $\partial f/\partial x^i$ as the covariant components of an underlying vector (the gradient vector defined in Sec. 4.3.2), which has perfectly well-defined contravariant and mixed components as well, found by using formulas like (2.53).

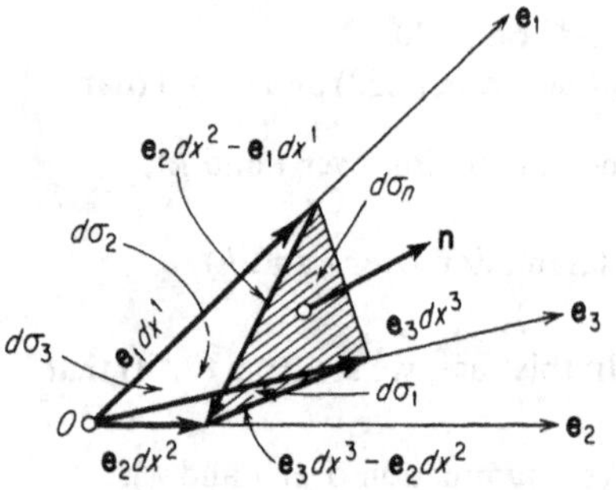

FIG. 2.13. The stress tensor in generalized coordinates.

Example 2 (The stress tensor in generalized coordinates.) Again let x^1, x^2, x^3 be generalized coordinates, and let $\mathbf{e}_1, \mathbf{e}_2, \mathbf{e}_3$ be the corresponding (right-handed) basis. Consider the elementary tetrahedron with edges $\mathbf{e}_1\,dx^1$, $\mathbf{e}_2\,dx^2$, $\mathbf{e}_3\,dx^3$ drawn from the origin O and "bottom" of area $d\sigma_n$. Let $\mathbf{n}$ be the unit exterior normal to $d\sigma_n$, and let the areas of the faces of the tetrahedron be $d\sigma_1, d\sigma_2, d\sigma_3$. Then, as shown in Fig. 2.13,

$$\begin{aligned} d\sigma_1 &= |\mathbf{e}_2 \times \mathbf{e}_3|\,dx^2\,dx^3, \\ d\sigma_2 &= |\mathbf{e}_3 \times \mathbf{e}_1|\,dx^3\,dx^1, \\ d\sigma_3 &= |\mathbf{e}_1 \times \mathbf{e}_2|\,dx^1\,dx^2, \end{aligned} \tag{2.55}$$

and

$$\begin{aligned} \mathbf{n}\,d\sigma_n &= (\mathbf{e}_2\,dx^2 - \mathbf{e}_1\,dx^1) \times (\mathbf{e}_3\,dx^3 - \mathbf{e}_2\,dx^2) \\ &= (\mathbf{e}_2 \times \mathbf{e}_3)\,dx^2\,dx^3 + (\mathbf{e}_3 \times \mathbf{e}_1)\,dx^3\,dx^1 + (\mathbf{e}_1 \times \mathbf{e}_2)\,dx^1\,dx^2. \end{aligned}$$

Introducing the vectors $\mathbf{e}^1$, $\mathbf{e}^2$, $\mathbf{e}^2$ of the reciprocal lattice, and taking account of (1.36), (1.54) and (2.55), we find that

$$\mathbf{e}_2 \times \mathbf{e}_3 = \mathbf{e}^1\sqrt{G},$$

say, and hence,

$$(\mathbf{e}_2 \times \mathbf{e}_3)\,dx^1\,dx^2 = \frac{\mathbf{e}^1}{|\mathbf{e}^1|}\,d\sigma_1.$$

It follows that

$$\mathbf{n}\,d\sigma_n = \sum_{i=1}^{3} \frac{\mathbf{e}^i}{|\mathbf{e}^i|}\,d\sigma_i.$$

Therefore

$$n_i\,d\sigma_n = \frac{d\sigma_i}{|\mathbf{e}^i|}, \tag{2.56}$$

where n_i are the covariant components of the vector $\mathbf{n}$.

Just as in Sec. 2.4.1, the stress vector $\mathbf{p}_n$ acting on $d\sigma_n$ equals

$$\mathbf{p}_n = \sum_{i=1}^{3} \mathbf{p}_i \, d\sigma_i \frac{1}{d\sigma_n},$$

or

$$\mathbf{p}_n = \sum_{i=1}^{3} \mathbf{p}_i \, |\mathbf{e}^i| \, n_i$$

after substituting from (2.56). The components of the stress vector $\mathbf{p}_n$ with respect to the basis $\mathbf{e}_1$, $\mathbf{e}_2$, $\mathbf{e}_3$ are of course contravariant [recall (1.41)], and in this sense the vector $\mathbf{p}_i \, |\mathbf{e}^i|$ itself is contravariant. Denoting the contravariant components of $\mathbf{p}_i \, |\mathbf{e}^i|$ by p^{ik}, so that

$$\mathbf{p}_i \, |\mathbf{e}^i| = \sum_{k=1}^{3} p^{ik} \mathbf{e}_k,$$

we have

$$\mathbf{p}_n = \sum_{i,k=1}^{3} p^{ik} n_i \mathbf{e}_k.$$

Therefore the contravariant components of the stress vector $\mathbf{p}_n$ are

$$p_n^k = p^{ik} n_i.$$

The nine quantities p^{ik} are the contravariant components of a single physical quantity, namely the stress tensor, and allow us to determine (in the given system) the stress at an arbitrary point acting on an arbitrary element of area characterized by the vector $\mathbf{n}$. The covariant and mixed components of p^{ik} are found in the usual way, e.g.,

$$p_{ik} = g_{il} g_{km} p^{lm}, \qquad p_i^{\cdot k} = g_{il} p^{lk}.$$

Remark. Just as the vector $\mathbf{A}$ with components A_i is denoted by the boldface letter $\mathbf{A}$ without subscripts, we can use $\mathbf{A}$ to denote the *tensor* with components A_{ik}, A_{ikl}, etc. The analogue of the representation

$$\mathbf{A} = A_i \mathbf{e}^i$$

is then

$$\mathbf{A} = A_{ik} \mathbf{e}^i \mathbf{e}^k, \qquad \mathbf{A} = A_{ikl} \mathbf{e}^i \mathbf{e}^k \mathbf{e}^l, \tag{2.57}$$

and so on. Given two vectors $\mathbf{A}$ and $\mathbf{B}$, the second-order tensor $\mathbf{C}$ with components $C_{ik} = A_i B_k$ (recall Example 1, p. 64) can then be written as

$$\mathbf{C} = \mathbf{AB} = A_i B_k \mathbf{e}^i \mathbf{e}^k = C_{ik} \mathbf{e}^i \mathbf{e}^k.$$

Here the quantity $\mathbf{AB}$, called a *dyad*, has no multiplication sign and is not to be confused with the scalar or vector product. Note that $\mathbf{AB} \neq \mathbf{BA}$, and in fact the matrix corresponding to $\mathbf{AB}$ (see Remark 2, p. 64) is the transpose of that corresponding to $\mathbf{BA}$. The natural generalization of (2.57) to tensors of higher order is

$$\mathbf{T} = T_{ik..}^{\cdot\cdot\cdot\cdot pq\cdot\cdot} \mathbf{e}^i \mathbf{e}^k \cdots \mathbf{e}_p \mathbf{e}_q \cdots$$

SOLVED PROBLEMS

Problem 1. Find a formula for the moment of inertia of a system of n particles of masses $m_1, m_2, \ldots, m_n$ about an axis u characterized by the unit vector $\mathbf{u}_0$.

First solution. Let $\mathbf{r}_k$ be the radius vector of the kth particle with respect to some origin O lying on u. Then $|\mathbf{r}_k \times \mathbf{u}_0|$ is the distance from the kth particle to the axis u, and hence the moment of inertia of the system of particles about u is given by

$$I_u = \sum_{k=1}^{n} m_k(\mathbf{r}_k \times \mathbf{u}_0)^2.$$

It follows from (1.27) and (1.30) that

$$\begin{aligned} I_u &= \sum_{k=1}^{n} m_k(\mathbf{r}_k \times \mathbf{u}_0) \cdot (\mathbf{r}_k \times \mathbf{u}_0) = \sum_{k=1}^{n} m_k\mathbf{r}_k \cdot [\mathbf{u}_0 \times (\mathbf{r}_k \times \mathbf{u}_0)] \\ &= \sum_{k=1}^{n} m_k\mathbf{r}_k \cdot [\mathbf{r}_k - \mathbf{u}_0(\mathbf{r}_k \cdot \mathbf{u}_0)] = \sum_{k=1}^{n} m_k[r_k^2 - (\mathbf{r}_k \cdot \mathbf{u}_0)^2], \end{aligned} \tag{2.58}$$

where $r_k = |\mathbf{r}_k|$. In a coordinate system K with origin O, we have

$$I_u = \sum_{k=1}^{n} m_k[x_l^{(k)}x_l^{(k)} - (x_l^{(k)}u_{0l})^2], \tag{2.59}$$

where

$$\mathbf{r}_k = x_1^{(k)}\mathbf{i}_1 + x_2^{(k)}\mathbf{i}_2 + x_3^{(k)}\mathbf{i}_3$$

and $\mathbf{u}_0$ has components u_{0l} $(l = 1, 2, 3)$.

Second solution. Let u be the x_1'-axis of a new coordinate system K'. Then

$$I_u = I_{11}' = \alpha_{1'r}\alpha_{1's}I_{rs},$$

where I_{rs} is the moment of inertia tensor about O in the system K. But

$$\alpha_{1'r} = \mathbf{i}_1' \cdot \mathbf{i}_r = \mathbf{u}_0 \cdot \mathbf{i}_r = u_{0r},$$

and hence

$$I_u = I_{rs}u_{0r}u_{0s}, \tag{2.60}$$

which gives (2.59) after substituting from (2.17).

Problem 2. Show that the kinetic energy of a rigid system of n particles of masses $m_1, m_2, \ldots, m_n$ rotating with instantaneous angular velocity $\boldsymbol{\omega}$ about a fixed point O equals

$$T = \tfrac{1}{2}I_\omega\omega^2,$$

where I_ω is the moment of inertia of the system about the instantaneous axis of rotation (the axis with the direction of $\boldsymbol{\omega}$).

Solution. Let $\mathbf{r}_k$ be the radius vector (with respect to O) and $\mathbf{v}_k$ the instantaneous velocity of the kth particle. Then, using Prob. 11, p. 45, we have

$$T = \frac{1}{2}\sum_{k=1}^{n} m_k v_k^2 = \frac{1}{2}\sum_{k=1}^{n} m_k(\boldsymbol{\omega} \times \mathbf{r}_k)^2 = \frac{1}{2}\sum_{k=1}^{n} m_k(\boldsymbol{\omega} \times \mathbf{r}_k) \cdot (\boldsymbol{\omega} \times \mathbf{r}_k)$$

$$= \frac{1}{2}\sum_{k=1}^{n} m_k \boldsymbol{\omega} \cdot [\mathbf{r}_k \times (\boldsymbol{\omega} \times \mathbf{r}_k)] = \frac{1}{2}\sum_{k=1}^{n} m_k \boldsymbol{\omega} \cdot [r_k^2 \boldsymbol{\omega} - \mathbf{r}_k(\mathbf{r}_k \cdot \boldsymbol{\omega})]$$

$$= \frac{1}{2}\sum_{k=1}^{n} m_k[r_k^2 \omega^2 - (\mathbf{r}_k \cdot \boldsymbol{\omega})^2].$$

Let

$$\boldsymbol{\omega}_0 = \frac{\boldsymbol{\omega}}{\omega}$$

be the unit vector characterizing the axis of rotation, and let I_ω be the moment of inertia about this axis. Then

$$T = \frac{1}{2}\omega^2 \sum_{k=1}^{n} m_k[r_k^2 - (\mathbf{r}_k \cdot \boldsymbol{\omega}_0)^2] = \frac{1}{2} I_\omega \omega^2$$

[recall (2.58)]. Using (2.60), we can also write

$$T = \frac{1}{2} I_\omega \omega^2 = \frac{1}{2} I_{rs} \omega_{0r} \omega_{0s} \omega^2 = \frac{1}{2} I_{rs} \omega_r \omega_s$$

in terms of the components of $\boldsymbol{\omega}$ and the moment of inertia tensor about O.

Problem 3. Let I_u be the moment of inertia of a system of n particles of masses $m_1, m_2, \ldots, m_n$ about an axis u through the center of mass of the system. Find the moment of inertia of the system about an axis v parallel to u.

Solution. Let $\mathbf{R}$ be the radius vector of the center of mass (which lies on u) with respect to some origin O on v. Suppose $\mathbf{r}_k$ is the radius vector of the kth particle with respect to O, while $\mathbf{r}'_k$ is its radius vector with respect to the center of mass. Then

$$\mathbf{r}_k = \mathbf{r}'_k + \mathbf{R},$$

and hence

$$I_v = \sum_{k=1}^{n} m_k(\mathbf{r}_k \times \mathbf{v}_0)^2 = \sum_{k=1}^{n} m_k[(\mathbf{r}'_k + \mathbf{R}) \times \mathbf{v}_0]^2$$

$$= \sum_{k=1}^{n} m_k(\mathbf{R} \times \mathbf{v}_0)^2 + \sum_{k=1}^{n} m_k(\mathbf{r}'_k \times \mathbf{v}_0)^2 + 2\sum_{k=1}^{n} m_k(\mathbf{R} \times \mathbf{v}_0) \cdot (\mathbf{r}'_k \times \mathbf{v}_0),$$

where $\mathbf{v}_0$ is a unit vector along v. But

$$\sum_{k=1}^{n} m_k(\mathbf{r}'_k \times \mathbf{v}_0)^2 = \sum_{k=1}^{n} m_k(\mathbf{r}'_k \times \mathbf{u}_0)^2 = I_u,$$

where $\mathbf{u}_0$ is a unit vector along u, since $\mathbf{v}_0 = \mathbf{u}_0$ (the axes are parallel). Moreover

$$\sum_{k=1}^{n} m_k(\mathbf{R} \times \mathbf{v}_0) \cdot (\mathbf{r}'_k \times \mathbf{v}_0) = (\mathbf{R} \times \mathbf{v}_0) \cdot \left[\left(\sum_{k=1}^{n} m_k \mathbf{r}'_k\right) \times \mathbf{v}_0\right] = 0,$$

since

$$\sum_{k=1}^{n} m_k \mathbf{r}'_k = 0,$$

by the definition of the center of mass. Therefore

$$I_v = I_u + M\,|\mathbf{R} \times \mathbf{v}_0|^2,$$

where M is the total mass of the system and $|\mathbf{R} \times \mathbf{v}_0|$ is the distance between the axes.

Problem 4. Find the most general linear function relating the viscous stress tensor $\hat{p}_{ik}$ to the rate of deformation tensor v_{ik} in an isotropic fluid.[11]

Solution. The most general linear function relating $\hat{p}_{ik}$ and v_{ik} is of the form

$$\hat{p}_{ik} = \eta_{iklm} v_{lm},$$

where η_{iklm} is a fourth-order tensor characterizing the properties of the fluid. Since the properties of the fluid must be the same in all directions, the components of η_{iklm} must be invariant under arbitrary rotations of the coordinate system. Such a tensor is said to be *isotropic*. It can be shown[12] that the most general isotropic tensor of order four is of the form

$$\eta_{iklm} = A\delta_{ik}\delta_{lm} + B\delta_{il}\delta_{km} + C\delta_{im}\delta_{kl},$$

where

$$\delta_{ij} = \begin{cases} 0 \text{ if } i \neq j, \\ 1 \text{ if } i = j \end{cases}$$

is the Kronecker delta. It follows that

$$\begin{aligned} \hat{p}_{ik} &= A\delta_{ik}\delta_{lm}v_{lm} + B\delta_{il}\delta_{km}v_{lm} + C\delta_{im}\delta_{kl}v_{lm} \\ &= A\delta_{ik}v_{ll} + Bv_{ik} + Cv_{ki}. \end{aligned}$$

Since $v_{ik} = v_{ki}$ [recall (2.22)], we have

$$\hat{p}_{ik} = 2\mu v_{ik} + \mu'\delta_{ik}v_{ll}, \tag{2.61}$$

where

$$B + C = 2\mu, \quad A = \mu'.$$

---

[11] The *viscous stress tensor* $\hat{p}_{ik}$ is the part of the stress tensor p_{ik} which vanishes when the fluid is at rest. The relation between p_{ik} and $\hat{p}_{ik}$ is $p_{ik} = -p\delta_{ik} + \hat{p}_{ik}$, where p is the hydrostatic pressure. See R. Aris, *op. cit.*, p. 107.

[12] *Ibid.*, pp. 33–34.

Formula (2.61) is basic in hydrodynamics. The constants μ and μ' (called *viscosity coefficients*) characterize the properties of the isotropic fluid.[13]

Problem 5. Show that an arbitrary second-order tensor can be represented as a sum of products of the components of three vectors, taken two at a time.

Solution. Introduce an orthogonal coordinate system (in general, nonrectangular) with orthonormal basis $\mathbf{e}_1, \mathbf{e}_2, \mathbf{e}_3$. Let T_{ik} be the covariant components (say) of the tensor in this system, and suppose the vector $\mathbf{e}_\alpha$ has covariant components $e_{\alpha i}$ and contravariant components e^i_α.[14] Consider the scalars

$$T_{(\alpha\beta)} = T_{ik} e^i_\alpha e^k_\beta. \tag{2.62}$$

Multiplying (2.62) by $e_{\alpha p} e_{\beta q}$ and summing over α and β, we obtain

$$\sum_{\alpha,\beta=1}^{3} T_{(\alpha\beta)} e_{\alpha p} e_{\beta q} = \sum_{i,k=1}^{3} T_{ik} \sum_{\alpha,\beta=1}^{3} e^i_\alpha e_{\alpha p} e^k_\beta e_{\beta q}. \tag{2.63}$$

But

$$\mathbf{e}_\alpha \cdot \mathbf{e}_\beta = \sum_{i=1}^{3} e^i_\alpha e_{\beta i} = \delta_{\alpha\beta}, \tag{2.64}$$

since the basis $\mathbf{e}_1, \mathbf{e}_2, \mathbf{e}_3$ is orthonormal. Multiplying (2.64) by e^α_β and summing over β, we find that

$$\sum_{\beta=1}^{3} \delta_{\alpha\beta} e^\alpha_\beta = \sum_{i=1}^{3} e^i_\alpha \sum_{\beta=1}^{3} e_{\beta i} e^\alpha_\beta$$

or

$$e^\alpha_\alpha = \sum_{i=1}^{3} e^i_\alpha \sum_{\beta=1}^{3} e_{\beta i} e^\alpha_\beta. \tag{2.65}$$

On the other hand, since $g_i^{\cdot\alpha} = 1$ if $i = \alpha$ and 0 otherwise,

$$e^\alpha_\alpha = \sum_{i=1}^{3} g_i^{\cdot\alpha} e^i_\alpha.$$

A comparison of this formula with (2.65) shows that

$$g_i^{\cdot\alpha} = \sum_{\beta=1}^{3} e_{\beta i} e^\alpha_\beta. \tag{2.66}$$

Therefore the right-hand side of (2.63) equals

$$\sum_{i,k=1}^{3} T_{ik} g^{\cdot i}_p g^{\cdot k}_q = T_{pq},$$

and hence

$$T_{pq} = \sum_{\alpha,\beta=1}^{3} T_{(\alpha\beta)} e_{\alpha p} e_{\beta q}, \tag{2.67}$$

---

[13] R. Aris, *op. cit.*, pp. 111–112.

[14] Do not think of $e_{\alpha i}$ and e^i_α as tensors. A somewhat "safer" but clumsier notation would be $e_{(\alpha)i}$ and $e^i_{(\alpha)}$.

which is a representation of the required form. Analogous representations for T^{ik} and $T_{i}^{\cdot k}$ are easily deduced from (2.67).

The natural generalization of (2.67) to the case of tensors of order n is

$$T_{i_1 i_2 \ldots i_n} = \sum_{\alpha_1, \alpha_2, \ldots, \alpha_n}^{3} T_{(\alpha_1 \alpha_2 \ldots \alpha_n)} e_{\alpha_1 i_1} e_{\alpha_2 i_2} \cdots e_{\alpha_n i_n}.$$

Problem 6. Find representations analogous to (2.66) for the components g_{ik} and g^{ik} of the metric tensor.

Solution. Using (1.48), we find that

$$g_{jk} e_{\alpha}^{j} = e_{\alpha k} \qquad (\alpha = 1, 2, 3).$$

Multiplying each of these formulas by $e_{\alpha i}$ and summing over α, we obtain

$$\sum_{j=1}^{3} g_{jk} \sum_{\alpha=1}^{3} e_{\alpha i} e_{\alpha}^{j} = \sum_{\alpha=1}^{3} e_{\alpha i} e_{\alpha k}.$$

It follows from (2.66) that

$$\sum_{j=1}^{3} g_{jk} g_{i}^{\cdot j} = \sum_{\alpha=1}^{3} e_{\alpha i} e_{\alpha k},$$

and hence

$$g_{ik} = \sum_{\alpha=1}^{3} e_{\alpha i} e_{\alpha k}.$$

Similarly, we find that

$$g^{ik} = \sum_{\alpha=1}^{3} e_{\alpha}^{i} e_{\alpha}^{k}.$$

Problem 7. Given a rectangular coordinate system K with orthonormal basis $\mathbf{i}_1$, $\mathbf{i}_2$, $\mathbf{i}_3$, consider the second-order tensor with components

$$\|A_{ik}\| = \|A^{ik}\| = \|A_{i}^{\cdot k}\| = \|A_{\cdot k}^{i}\| = \begin{Vmatrix} 2 & 1 & 3 \\ 2 & 3 & 4 \\ 1 & 2 & 1 \end{Vmatrix}.$$

Let K' be a nev coordinate system with basis vectors

$$\begin{aligned} \mathbf{e}_1 &= \mathbf{i}_1, \\ \mathbf{e}_2 &= \mathbf{i}_1 + \mathbf{i}_2, \\ \mathbf{e}_3 &= \mathbf{i}_1 + \mathbf{i}_2 + \mathbf{i}_3. \end{aligned} \tag{2.68}$$

Express the ccvariant, contravariant and mixed components of the given tensor in the system K'.

Solution. According to (2.52),

$$A'_{ik} = \alpha_{i'}^{l} \alpha_{k'}^{m} A_{lm},$$

where $\alpha_{i'}^{l}$, $\alpha_{k'}^{m}$, are the coefficients of the direct transformation (2.68), i.e.,

$$\alpha_{1'}^{1} = 1, \quad \alpha_{1'}^{2} = 0, \quad \alpha_{1'}^{3} = 0,$$
$$\alpha_{2'}^{1} = 1, \quad \alpha_{2'}^{2} = 1, \quad \alpha_{2'}^{3} = 0,$$
$$\alpha_{3'}^{1} = 1, \quad \alpha_{3'}^{2} = 1, \quad \alpha_{3'}^{3} = 1,$$

and hence

$$\|A'_{ik}\| = \left\| \begin{matrix} 2 & 3 & 6 \\ 4 & 8 & 15 \\ 5 & 11 & 19 \end{matrix} \right\|.$$

To find A'^{ik}, $A'^{.k}_{i}$ and $A'^{i}_{.k}$, we use the formulas

$$A'^{ik} = g^{il}g^{km}A'_{lm},$$
$$A'^{.k}_{i} = g^{kl}A'_{il},$$
$$A'^{i}_{.k} = g^{il}A'_{lk}$$

[see (2.53)], after first noting that

$$\|g_{ik}\| = \|\mathbf{e}_i \cdot \mathbf{e}_k\| = \left\| \begin{matrix} 1 & 1 & 1 \\ 1 & 2 & 2 \\ 1 & 2 & 3 \end{matrix} \right\|,$$

$$\|g^{ik}\| = \left\| \begin{matrix} 2 & -1 & 0 \\ -1 & 2 & -1 \\ 0 & -1 & 1 \end{matrix} \right\|$$

[see (1.47) and (1.52)]. As a result, we obtain

$$\|A'^{ik}\| = \left\| \begin{matrix} 2 & -1 & -1 \\ 0 & -2 & 3 \\ -1 & 1 & 1 \end{matrix} \right\|,$$

$$\|A'^{.k}_{i}\| = \left\| \begin{matrix} 1 & -2 & 3 \\ 0 & -3 & 7 \\ -1 & -2 & 8 \end{matrix} \right\|,$$

$$\|A'^{i}_{.k}\| = \left\| \begin{matrix} 0 & -2 & -3 \\ 1 & 2 & 5 \\ 1 & 3 & 4 \end{matrix} \right\|.$$

EXERCISES

1. Given a rectangular coordinate system K, let K' be the coordinate system obtained from K by rotating K first through the angle $\pi/6$ about the x_3-axis and then through $\pi/2$ about the x_1'-axis so that the x_2' and x_3-axes coincide (see Fig. 2.14). Find

a) The components of the vectors

$$\mathbf{A} = \mathbf{i}_1 + 2\mathbf{i}_2 + 3\mathbf{i}_3,$$

$$\mathbf{B} = 4\mathbf{i}_1 + 5\mathbf{i}_2 + 6\mathbf{i}_3$$

in the system K';

b) The tangential and normal stresses on elements of area perpendicular to the axes of K' if the stress tensor in the system K is of the form

$$\|p_{ik}\| = \begin{Vmatrix} p_1 & 0 & 0 \\ 0 & p_2 & 0 \\ 0 & 0 & p_3 \end{Vmatrix};$$

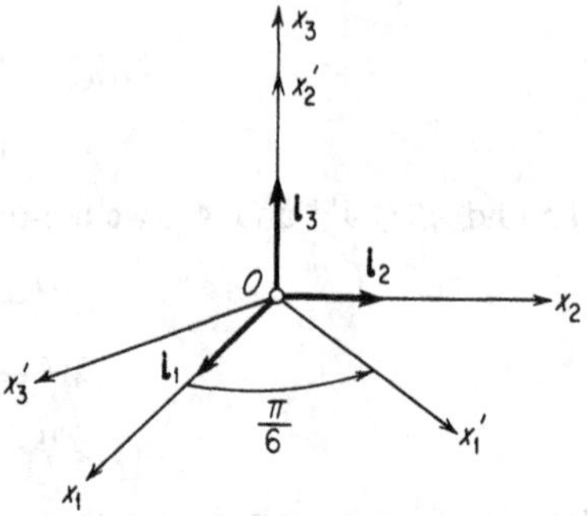

FIG. 2.14. Illustrating Exercise 1.

c) The stress on an element of area passing through the bisector of the first quadrant of the x_1x_2-plane and making angle $\pi/4$ with the x_3-axis.

2. The moment of inertia tensor of a right circular cylinder with respect to axes passing through its center of mass (the x_3-axis is parallel to the generators) is of the form

$$\|I_{ik}\| = \begin{Vmatrix} I_0 & 0 & 0 \\ 0 & I_0 & 0 \\ 0 & 0 & I_1 \end{Vmatrix}.$$

Find the moment of inertia of the cylinder about the bisectors of the angles between the various coordinate axes.

3. What are the analogues of formulas (2.37) and (2.38) for tensors of order 3? Of order n?

4. If A_{ikl} is a covariant tensor of order 3 and B^{pqmn} is a contravariant tensor of order 4, prove that $A_{ikl}B^{klmn}$ is a mixed tensor of order 3 (with one covariant and two contravariant indices).

5. If v_i is a covariant vector and x^k are generalized coordinates, prove that the quantities

$$\frac{\partial v_i}{\partial x^k}$$

form a second-order tensor.

6. Let q^1, q^2, q^3 be coordinates related to rectangular coordinates x_1, x_2, x_3 with orthonormal basis $\mathbf{i}_1, \mathbf{i}_2, \mathbf{i}_3$ by the formulas

$$q^1 = x_1 + x_2, \quad q^2 = x_1 - x_2, \quad q^3 = 2x_3.$$

a) Prove that q^1, q^2, q^3 are themselves rectangular coordinates;
b) Find the corresponding basis vectors;
c) Find the metric tensor g_{ik};
d) Find the covariant and contravariant components of the vectors

$$2\mathbf{i}_1, \quad A = \mathbf{i}_1 + \mathbf{i}_2, \quad B = 2\mathbf{i}_1 - 3\mathbf{i}_3;$$

e) Calculate $G = \det \|g_{ik}\|$;
f) Find the components of $\mathbf{A} \times \mathbf{B}$ in both coordinate systems.

7. Show that the vectors with contravariant components

$$(1, 0, 0), \quad \left(0, \frac{1}{R}, 0\right), \quad \left(0, 0, \frac{1}{R \sin \theta}\right)$$

in a spherical coordinate system are perpendicular to each other.

8. Given a system of rectangular coordinates x_1, x_2 with orthonormal basis $\mathbf{i}_1, \mathbf{i}_2$, the coordinates q^1, q^2 defined by the formulas

$$x_1 = \frac{a \sinh q^1}{\cosh q^1 + \cos q^2}, \qquad x_2 = \frac{a \sin q^2}{\cosh q^1 + \cos q^2} \qquad (a = \text{const})$$

are called *bipolar coordinates*. Find
a) The basis vectors $\mathbf{e}_1, \mathbf{e}_2$ (are they orthogonal?);
b) The metric tensor g_{ik};
c) The covariant and contravariant components of $\mathbf{i}_1, \mathbf{i}_2$ in bipolar coordinates.

9. Two tensors have components

$$A_1 = x_1 x_2, \quad A_2 = 2x_2 - x_3^2, \quad A_3 = x_1 x_3$$

and

$$\|T_{ik}\| = \begin{Vmatrix} x_1 & x_2 x_1 & 0 \\ x_1 x_2 & x_2 & 1 \\ 0 & 1 & x_3^2 \end{Vmatrix}$$

in rectangular coordinates. Find the covariant, contravariant and physical components of these tensors in cylindrical and spherical coordinates.

10. Show that the angles $\theta_{12}, \theta_{23}, \theta_{13}$ between the coordinate axes of a generalized coordinate system with metric tensor g_{ik} are given by the formulas

$$\cos \theta_{12} = \frac{g_{12}}{\sqrt{g_{11} g_{22}}}, \quad \cos \theta_{23} = \frac{g_{23}}{\sqrt{g_{22} g_{33}}}, \quad \cos \theta_{31} = \frac{g_{31}}{\sqrt{g_{33} g_{11}}}.$$

Find the angles between the vector with components A^i and the coordinate axes.

11. Given a scalar function $\Phi = \Phi(x^1, x^2, x^3)$, do the quantities

$$\frac{\partial^2 \Phi}{\partial x^i \partial x^k}$$

form a tensor?

12. Solve Prob. 7, p. 98 if K' has basis vectors

$$\mathbf{e}_1 = -\mathbf{i}_1 + \mathbf{i}_3,$$

$$\mathbf{e}_2 = \mathbf{i}_2 + \mathbf{i}_3,$$

$$\mathbf{e}_3 = 2\mathbf{i}_3,$$

instead of (2.68).

13. Given two generalized coordinate systems K and K', with local bases $\mathbf{e}_1, \mathbf{e}_2, \mathbf{e}_3$ and $\mathbf{e}_{1'}, \mathbf{e}_{2'}, \mathbf{e}_{3'}$, respectively, let $\alpha^k_{i'}$ be the coefficients of the direct transformation and $\alpha^{k'}_i$ the coefficients of the inverse transformation, so that

$$\alpha^k_{i'} = \mathbf{e}_{i'} \cdot \mathbf{e}^k, \qquad \alpha^{k'}_i = \mathbf{e}_i \cdot \mathbf{e}^{\prime k}$$

[cf. (1.45)]. Prove that

$$g^k_i = \alpha^{l'}_i \alpha^k_{l'}.$$

3

TENSOR ALGEBRA

3.1. Addition of Tensors

Let A_{ik} and B_{ik} be the components of two second-order (Cartesian) tensors, and let

$$C_{ik} = A_{ik} + B_{ik}.$$

Then the numbers C_{ik} are themselves the components of a second-order tensor, called the *sum* of the tensors with components A_{ik} and B_{ik}. In fact, A_{ik} and B_{ik} transform according to the formulas

$$\begin{aligned} A'_{ik} &= \alpha_{i'l}\alpha_{k'm}A_{lm}, \\ B'_{ik} &= \alpha_{i'l}\alpha_{k'm}B_{lm}, \end{aligned} \tag{3.1}$$

and hence C_{ik} transforms in the same way:

$$C'_{ik} = A'_{ik} + B'_{ik} = \alpha_{i'l}\alpha_{k'm}(A_{lm} + B_{lm}) = \alpha_{i'l}\alpha_{k'm}C_{lm}.$$

Addition of any number of tensors of arbitrary order is defined similarly, i.e., the sum of two or more tensors of the same order is the tensor whose components are the sums of the corresponding components of the summands. Note that tensors of different orders cannot be added. Subtraction of tensors of the same order is defined in the obvious way.

In the case of generalized coordinates, the tensors must have not only the same order but also the same *structure*, i.e., the same numbers of covariant and contravariant indices in the same places. For example, the sums

$$C^{ik} = A^{ik} + B^{ik},$$

$$C_{i}^{\cdot k} = A_{i}^{\cdot k} + B_{i}^{\cdot k},$$

$$C^{ik}_{\cdot\cdot l} = A^{ik}_{\cdot\cdot l} + B^{ik}_{\cdot\cdot l}$$

all make sense, and each expression on the left is a tensor of the same order and structure as the terms on the right. This is a consequence of the fact that any transformation law for a tensor is homogeneous in the transformation coefficients and linear in the components of the tensors.

3.2. Multiplication of Tensors

Again let A_{ik} and B_{ik} be the components of two second-order tensors, but this time consider all possible products of the form

$$C_{iklm} = A_{ik}B_{lm}.$$

Then the numbers C_{iklm} are the components of a fourth-order tensor, called the (*outer*) *product* of the tensors with components A_{ik} and B_{ik}. In fact, it follows from (3.1) that

$$C'_{iklm} = A'_{ik}B'_{lm} = \alpha_{i'n}\alpha_{k'p}\alpha_{l'r}\alpha_{m's}A_{np}B_{rs} = \alpha_{i'n}\alpha_{k'p}\alpha_{l'r}\alpha_{m's}C_{nprs}.$$

It is easy to see that tensor multiplication is noncommutative, e.g.,

$$C_{iklm} = A_{ik}B_{lm} \neq C_{lmik} = A_{lm}B_{ik}.$$

Multiplication of any number of tensors of arbitrary order is defined similarly, i.e., the product of two or more tensors is the tensor whose components are the products of the components of the factors. The order of a tensor product is clearly the sum of the orders of the factors.

In the case of generalized coordinates, some of the indices may be covariant while others are contravariant. For example, the product of the tensors with components $A^{i}_{.kl}$ and B^{ik} is the tensor with components

$$C^{i..mn}_{.kl} = A^{i}_{.kl}B^{mn},$$

where the fact that $C^{i..mn}_{.kl}$ is a fifth-order tensor of the indicated structure is an immediate consequence of the transformation laws of $A^{i}_{.kl}$ and B^{ik}. Note that tensors of arbitrary order and structure can be multiplied (but not added).

3.3. Contraction of Tensors

The operation of summing a tensor of order n ($n \geqslant 2$) over two of its indices is called *contraction*. For example, contraction of the first and second indices of a third-order tensor A_{ikl} gives the quantity

$$A_{iil} = \sum_{i=1}^{3} A_{iil} = A_{11l} + A_{22l} + A_{33l} \qquad (l = 1, 2, 3).$$

The two other possible contractions of A_{ikl} are A_{iki} and A_{ill}. Each such contraction is a first-order tensor, i.e., a vector. In fact, being a third-order tensor, A_{ikl} transforms according to the formula

$$A'_{ikl} = \alpha_{i'm}\alpha_{k'n}\alpha_{l'r}A_{mnr}.$$

Hence, setting $k = i$ and summing over i, we obtain

$$A'_{iil} = \alpha_{i'm}\alpha_{i'n}\alpha_{l'r}A_{mnr} = \delta_{mn}\alpha_{l'r}A_{mnr} = \alpha_{l'r}A_{mmr}$$

[recall the orthogonality conditions (1.62)], i.e., the quantities A'_{iil} transform like a vector, as asserted.

More generally, it is clear that contraction of a tensor of order n ($n \geqslant 2$) leads to a tensor of order $n - 2$. This tensor of order $n - 2$ can then be contracted again (provided that $n \geqslant 4$), giving a tensor of order $n - 4$, and so on, until we obtain a tensor of order less than 2. In fact, repeated contraction of a tensor of order n eventually gives a scalar if n is even and a vector if n is odd.

The result of multiplying two or more tensors and then contracting the product with respect to indices belonging to different factors is often called an *inner product* of the given tensors. For example, the expressions $A_{ik}B_k$ and $\lambda_{iklm}B_{lm}$ are both inner products, and so is the scalar product A_iB_i of two vectors **A** and **B**.

In the case of generalized coordinates, it is important to note that contraction (like summation itself) can be performed only on pairs of indices in different positions (recall Remark 2, p. 34), i.e., one contracted index must be covariant and the other contravariant. Otherwise, the result of contraction will not be a tensor. For example, suppose we contract the tensor $A_i^{\cdot kl}$ in the indices i and k. Then $A_i^{\cdot il}$ is a tensor (in fact, a contravariant vector), since it follows from (2.52) and (1.14) that

$$A_i'^{\cdot il} = \alpha_{i'}^m \alpha_n^{i'} \alpha_r^{l'} A_m^{\cdot nr} = \alpha_r^{l'} A_n^{\cdot nr}.$$

However, contracting $A_i^{\cdot kl}$ in the indices k and l gives a quantity whose transformation law

$$A_i'^{\cdot kk} = \alpha_{i'}^m \alpha_n^{k'} \alpha_r^{k'} A_m^{\cdot nr}$$

is not that of a vector. Similarly, in forming inner products in generalized coordinates, we can only sum over indices in different positions, obtaining expressions like A^iB_i, $A_{ik}B^k$, $\lambda_{ik}^{\cdot\cdot lm}B_{lm}$, $\lambda_{ik\cdot m}^{\cdot\cdot l}B_l^{\cdot m}$, etc.

3.4. Symmetry Properties of Tensors

3.4.1. Symmetric and antisymmetric tensors. A tensor $S_{ikl\ldots}$ (of order 2 or higher) is said to be *symmetric* in the indices i and k (say) if

$$S_{ikl\ldots} = S_{kil\ldots},$$

i.e., if interchanging i and k has no effect on the corresponding components. Thus

$$S_{12l\ldots} = S_{21l\ldots}, \qquad S_{23l\ldots} = S_{32l,\ldots},$$

and so on.

A tensor $A_{ikl\ldots}$ (of order 2 or higher) is said to be *antisymmetric* in the indices i and k (say) if

$$A_{ikl\ldots} = -A_{kil\ldots}, \tag{3.2}$$

i.e., if interchanging i and k changes the sign of the corresponding components. Thus

$$A_{12l\ldots} = -A_{21l\ldots}, \qquad A_{23l\ldots} = -A_{32l\ldots},$$

and so on. If $A_{ikl\ldots}$ is antisymmetric in a pair of indices, then the components obtained by equating these indices must vanish. For example, (3.2) implies $A_{11l\ldots} = -A_{11l\ldots}$ and hence $A_{11l\ldots} = 0$.

Example 1. In a system of rectangular coordinates x_1, x_2, x_3 with orthonormal basis $\mathbf{i}_1, \mathbf{i}_2, \mathbf{i}_3$, the element of arc length is given by

$$(ds)^2 = \delta_{jk}\, dx_j\, dx_k$$

in terms of the symmetric second-order tensor $\delta_{jk} = \mathbf{i}_j \cdot \mathbf{i}_k$ called the Kronecker delta (see p. 39) or the unit tensor (see p. 70).

Example 2. Given two vectors $\mathbf{A}$ and $\mathbf{B}$, with components A_i and B_i, the second-order tensor with components

$$C_{ik} = A_iB_k - A_kB_i$$

is antisymmetric.

A tensor which is symmetric (or antisymmetric) in one coordinate system remains symmetric (or antisymmetric) in any other coordinate system. In fact, if T_{ik} is symmetric in a given system, i.e., if $T_{ik} = T_{ki}$, then

$$T'_{ik} = \alpha_{i'l}\alpha_{k'm}T_{lm} = \alpha_{i'l}\alpha_{k'm}T_{ml} = \alpha_{k'm}\alpha_{i'l}T_{ml} = T'_{ki},$$

and similarly for antisymmetry and tensors of higher order.

A symmetric second-order tensor S_{ik} has a matrix of the form

$$\|S_{ik}\| = \begin{Vmatrix} S_{11} & S_{12} & S_{13} \\ S_{12} & S_{22} & S_{23} \\ S_{13} & S_{23} & S_{33} \end{Vmatrix},$$

while an antisymmetric second-order tensor A_{ik} has a matrix of the form

$$\|A_{ik}\| = \begin{Vmatrix} 0 & A_{12} & A_{13} \\ -A_{12} & 0 & A_{23} \\ -A_{13} & -A_{23} & 0 \end{Vmatrix}.$$

Thus a symmetric second-order tensor has 6 independent components, while an antisymmetric second-order tensor has only 3 independent components.

Any second-order tensor T_{ik} can be represented as a sum of a symmetric tensor and an antisymmetric tensor. In fact, let

$$T_{ik} = S_{ik} + A_{ik},$$

where

$$S_{ik} = \tfrac{1}{2}(T_{ik} + T_{ki}), \qquad A_{ik} = \tfrac{1}{2}(T_{ik} - T_{ki}).$$

Then obviously S_{ik} is symmetric and A_{ik} is antisymmetric. S_{ik} is called the *symmetric part* of T_{ik} and A_{ik} is called the *antisymmetric part* of T_{ik}.

Remark. The operation leading from an arbitrary tensor with components T_{ik} to the tensor with components $T_{ik} + T_{ki}$ is called *symmetrization*, while the operation leading to the tensor with components $T_{ik} - T_{ki}$ is called *antisymmetrization.*

In generalized coordinates, the concepts of symmetry and antisymmetry apply only to pairs of indices in the same positions. Thus $A_{ik}^{\cdot\cdot l}$ is symmetric in i and k if $A_{ik}^{\cdot\cdot l} = A_{ki}^{\cdot\cdot l}$, while $B^{ik}_{\cdot\cdot l}$ is antisymmetric in i and k if $B^{ik}_{\cdot\cdot l} = -B^{ki}_{\cdot\cdot l}$.

Example. In a system of generalized coordinates x^1, x^2, x^3 with basis $\mathbf{e}_1, \mathbf{e}_2, \mathbf{e}_3$, the element of arc length is given by

$$(ds)^2 = g_{jk}dx^j dx^k$$

in terms of the symmetric second-order tensor $g_{jk} = \mathbf{e}_j \cdot \mathbf{e}_k$, called the metric tensor (see Sec. 2.9.2).

3.4.2. Equivalence of an antisymmetric second-order tensor to an axial vector. Let A_{jk} be an antisymmetric second-order tensor. Then the transformation law of A_{jk} is

$$\begin{aligned} A'_{jk} = \alpha_{j'l}\alpha_{k'm}A_{lm} &= \alpha_{j'1}\alpha_{k'2}A_{12} + \alpha_{j'2}\alpha_{k'1}A_{21} + \alpha_{j'1}\alpha_{k'3}A_{13} \\ &\quad + \alpha_{j'3}\alpha_{k'1}A_{31} + \alpha_{j'2}\alpha_{k'3}A_{23} + \alpha_{j'3}\alpha_{k'2}A_{32} \end{aligned} \tag{3.3}$$

$(A_{11} = A_{22} = A_{33} = 0)$. Since $A_{lm} = -A_{ml}$, we can write (3.3) in the form

$$A'_{jk} = (\alpha_{j'l}\alpha_{k'm} - \alpha_{j'm}\alpha_{k'l})A_{lm}, \tag{3.4}$$

where the indices l and m are restricted to the values 1, 2 or 2, 3 or 3, 1. Suppose we introduce the notation

$$A_{12} = -A_{21} \equiv A_3, \quad A_{23} = -A_{32} \equiv A_1, \quad A_{31} = -A_{13} \equiv A_2,$$

or more concisely,

$$A_{lm} \equiv A_n$$

where l, m, n is a cyclic permutation of the numbers 1, 2, 3 and similarly

$$A'_{jk} \equiv A'_r$$

where j, k, r is also a cyclic permutation of 1, 2, 3. Then (3.4) can be written as

$$A'_r = \sum_{l,m,n} (\alpha_{j'l}\alpha_{k'm} - \alpha_{j'm}\alpha_{k'l})A_n, \tag{3.5}$$

where the indices r, j, k and l, m, n are both cyclic permutations of 1, 2, 3.

To simplify (3.5) further, we first expand the orthonormal basis $\mathbf{i}'_1, \mathbf{i}'_2, \mathbf{i}'_3$ of the new coordinate system K' with respect to the orthonormal basis $\mathbf{i}_1, \mathbf{i}_2, \mathbf{i}_3$ of the old coordinate system K:

$$\mathbf{i}'_r = \alpha_{r'l}\mathbf{i}_l.$$

Then we calculate the vector products $\mathbf{i}'_j \times \mathbf{i}'_k$, obtaining

$$\mathbf{i}'_j \times \mathbf{i}'_k = \alpha_{j'l}\alpha_{k'm}(\mathbf{i}_l \times \mathbf{i}_m), \tag{3.6}$$

where the vector products are both taken in some underlying right-handed coordinate system (say K itself) and the sum on the right is over all values of l and m. Forming the scalar product of (3.6) with $\mathbf{i}_n$, we find that

$$(\mathbf{i}'_j \times \mathbf{i}'_k) \cdot \mathbf{i}_n = \alpha_{j'l}\alpha_{k'm}(\mathbf{i}_l \times \mathbf{i}_m) \cdot \mathbf{i}_n,$$

where in the right-hand side only two terms differing in sign survive for each fixed value of n. Specifically, if the old system K is right-handed, then

$$(\mathbf{i}'_j \times \mathbf{i}'_k) \cdot \mathbf{i}_n = \alpha_{j'l}\alpha_{k'm} - \alpha_{j'm}\alpha_{k'l}, \tag{3.7}$$

where l, m, n is a cyclic permutation of 1, 2, 3. Moreover, if the new system K' is right-handed, like K itself, then

$$\mathbf{i}'_j \times \mathbf{i}'_k = \mathbf{i}'_r,$$

while if K' is left-handed,

$$\mathbf{i}'_j \times \mathbf{i}'_k = -\mathbf{i}'_r,$$

where in both formulas j, k, r is a cyclic permutation of 1, 2, 3. It follows that

$$(\mathbf{i}'_j \times \mathbf{i}'_k) \cdot \mathbf{i}_n = \mathbf{i}'_r \cdot \mathbf{i}_n = \alpha_{r'n}$$

if K' is right-handed (like K), while

$$(\mathbf{i}'_j \times \mathbf{i}'_k) \cdot \mathbf{i}_n = -\mathbf{i}'_r \cdot \mathbf{i}_n = -\alpha_{r'n}$$

if K' is left-handed. Therefore formula (3.7) finally becomes

$$\alpha_{r'n} = \pm(\alpha_{j'l}\alpha_{k'm} - \alpha_{j'm}\alpha_{k'l}), \tag{3.8}$$

where j, k, r and n, l, m are both cyclic permutations of 1, 2, 3 and we choose the plus sign if K' is right-handed and the minus sign if K' is left-handed.

Returning now to the transformation law (3.5) and using (3.8), we find that

$$A'_r = \alpha_{r'n}A_n \tag{3.9}$$

under transformations from a right-handed coordinate system to another right-handed system, while

$$A'_r = -\alpha_{r'n} A_n \tag{3.10}$$

under transformations from a right-handed system to a left-handed system. It can be shown (see Sec. 3.7.1) that the law (3.9) also governs transformations from a left-handed coordinate system to another left-handed system, while (3.10) governs transformations from a left-handed system to a right handed system. Vectors transforming according to formulas (3.9) and (3.10) are called *axial* vectors, and have already appeared on p. 18. Thus we have finally shown that *the components of an antisymmetric second-order tensor transform like the components of an axial vector.*

Axial vectors are a special case of a class of tensors, called *pseudotensors,* whose components change sign when the "handedness" of the coordinate system is changed. Pseudotensors will be discussed in detail in Sec. 3.7.

3.5. Reduction of Tensors to Principal Axes

3.5.1. Statement of the problem. We now consider a problem of great physical importance. Given a fixed second-order tensor with components T_{ik} and any vector **A** with components A_i, we form the inner product

$$T_{ik}A_k = B_i,$$

thereby obtaining a new vector **B** with components B_i. In general, the vector **B** differs from **A** in both direction and magnitude, i.e., the operation $T_{ik}A_k$ both rotates **B** and changes its length. Suppose we pose the problem of finding all vectors **A** which are not rotated by inner multiplication with T_{ik}, i.e., all vectors **A** such that

$$T_{ik}A_k = \lambda A_i, \tag{3.11}$$

where λ is a scalar (the length of **A** is changed if $\lambda \neq \pm 1$). Such vectors, if they exist, are called the *characteristic vectors* or *eigenvectors* of the tensor T_{ik}, and their directions are called the *characteristic* or *principal directions* of T_{ik}. Moreover, the axes determined by the principal directions are called the *principal axes* of T_{ik}. The problem of finding the principal axes of T_{ik} is often called the problem of reducing T_{ik} to principal axes (more exactly, to principal axis form).

The values of the components T_{ik} in the coordinate system determined by the principal axes are called the *characteristic values* or *eigenvalues* of the corresponding tensor. It will turn out that these are just the values of λ for which equation (3.11) has solutions.

We now give two examples illustrating the physical meaning of these concepts.

Example 1. If p_{ik} is the stress tensor, then the stress $\mathbf{p}_n$ on the element of area with unit normal $\mathbf{n}$ has components

$$p_{nk} = p_{ik}n_i$$

(see p. 67), where in general $\mathbf{p}_n$ is not parallel to $\mathbf{n}$, i.e., in general there are tangential as well as normal stresses on any element of area. Suppose we are interested in finding elements of area on which there are only normal stresses, with no tangential stresses at all. For such elements of area, $\mathbf{p}_n$ is parallel to $\mathbf{n}$ and hence

$$\mathbf{p}_n = \mathbf{p}_i n_i = \lambda \mathbf{n}$$

or

$$p_{ik}n_i = \lambda n_k.$$

In other words, the normals to these elements of area lie along the principal directions of the tensor p_{ik}.

Example 2. The displacement vector **D** in a dielectric medium is a linear vector function (recall Example 3, p. 65) of the electric field **E**, i.e.,

$$D_i = \varepsilon_{ik}E_k$$

where ε_{ik} is the *dielectric tensor*. In general, **D** and **E** have different directions. However, for special choices of **E**, determined by the solutions of the equation

$$\varepsilon_{ik}E_k = \lambda E_i,$$

the directions of **D** and **E** coincide.

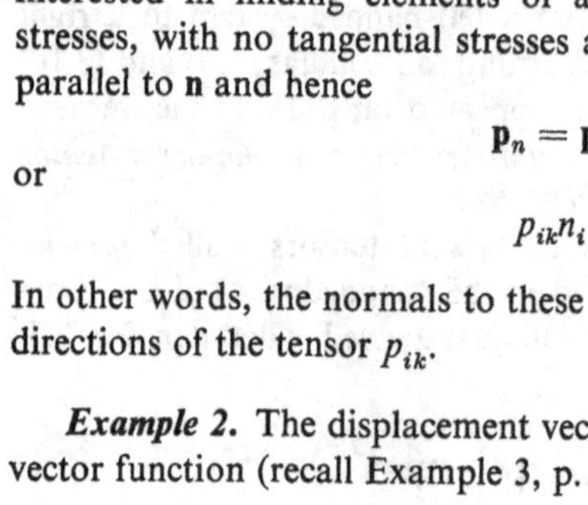

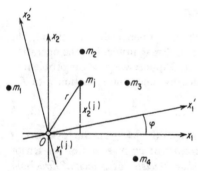

FIG. 3.1. The principal axes of the moment of inertia tensor of a plane system of particles.

3.5.2. The two-dimensional case. Before studying the problem of reducing three-dimensional tensors to principal axes, we consider the simpler problem obtained when the number of dimensions is only two. Thus, to be explicit, consider a system of n particles of masses $m_1, m_2, \ldots, m_n$ distributed in a plane, and let I_{ik} be the moment of inertia tensor of the system with respect to a system of rectangular coordinates x_1, x_2 with origin O (see Fig. 3.1). Being two-dimensional, the tensor I_{ik} has only four components. Its matrix is of the form

$$\|I_{ik}\| = \begin{Vmatrix} I_{11} & I_{12} \\ I_{21} & I_{22} \end{Vmatrix},$$

where

$$I_{11} = \sum_{j=1}^{n} m_j (x_2^{(j)})^2,$$

$$I_{22} = \sum_{j=1}^{n} m_j (x_1^{(j)})^2,$$

$$I_{12} = I_{21} = -\sum_{j=1}^{n} m_j x_1^{(j)} x_2^{(j)}.$$

If the vector **A** lies along a principal axis of the tensor I_{ik}, then its components must satisfy the homogeneous system

$$\begin{aligned} I_{11}A_1 + I_{21}A_2 &= \lambda A_1, \\ I_{12}A_1 + I_{22}A_2 &= \lambda A_2 \end{aligned} \tag{3.12}$$

or

$$\begin{aligned} (I_{11} - \lambda)A_1 + I_{12}A_2 &= 0, \\ I_{12}A_1 + (I_{22} - \lambda)A_2 &= 0 \end{aligned} \tag{3.13}$$

[cf. (3.11)]. The system (3.13) has a nontrivial (i.e., nonzero) solution if and only if

$$\begin{vmatrix} I_{11} - \lambda & I_{12} \\ I_{12} & I_{22} - \lambda \end{vmatrix} = 0$$

or

$$\lambda^2 - \lambda(I_{11} + I_{22}) + I_{11}I_{22} - I_{12}^2 = 0. \tag{3.14}$$

Therefore (3.13) has a nontrivial solution if and only if $\lambda = \lambda_1$ or $\lambda = \lambda_2$, where

$$\begin{aligned} \lambda_1 &= \frac{I_{11} + I_{22}}{2} + \sqrt{\left(\frac{I_{11} - I_{22}}{2}\right)^2 + I_{12}^2}, \\ \lambda_2 &= \frac{I_{11} + I_{22}}{2} - \sqrt{\left(\frac{I_{11} - I_{22}}{2}\right)^2 + I_{12}^2} \end{aligned} \tag{3.15}$$

are the roots of equation (3.14).

If $I_{12} = 0$, it follows from (3.12) that the original x_1 and x_2-axes are the principal axes, since then **A** is not rotated by inner multiplication with I_{ik}. Thus suppose $I_{12} \neq 0$. Then $\lambda_1 \neq \lambda_2$ and there are two distinct principal axes, determined by vectors $\mathbf{A}^{(1)}$ and $\mathbf{A}^{(2)}$. Using (3.13), we find that the directions of $\mathbf{A}^{(1)}$ and $\mathbf{A}^{(2)}$ have slopes

$$\tan \varphi_1 = \frac{A_2^{(1)}}{A_1^{(1)}} = \frac{\lambda_1 - I_{11}}{I_{12}} = \frac{I_{12}}{\lambda_1 - I_{22}},$$

$$\tan \varphi_2 = \frac{A_2^{(2)}}{A_1^{(2)}} = \frac{\lambda_2 - I_{11}}{I_{12}} = \frac{I_{12}}{\lambda_2 - I_{22}},$$

where φ_1 and φ_2 are the angles between the x_1-axis and the principal axes of the tensor I_{ik}. Therefore

$$\tan 2\varphi_1 = \frac{2\tan\varphi_1}{1-\tan^2\varphi_1} = \frac{2\dfrac{\lambda_1 - I_{11}}{I_{12}}}{1-\left(\dfrac{\lambda_1 - I_{11}}{I_{12}}\right)^2}$$

$$= \frac{2I_{12}(\lambda_1 - I_{11})}{I_{12}^2 - (\lambda_1 - I_{11})^2},$$

and similarly for $\tan\varphi_2$ with λ_2 instead of λ_1. Substituting from (3.15) and carrying out some elementary calculations, we find that

$$\tan 2\varphi_1 = \tan 2\varphi_2 = \frac{2I_{12}}{I_{11} - I_{22}}.$$

Hence

$$\varphi_2 = \varphi_1 + \frac{\pi}{2},$$

i.e., the principal axes are perpendicular to each other.

In the coordinate system determined by the principal axes, the tensor I_{ik} has components

$$I'_{11} = \lambda_1,$$
$$I'_{22} = \lambda_2, \qquad (3.16)$$
$$I'_{12} = 0.$$

Rather than verify (3.16) by direct calculation, we need only write (3.12) in the principal axis (primed) system, obtaining

$$\lambda_1 A_1^{(1)'} = I'_{11}A_1^{(1)'} + I'_{12}A_2^{(1)'},$$
$$\lambda_1 A_2^{(1)'} = I'_{12}A_1^{(1)'} + I'_{22}A_2^{(1)'},$$

and

$$\lambda_2 A_1^{(2)'} = I'_{11}A_1^{(2)'} + I'_{12}A_2^{(2)'},$$
$$\lambda_2 A_2^{(2)'} = I'_{12}A_1^{(2)'} + I'_{22}A_2^{(2)'},$$

which immediately give (3.16) since $A_1^{(1)'} \neq 0$, $A_2^{(1)'} = 0$ and $A_1^{(2)'} = 0$, $A_2^{(2)'} \neq 0$.

The principal axes of I_{ik} can be found more simply if it is known in advance that the principal axes are perpendicular and that I'_{ik} vanishes in the principal axis system. Thus, introducing new rectangular coordinates x'_1 and x'_2, we have

$$I'_{12} = \alpha_{1'k}\alpha_{2'l}I_{kl},$$

where all the cosines $\alpha_{1'k}$, $\alpha_{2'l}$ can be expressed in terms of one parameter φ, the angle between the new x_1'-axis and the old x_1-axis (see Fig. 3.1). In fact,

$$\alpha_{1'1} = \cos\varphi, \qquad \alpha_{1'2} = \sin\varphi,$$

$$\alpha_{2'1} = -\sin\varphi, \qquad \alpha_{2'2} = \cos\varphi,$$

and therefore

$$\begin{aligned} I'_{12} &= \alpha_{1'1}\alpha_{2'1}I_{11} + \alpha_{1'1}\alpha_{2'2}I_{12} + \alpha_{1'2}\alpha_{2'1}I_{21} + \alpha_{1'2}\alpha_{2'2}I_{22} \\ &= -I_{11}\cos\varphi\sin\varphi + I_{12}\cos^2\varphi + I_{21}\sin^2\varphi + I_{22}\cos\varphi\sin\varphi \\ &= \frac{I_{22}-I_{11}}{2}\sin 2\varphi + I_{12}\cos 2\varphi. \end{aligned}$$

Hence, to make I'_{12} vanish, we must set

$$\tan 2\varphi = \frac{2I_{12}}{I_{11}-I_{22}},$$

thereby uniquely determining the positions of the principal axes of the tensor I_{ik}. Moreover, the characteristic value I'_{11} is given by

$$\begin{aligned} I'_{11} &= \alpha_{1'k}\alpha_{1'l}I_{kl} = I_{11}\cos^2\varphi + I_{12}\sin 2\varphi + I_{22}\sin^2\varphi \\ &= I_{11}\frac{1+\cos 2\varphi}{2} + I_{22}\frac{1-\cos 2\varphi}{2} + I_{12}\sin 2\varphi \\ &= \frac{I_{11}+I_{22}}{2} + \frac{I_{11}-I_{22}}{2}\cos 2\varphi + I_{12}\sin 2\varphi \\ &= \frac{I_{11}+I_{22}}{2} + \sqrt{\left(\frac{I_{11}-I_{22}}{2}\right)^2 + I_{12}^2} \qquad (= \lambda_1), \end{aligned}$$

and similarly

$$I'_{22} = \alpha_{2'k}\alpha_{2'l}I_{kl} = \frac{I_{11}+I_{22}}{2} - \sqrt{\left(\frac{I_{11}-I_{22}}{2}\right)^2 + I_{12}^2} \qquad (= \lambda_2).$$

Thus the moment of inertia tensor in the principal axis system is

$$\|I'_{ik}\| = \begin{Vmatrix} I'_{11} & 0 \\ 0 & I'_{22} \end{Vmatrix}.$$

3.5.3. The three-dimensional case. We now consider the problem of reducing a three-dimensional tensor T_{ik} to principal axes. According to

(3.11), the components A_i of the vectors **A** determining the principal axes of T_{ik} satisfy the homogeneous system

$$T_{ik}A_k - \lambda A_i = (T_{ik} - \lambda\delta_{ik})A_k = 0,$$

or

$$\begin{aligned} (T_{11} - \lambda)A_1 + T_{12}A_2 + T_{13}A_3 &= 0, \\ T_{21}A_1 + (T_{22} - \lambda)A_2 + T_{23}A_3 &= 0, \\ T_{31}A_1 + T_{32}A_2 + (T_{33} - \lambda)A_3 &= 0, \end{aligned} \tag{3.17}$$

when written out in full. The system (3.17) has a nontrivial solution if and only if its determinant vanishes:

$$\begin{vmatrix} T_{11} - \lambda & T_{12} & T_{13} \\ T_{21} & T_{22} - \lambda & T_{23} \\ T_{31} & T_{32} & T_{33} - \lambda \end{vmatrix} = 0. \tag{3.18}$$

The equation (3.18), called the *characteristic equation* of the tensor T_{ik}, is clearly cubic in λ.

Remark. In the case of generalized coordinates, there are various ways of reducing a tensor to principal axes, depending on how the components are chosen. For example, choosing covariant components, we have

$$T_{ik}A^k = \lambda A_i \tag{3.19}$$

instead of (3.11). Since $A_i = g_{ik}A^k$, (3.19) implies

$$(T_{ik} - \lambda g_{ik})A^k = 0. \tag{3.20}$$

However, to bring (3.20) into a form resembling (3.17), we must use mixed components of the tensor. In fact, replacing the dummy index i by l in (3.20), multiplying by g^{il} and summing over l, we obtain

$$(T^i_{.k} - \lambda g^i_{.k})A^k = 0$$

[recall (2.53)]. This gives the following characteristic equation for determining the eigenvalues of the tensor:

$$\begin{vmatrix} T^1_{.1} - \lambda & T^1_{.2} & T^1_{.3} \\ T^2_{.1} & T^2_{.2} - \lambda & T^2_{.3} \\ T^3_{.1} & T^3_{.2} & T^3_{.3} - \lambda \end{vmatrix} = 0. \tag{3.21}$$

Henceforth we shall assume that T_{ik} is a symmetric Cartesian tensor, i.e., a tensor written in rectangular components whose components satisfy the condition $T_{ik} = T_{ki}$. In this case, the roots of the characteristic equation

(3.18) are all real. To see this, let λ be any root of (3.18) and let A_i be the corresponding eigenvector, so that

$$T_{ik}A_k = \lambda A_i.$$

Multiplying this identity by $\bar{A}_i$ (the overbar denotes the complex conjugate) and summing over i, we obtain

$$T_{ik}A_k\bar{A}_i = \lambda A_i\bar{A}_i. \tag{3.22}$$

The left-hand side of (3.22) is real. In fact, since $T_{ik} = T_{ki}$,

$$\begin{aligned} T_{ik}A_k\bar{A}_i &= \tfrac{1}{2}(T_{ik}A_k\bar{A}_i + T_{ik}A_k\bar{A}_i) \\ &= \tfrac{1}{2}(T_{ik}A_k\bar{A}_i + T_{ki}\bar{A}_kA_i) \\ &= \tfrac{1}{2}T_{ik}(A_k\bar{A}_i + \bar{A}_kA_i), \end{aligned}$$

where the last expression is the product of two real numbers (T_{ik} is real by hypothesis and $A_k\bar{A}_i + \bar{A}_kA_i$ is real by inspection). But $A_i\bar{A}_i$ is obviously real, and hence (3.22) implies that λ itself is real.

Suppose now that the eigenvalues $\lambda_1, \lambda_2, \lambda_3$ are all distinct (the case of multiple eigenvalues will be treated later). Then each eigenvalue λ_r is characterized by a set of numbers $A_1^{(r)}, A_2^{(r)}, A_3^{(r)}$ (the components of the eigenvector $\mathbf{A}^{(r)}$) determined to within a constant factor by the system of equations

$$(T_{ik} - \lambda_r\delta_{ik})A_k^{(r)} = 0.$$

For example,

$$\begin{aligned} (T_{11} - \lambda_1)A_1^{(1)} + T_{12}A_2^{(1)} + T_{13}A_3^{(1)} &= 0, \\ T_{21}A_1^{(1)} + (T_{22} - \lambda_1)A_2^{(1)} + T_{23}A_3^{(1)} &= 0, \\ T_{31}A_1^{(1)} + T_{32}A_2^{(1)} + (T_{33} - \lambda_1)A_3^{(1)} &= 0, \end{aligned}$$

which implies

$$\frac{A_1^{(1)}}{\begin{vmatrix} T_{22} - \lambda_1 & T_{23} \\ T_{32} & T_{33} - \lambda_1 \end{vmatrix}} = \frac{A_2^{(1)}}{\begin{vmatrix} T_{23} & T_{21} \\ T_{33} - \lambda_1 & T_{31} \end{vmatrix}} = \frac{A_3^{(1)}}{\begin{vmatrix} T_{21} & T_{22} - \lambda_1 \\ T_{31} & T_{32} \end{vmatrix}}. \tag{3.23}$$

The relations (3.23) uniquely determine the direction (but not the magnitude) of the vector $\mathbf{A}^{(1)}$, i.e., the direction of one of the principal axes of the tensor T_{ik}. The other two principal directions are determined by the vectors $\mathbf{A}^{(2)}$ and $\mathbf{A}^{(3)}$.

It is easy to see that the principal axes of the tensor T_{ik} are perpendicular. In fact, let $\mathbf{A}^{(r)}$ and $\mathbf{A}^{(s)}$ be the eigenvectors corresponding to the eigenvalues λ_r and λ_s ($r \neq s$), respectively, so that

$$T_{ik}A_k^{(r)} = \lambda_r A_i^{(r)}, \tag{3.24}$$

$$T_{ik}A_k^{(s)} = \lambda_s A_i^{(s)}. \tag{3.25}$$

Multiplying (3.24) by $A_i^{(s)}$ and (3.25) by $A_i^{(r)}$, summing over i and subtracting the second equation from the first, we obtain

$$T_{ik}A_k^{(r)}A_i^{(s)} - T_{ik}A_k^{(s)}A_i^{(r)} = (\lambda_r - \lambda_s)A_i^{(r)}A_i^{(s)}.$$

Since $T_{ik} = T_{ki}$, it follows that

$$0 = (\lambda_r - \lambda_s)A_i^{(r)}A_i^{(s)},$$

and hence

$$A_i^{(r)}A_i^{(s)} = \mathbf{A}^{(r)} \cdot \mathbf{A}^{(s)} = 0$$

since $\lambda_r \neq \lambda_s$ (the eigenvalues are distinct, by hypothesis). In other words, $\mathbf{A}^{(r)}$ is perpendicular to $\mathbf{A}^{(s)}$, as asserted.

There is another way of proving the existence of three perpendicular principal axes which works even when the characteristic equation (3.18) has multiple roots. Let λ_1 be a root of (3.18) and let $\mathbf{A}^{(1)}$ be the corresponding eigenvector describing a principal axis of the tensor T_{ik}, so that

$$T_{ik}A_k^{(1)} = \lambda_1 A_i^{(1)}.$$

Let M be the plane through the initial point of $\mathbf{A}^{(1)}$ perpendicular to $\mathbf{A}^{(1)}$. Then T_{ik} carries every vector in M into another vector in M. In fact, let $\mathbf{P}$ be a vector in M, so that

$$\mathbf{P} \cdot \mathbf{A}^{(1)} = P_i A_i^{(1)} = 0.$$

Then

$$T_{ik}A_k^{(1)}P_i = \lambda_1 A_i^{(1)}P_i = 0,$$

i.e., the vector $\mathbf{Q}$ with components $Q_k = T_{ik}P_i$ is perpendicular to $\mathbf{A}^{(1)}$ (since $Q_k A_k^{(1)} = 0$) and hence lies in the plane M (see Fig. 3.2).

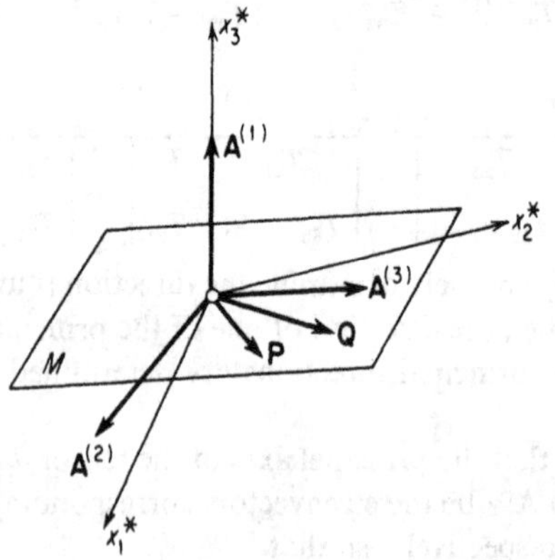

FIG. 3.2. Illustrating the three principal axes of a symmetric tensor. The existence of such axes does not depend on the multiplicity of the roots of the characteristic equation.

Now consider a system of rectangular coordinates x_1^*, x_2^*, x_3^* such that the x_3^*-axis lies along $\mathbf{A}^{(1)}$, while the other two axes lie in the plane M, as in Fig. 3.2. In this coordinate system, a vector $\mathbf{A}$ in the plane M has components A_1^*, A_2^*, 0, and the effect of the tensor on $\mathbf{A}$ is described by a matrix of the form

$$\|T_{ik}^*\| = \left\| \begin{matrix} T_{11}^* & T_{12}^* & 0 \\ T_{21}^* & T_{22}^* & 0 \\ 0 & 0 & 1 \end{matrix} \right\| \qquad (T_{12}^* = T_{21}^*).$$

Suppose we look for vectors $\mathbf{A}$ in the plane M which are not rotated by T_{ik}, but only have their lengths changed. Then we get *two* equations

$$T_{ik}^* A_k^* = \lambda A_i^* \qquad (i = 1, 2)$$

(the third equation $T_{3k}^* A_k^* = A_3^*$ reduces to the trivial identity $0 = 0$), and a corresponding characteristic equation

$$\begin{vmatrix} T_{11}^* - \lambda & T_{12}^* \\ T_{21}^* & T_{22}^* - \lambda \end{vmatrix} = 0$$

which is quadratic instead of cubic in λ. Let λ_2 be a root of this equation, and let $\mathbf{A}^{(2)}$ be the corresponding vector, so that

$$T_{ik}^* A_k^{(2)*} = \lambda_2 A_i^{(2)*}$$

Then $\mathbf{A}^{(2)}$ determines a second principal axis of the tensor. Moreover, $\mathbf{A}^{(2)}$ is perpendicular to $\mathbf{A}^{(1)}$, since it lies in the plane M.

Finally, let $\mathbf{A}^{(3)}$ be a vector perpendicular to both $\mathbf{A}^{(1)}$ and $\mathbf{A}^{(2)}$. Then the vector with components $T_{ik}^* A_i^{(3)*}$ is perpendicular to $\mathbf{A}^{(1)}$ (since it lies in the plane M) and also perpendicular to $\mathbf{A}^{(2)}$ (since the orthogonality of $\mathbf{A}^{(2)}$ and $\mathbf{A}^{(3)}$ implies $T_{ik}^* A_i^{(3)*} A_k^{(2)*} = \lambda_2 A_i^{(2)*} A_i^{(3)*} = 0$). Therefore the vector with components $T_{ik}^* A_i^{(3)*}$ is collinear with $\mathbf{A}^{(3)}$, i.e.,

$$T_{ik}^* A_i^{(3)*} = \lambda_3 A_k^{(3)*}$$

for some constant λ_3. But then

$$T_{ki}^* A_i^{(3)} = \lambda_3 A_k^{(3)}$$

since $T_{ik}^* = T_{ki}^*$, i.e., $\mathbf{A}^{(3)}$ determines a principal axis of our tensor. Thus, since it was nowhere stipulated that the numbers $\lambda_1, \lambda_2, \lambda_3$ be distinct, we have finally proved that a symmetric tensor has three perpendicular principal axes regardless of the multiplicity of the roots of the characteristic equation (3.18).

Suppose the tensor T_{ik} is originally written in a system of rectangular coordinates x_1, x_2, x_3. Then the principal axes of T_{ik} determine a new system of rectangular coordinates x'_1, x'_2, x'_3 with a corresponding orthonormal basis

$$\mathbf{n}^{(r)} = \frac{\mathbf{A}^{(r)}}{|\mathbf{A}^{(r)}|} \qquad (r = 1, 2, 3)$$

(see Fig. 3.3). In the old coordinates we have

$$T_{ik}n_k^{(r)} = \lambda_k n_i^{(r)} \qquad (i, r = 1, 2, 3),$$

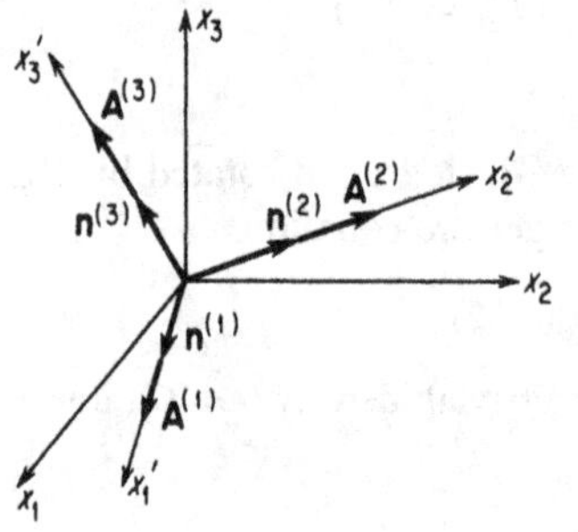

FIG. 3.3. The principal axes of a symmetric tensor.

where $n_i^{(r)}$ is the ith component of $\mathbf{n}^{(r)}$. Similarly

$$T'_{ik}n_k^{(r)'} = \lambda_r n_i^{(r)'} \qquad (i, r = 1, 2, 3) \quad (3.26)$$

in the new coordinates. Since $n_1^{(1)'} = n_2^{(2)'} = n_3^{(3)'} = 1$ and $n_i^{(r)'} = 0$ if $i \neq r$, the nine equations (3.26) reduce to

$$T'_{11} = 1, \quad T'_{12} = 0, \quad T'_{13} = 0,$$

$$T'_{21} = 0, \quad T'_{22} = 1, \quad T'_{23} = 0,$$

$$T'_{31} = 0, \quad T'_{32} = 0, \quad T'_{33} = 1.$$

Therefore in the principal axis system our tensor has a diagonal matrix of the form

$$\|T'_{ik}\| = \begin{Vmatrix} \lambda_1 & 0 & 0 \\ 0 & \lambda_2 & 0 \\ 0 & 0 & \lambda_3 \end{Vmatrix},$$

with the eigenvalues of the tensor (the roots $\lambda_1, \lambda_2, \lambda_3$ of its characteristic equation) along the main diagonal and zeros everywhere off the main diagonal.

3.5.4. The tensor ellipsoid. Suppose the tensor T_{ik} carries a vector $\mathbf{P}$ into a vector $\mathbf{Q}$, so that

$$Q_i = T_{ik}P_k. \qquad (3.27)$$

In the principal axis (primed) system, (3.27) becomes

$$Q'_i = T'_{ik}P'_k$$

or

$$\begin{aligned} Q'_1 &= T'_{11}P'_1 = \lambda_1 P'_1, \\ Q'_2 &= T'_{22}P'_2 = \lambda_2 P'_2, \\ Q'_3 &= T'_{33}P'_3 = \lambda_3 P'_3, \end{aligned} \qquad (3.28)$$

when written out in full. If the eigenvalues $\lambda_1, \lambda_2, \lambda_3$ are all distinct, then T_{ik} has three unique principal directions (those determined by $\lambda_1, \lambda_2, \lambda_3$). In fact, it follows from (3.28) that the effect of applying T_{ik} to **P** is to rotate **P** as well as change its length unless **P** lies along one of the principal axes.

Next suppose $\lambda_1 = \lambda_2 \neq \lambda_3$, so that the characteristic equation has one simple root and one double root. Then (3.28) implies

$$Q'_1 = \lambda_1 P'_1,$$
$$Q'_2 = \lambda_1 P'_2,$$
$$Q'_3 = \lambda_3 P'_3,$$

and hence the effect of applying T_{ik} to any vector **P** in the $x'_1x'_2$-plane is merely to change the length of **P** but not to rotate **P**. In other words, the whole $x'_1x'_2$-plane is a *characteristic plane* in the sense that every direction in this plane is a principal direction. Thus if one principal direction has been determined (corresponding to the eigenvalue λ_3, say) and if the other two eigenvalues coincide, then any two directions perpendicular to each other and to the first direction can serve as principal axes of the tensor.

Finally, suppose the characteristic equation has a triple root $\lambda_1 = \lambda_2 = \lambda_3 = \lambda$. Then (3.28) becomes

$$Q'_1 = \lambda P'_1,$$
$$Q'_2 = \lambda P'_2,$$
$$Q'_3 = \lambda P'_3,$$

i.e., the effect of applying T_{ik} to any vector at all is now to change its length (if $\lambda \neq 1$) without rotating it. In other words, in this case every direction is a principal direction of the tensor T_{ik}, which is said to be isotropic (as in Prob. 4, p. 96).

There is a one-to-one correspondence between vectors **A** and planes of the form

$$\mathbf{A} \cdot \mathbf{r} = A_i x^i = 1.$$

In the same way, there is a one-to-one correspondence between *symmetric* tensors T_{ik} and quadric surfaces of the form

$$T_{ik}x^ix^k = 1 \tag{3.29}$$

(recall Example 2, p. 64).[1] The principal axes of this surface are clearly the same as the principal axes of the tensor T_{ik}. In the coordinates x'_1, x'_2, x'_3 corresponding to the principal axes, equation (3.29) takes the form

$$T'_{11}(x'_1)^2 + T'_{22}(x'_2)^2 + T'_{33}(x'_3)^2 = \lambda_1(x'_1)^2 + \lambda_2(x'_2)^2 + \lambda_3(x'_3)^2 = 1$$

[1] The correspondence is no longer one-to-one if T_{ik} is not symmetric. In fact, let $T_{ik}^{(1)}$ and $T_{ik}^{(2)}$ be two tensors with the same symmetric part S_{ik} but with different antisymmetric parts $A_{ik}^{(1)}$ and $A_{ik}^{(2)}$ (see p. 107). Then $T_{ik}^{(1)}$ and $T_{ik}^{(2)}$ have the same quadric surface, since

$$T_{ik}^{(1)}x^ix^k = S_{ik}x^ix^k + A_{ik}^{(1)}x^ix^k = S_{ik}x^ix^k = 1,$$
$$T_{ik}^{(2)}x^ix^k = S_{ik}x^ix^k + A_{ik}^{(2)}x^ix^k = S_{ik}x^ix^k = 1.$$

or

$$\frac{(x_1')^2}{\dfrac{1}{\lambda_1}}+\frac{(x_2')^2}{\dfrac{1}{\lambda_2}}+\frac{(x_3')^2}{\dfrac{1}{\lambda_3}}=1.$$

If λ_1, λ_2, λ_3 are all positive (the case of greatest practical importance), the surface (3.29) is an ellipsoid, called the *tensor ellipsoid*, with semiaxes of length

$$\frac{1}{\sqrt{\lambda_1}},\quad \frac{1}{\sqrt{\lambda_2}},\quad \frac{1}{\sqrt{\lambda_3}}.$$

The tensor ellipsoid is an ellipsoid of revolution if $\lambda_1 = \lambda_2$ and a sphere if $\lambda_1 = \lambda_2 = \lambda_3$ (see Fig. 3.4).

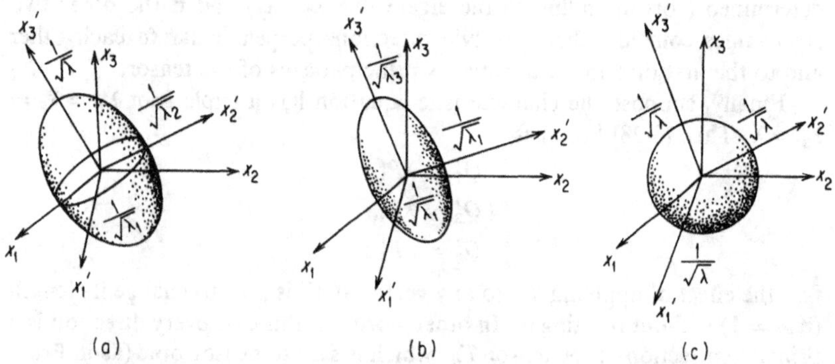

FIG. 3.4. Illustrating the tensor ellipsoid of a symmetric tensor T_{ik}. The principal axes of the ellipsoid coincide with those of the tensor.

(a) The case of distinct eigenvalues $\lambda_1, \lambda_2, \lambda_3$: the tensor has a matrix of the form

$$\|T'_{ik}\| = \begin{Vmatrix} \lambda_1 & 0 & 0 \\ 0 & \lambda_2 & 0 \\ 0 & 0 & \lambda_3 \end{Vmatrix}$$

in principal axes;

(b) The case of a double eigenvalue ($\lambda_1 = \lambda_2 \neq \lambda_3$): the tensor has a matrix of the form

$$\|T'_{ik}\| = \begin{Vmatrix} \lambda_1 & 0 & 0 \\ 0 & \lambda_1 & 0 \\ 0 & 0 & \lambda_3 \end{Vmatrix}$$

in principal axes;

(c) The case of a triple eigenvalue ($\lambda_1 = \lambda_2 = \lambda_3 = \lambda$): the tensor is isotropic and has a matrix of the form

$$\|T_{ik}\| = \lambda \begin{Vmatrix} 1 & 0 & 0 \\ 0 & 1 & 0 \\ 0 & 0 & 1 \end{Vmatrix}$$

in any rectangular coordinate system.

3.6. Invariants of a Tensor

Given a vector **A**, let A_i be the components of **A** in one rectangular coordinate system K and let A_i' be its components in another rectangular coordinate system K'. Although the components of **A** change in going from K to K', we can easily form an expression which does not change under this transformation, namely the square of the length of **A**:

$$A_iA_i = (A_1)^2 + (A_2)^2 + (A_3)^2 = (A_1')^2 + (A_2')^2 + (A_3')^2 = A_i'A_i'.$$

An expression like A_iA_i which does not change under transformations from one coordinate system to another is called an *invariant* of the vector **A**. It is easy to see that tensors of any order also have invariants.

For example, given a second-order tensor T_{ik}, consider the characteristic equation

$$\begin{vmatrix} T_{11}-\lambda & T_{12} & T_{13} \\ T_{21} & T_{22}-\lambda & T_{23} \\ T_{31} & T_{32} & T_{33}-\lambda \end{vmatrix} = 0,$$

or

$$\lambda^3 - \lambda^2(T_{11}+T_{22}+T_{33}) + \lambda\left(\begin{vmatrix} T_{22} & T_{32} \\ T_{23} & T_{33} \end{vmatrix} + \begin{vmatrix} T_{11} & T_{21} \\ T_{12} & T_{22} \end{vmatrix} + \begin{vmatrix} T_{11} & T_{31} \\ T_{13} & T_{33} \end{vmatrix}\right) - \begin{vmatrix} T_{11} & T_{12} & T_{13} \\ T_{21} & T_{22} & T_{23} \\ T_{31} & T_{32} & T_{33} \end{vmatrix} = 0 \tag{3.30}$$

after expanding the determinant. The numbers λ, λ^2, λ^3, being scalars (think of their geometric meaning!), are independent of the choice of the coordinate system, and hence so are the coefficients in (3.30). Therefore the quantities

$$\begin{aligned} I_1 &= T_{11}+T_{22}+T_{33}, \\ I_2 &= \begin{vmatrix} T_{22} & T_{32} \\ T_{23} & T_{33} \end{vmatrix} + \begin{vmatrix} T_{11} & T_{21} \\ T_{12} & T_{22} \end{vmatrix} + \begin{vmatrix} T_{11} & T_{31} \\ T_{13} & T_{33} \end{vmatrix}, \\ I_3 &= \begin{vmatrix} T_{11} & T_{12} & T_{13} \\ T_{21} & T_{22} & T_{23} \\ T_{31} & T_{32} & T_{33} \end{vmatrix} \end{aligned} \tag{3.31}$$

are all invariants of the tensor T_{ik}. Using (3.31), we can form infinitely many other invariants, e.g.

$$I_1^2 = (T_{ii})^2,$$

$$I_1^2 - 2I_2 = T_{ik}T_{ik},$$

and so on.

Tensors for which the invariant I_1 vanishes are called *deviators.* Any tensor T_{ik} can be written as the sum of a deviator and an isotropic tensor. In fact,

$$T_{ik} = T_{ik} - \tfrac{1}{3}T_{ll}\delta_{ik} + \tfrac{1}{3}T_{ll}\delta_{ik} = D_{ik} + \tfrac{1}{3}\delta_{ik}T_{ll},$$

where $\frac{1}{3}\delta_{ik}T_{ll}$ is clearly isotropic and D_{ik} is a deviator since

$$D_{ii} = D_{11} + D_{22} + D_{33} = T_{ii} - \tfrac{1}{3}T_{ll} \cdot 3 = 0.$$

3.6.1. A test for tensor character. Suppose T_{ik} $(i, k = 1, 2, 3)$ is such that the quantity

$$T_{ik}A_iB_k$$

is the same for all choices of the vectors **A** and **B** with components A_i and B_i. Then T_{ik} is a second-order tensor. In fact, introducing a new (primed) coordinate system, we have

$$T'_{ik}A'_iB'_k = T_{lm}A_lB_m = T_{lm}\alpha_{i'l}\alpha_{k'm}A'_iB'_k$$

or

$$(T'_{ik} - \alpha_{i'l}\alpha_{k'm}T_{lm})A'_iB'_k = 0.$$

But then

$$T'_{ik} - \alpha_{i'l}\alpha_{k'm}T_{lm} = 0$$

since **A** and **B** are arbitrary, i.e., T_{ik} is a second-order tensor, as asserted. As an exercise, the reader should state and prove the analogous criterion for $T_{i_1i_2\ldots i_n}$ to be a tensor of order n.

3.7. Pseudotensors

In Sec. 1.4.2 we distinguished between polar vectors whose direction is independent of the handedness of any underlying rectangular coordinate system K and axial vectors whose direction reverses if the handedness of K is changed. Correspondingly, it will be recalled from Sec. 3.4.2 that the transformation law of an axial vector differs from that of a polar vector by the presence of a minus sign in the case where the transformation changes the handedness of the underlying coordinate system. We now discuss a class of mathematical objects called *pseudotensors,* which generalize axial vectors in the same way that ordinary tensors generalize polar vectors.

3.7.1. Proper and improper transformations. According to Prob. 1, p. 38, an *orthogonal transformation,* i.e., the transformation from one rectangular coordinate system to another is described by the formulas

$$x'_i = \alpha_{i'k}x_k + x_{0i} \quad \text{(direct transformation)},$$

$$x_i = \alpha_{k'i}x'_k + x'_{0i} \quad \text{(inverse transformation)},$$

$$\left.\begin{aligned} \alpha_{l'i}\alpha_{l'k} &= \delta_{ik}, \\ \alpha_{i'l}\alpha_{k'l} &= \delta'_{ik} \end{aligned}\right\} \quad \text{(orthogonality conditions)},$$

where the old and new systems need not have the same handedness. The determinant of such a transformation, i.e., the quantity

$$\det \|\alpha_{i'k}\| \equiv \Delta = \begin{vmatrix} \alpha_{1'1} & \alpha_{1'2} & \alpha_{1'3} \\ \alpha_{2'1} & \alpha_{2'2} & \alpha_{2'3} \\ \alpha_{3'1} & \alpha_{3'2} & \alpha_{3'3} \end{vmatrix} \tag{3.32}$$

can only take one of the two values -1, $+1$. In fact, it follows from the orthogonality conditions and the law for multiplication of determinants[2] that

$$\det \|\delta'_{ik}\| = \det \left(\sum_{l=1}^{3} \alpha_{i'l}\alpha_{k'l} \right) = \det \|\alpha_{i'l}\| \cdot \det \|\alpha_{k'l}\|$$

$$= (\det \|\alpha_{i'k}\|)^2 = \Delta^2.$$

But

$$\det \|\delta'_{ik}\| = \begin{vmatrix} 1 & 0 & 0 \\ 0 & 1 & 0 \\ 0 & 0 & 1 \end{vmatrix} = 1,$$

and hence

$$\Delta^2 = 1 \quad \text{or} \quad \Delta = \pm 1.$$

We can now divide all orthogonal transformations into two classes: *proper transformations* for which $\Delta = +1$ and *improper transformations* for which $\Delta = -1$. Suppose a given orthogonal transformation carries the old system K into a new system K'. Then K' has the same handedness as K if the transformation is proper and different handedness if the transformation is improper. In fact, let K and K' have orthonormal bases $\mathbf{i}_1, \mathbf{i}_2, \mathbf{i}_3$ and $\mathbf{i}'_1, \mathbf{i}'_2, \mathbf{i}'_3$, respectively. Then a little algebra shows that

$$(\mathbf{i}'_1 \times \mathbf{i}'_2) \cdot \mathbf{i}'_3 = \alpha_{1'j}\alpha_{2'k}\alpha_{3'l}(\mathbf{i}_j \times \mathbf{i}_k) \cdot \mathbf{i}_l = (\mathbf{i}_1 \times \mathbf{i}_2) \cdot \mathbf{i}_3 \Delta, \tag{3.33}$$

where the vector products are both taken in some underlying right-handed coordinate system. But it will be recalled from p. 21 that a basis $\mathbf{i}_1, \mathbf{i}_2, \mathbf{i}_3$ is said to be right-handed if $(\mathbf{i}_1 \times \mathbf{i}_2) \cdot \mathbf{i}_3 > 0$ and left-handed if $(\mathbf{i}_1 \times \mathbf{i}_2) \cdot \mathbf{i}_3 < 0$, and similarly for $\mathbf{i}'_1, \mathbf{i}'_2, \mathbf{i}'_3$. It follows from (3.33) that the bases $\mathbf{i}_1, \mathbf{i}_2, \mathbf{i}_3$ and $\mathbf{i}'_1, \mathbf{i}'_2, \mathbf{i}'_3$ have the same handedness if and only if $\Delta = +1$ and different handedness if and only if $\Delta = -1$.

Example 1. The transformation

$$x'_1 = x_2, \quad x'_2 = -x_1, \quad x'_3 = x_3,$$

---

[2] See e.g., G. E. Shilov, *An Introduction to the Theory of Linear Spaces* (translated by R. A. Silverman), Prentice-Hall, Inc., Englewood Cliffs, N. J. (1961), p. 81.

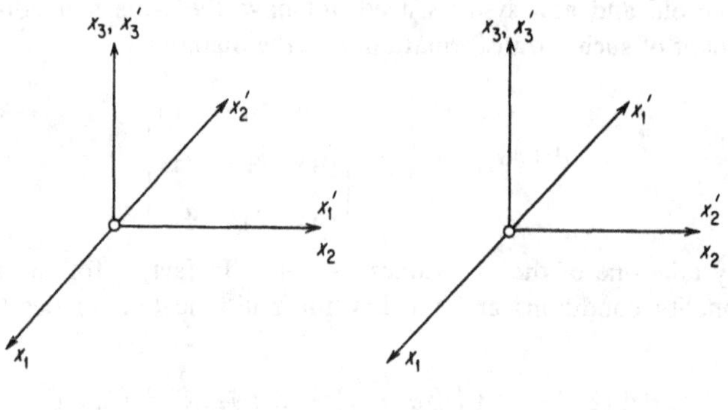

(a) (b)

FIG. 3.5. Illustrating proper and improper orthogonal transformations.

corresponding to rotation through 90° about the x_3-axis [see Fig. 3.5(a)], is a proper transformation.

***Example* 2.** The transformation

$$x_1' = -x_1, \quad x_2' = x_2, \quad x_3' = x_3,$$

corresponding to reflection in the x_2x_3-plane [see Fig. 3.5(b)], is an improper transformation.

3.7.2. Definition of a pseudotensor. By a *pseudotensor of order n* is meant a quantity specified by 3^n real numbers (the *components* of the pseudotensor) which transform under changes of coordinate system according to the law

$$A'_{i_1i_2\ldots i_n} = \alpha_{i_1'k_1}\alpha_{i_2'k_2}\cdots\alpha_{i_n'k_n}A_{k_1k_2\ldots k_n}\Delta, \tag{3.34}$$

where $A_{k_1k_2\ldots k_n}$, $A'_{i_1i_2\ldots i_n}$ are the components of the pseudotensor in the old and new coordinate systems K and K', respectively, $\alpha_{i_1'k_1}$ is the cosine of the angle between the i_1st axis of K' and the k_1st axis of K (similarly for $\alpha_{i_2'k_2}, \ldots, \alpha_{i_n'k_n}$), and Δ is the determinant (3.32). The transformation law (3.32) differs from the law (2.30) for an ordinary tensor only by the presence of the factor Δ. Thus pseudotensors behave the same way as ordinary tensors under proper transformations, but the transformation laws differ in sign if the transformations are improper.

The following algebraic properties of pseudotensors are easily verified:

1) The sum of two pseudotensors of order n is a pseudotensor of order n;
2) The product of two pseudotensors of orders m and n is an ordinary tensor of order $m + n$;
3) The product of a pseudotensor of order m and a tensor of order n is a pseudotensor of order $m + n$;
4) Contraction of a pseudotensor gives a pseudotensor of lower order.

Example. Let $\mathbf{C} = \mathbf{A} \times \mathbf{B}$ be the vector product of two vectors **A** and **B**. Then

$$C_i = A_j B_k - A_k B_j \tag{3.35}$$

in some rectangular coordinate system K, where i, j, k is a cyclic permutation of the numbers 1, 2, 3 [recall formula (1.23)]. In another rectangular coordinate system K', we have

$$A'_j = \alpha_{j'l} A_l, \qquad B'_k = \alpha_{k'm} B_m,$$

and hence

$$C'_i = \alpha_{j'l}\alpha_{k'm} A_l B_m - \alpha_{k'l}\alpha_{j'm} A_l B_m$$
$$= (\alpha_{j'l}\alpha_{k'm} - \alpha_{j'm}\alpha_{k'l}) A_l B_m,$$

where the sum is over all values of l and m (the terms with $l = m$ vanish). We can also write C'_1 in the form

$$C'_i = (\alpha_{j'l}\alpha_{k'm} - \alpha_{j'm}\alpha_{k'l})(A_l B_m - A_m B_l), \tag{3.36}$$

where the indices l and m are now restricted to the values 1,2 or 2,3 or 3,1. But according to (3.8),

$$\alpha_{j'l}\alpha_{k'm} - \alpha_{j'm}\alpha_{k'l} = \pm\alpha_{i'n},$$

where j, k, i and l, m, n are both cyclic permutations of 1, 2, 3, and we choose the plus sign if K and K' have the same handedness and the minus sign if K and K' have different handedness. In other words,

$$\alpha_{j'l}\alpha_{k'm} - \alpha_{j'm}\alpha_{k'l} = \alpha_{i'n}\Delta, \tag{3.37}$$

where Δ is the determinant of the transformation from K to K'. It follows from (3.35)–(3.37) that

$$C'_i = \alpha_{i'n} C_n \Delta,$$

i.e., that C is a pseudovector. Note that the concepts of a pseudovector and an axial vector are equivalent (why?).

3.7.3. The pseudotensor ε_{jkl}. Given a rectangular coordinate system K with orthonormal basis $\mathbf{i}_1, \mathbf{i}_2, \mathbf{i}_3$, let

$$\varepsilon_{jkl} = (\mathbf{i}_j \times \mathbf{i}_k) \cdot \mathbf{i}_l.$$

Then

$$\varepsilon_{jkl} = \begin{cases} +1 & \text{if } j, k, l \text{ is a cyclic permutation of } 1, 2, 3, \\ -1 & \text{if } j, k, l \text{ is a cyclic permutation of } 2, 1, 3, \\ 0 & \text{otherwise,} \end{cases} \tag{3.38}$$

so that the only nonzero components of ε_{jkl} are

$$\varepsilon_{123} = \varepsilon_{231} = \varepsilon_{312} = 1,$$

$$\varepsilon_{132} = \varepsilon_{321} = \varepsilon_{213} = -1.$$

The quantity ε_{jkl} is a pseudotensor of order 3 (called the *unit pseudotensor*). To see this, we first note that

$$\varepsilon'_{jkl} = (\mathbf{i}'_j \times \mathbf{i}'_k) \cdot \mathbf{i}'_l,$$

where $\mathbf{i}'_1$, $\mathbf{i}'_2$, $\mathbf{i}'_3$ is the orthonormal basis of a new rectangular coordinate system K'. Therefore

$$\varepsilon'_{jkl} = \alpha_{j'm}\alpha_{k'n}\alpha_{l'p}(\mathbf{i}_m \times \mathbf{i}_n) \cdot \mathbf{i}_p,$$

where the vector products are both taken in the system K'. But the vector product in K' is the same as that in K if K and K' have the same handedness and the negative of that in K if K and K' have different handedness. It follows that

$$\varepsilon'_{jkl} = \alpha_{j'm}\alpha_{k'n}\alpha_{l'p}\varepsilon_{mnp}\Delta,$$

where Δ is the determinant of the transformation from K to K', i.e., ε_{jkl} is a pseudotensor of order 3, as asserted. Note that (3.38) remains true if ε_{jkl} is replaced by ε'_{jkl}, i.e., ε_{jkl} has the same components in every rectangular coordinate system.

Example 1. The formula

$$C_j = \varepsilon_{jkl}A_kB_l.$$

gives the components of the vector product $\mathbf{C} = \mathbf{A} \times \mathbf{B}$. In fact,

$$\begin{aligned} C_1 = \varepsilon_{1kl}A_kB_l &= \varepsilon_{123}A_2B_3 + \varepsilon_{132}A_3B_2 \\ &= A_2B_3 - A_3B_2, \end{aligned}$$

and so on.

Example 2. Let φ be a scalar, T_{jk} a second-order tensor and T_{jkl} a third-order tensor. Then the quantities

$$\varepsilon_{jkl}\varphi, \quad \varepsilon_{jkl}T_{kl}, \quad \varepsilon_{jkl}T_{jkl}$$

are a pseudotensor of order 3, a pseudovector and a pseudoscalar, respectively (verify this assertion).

SOLVED PROBLEMS

Problem 1. Let

$$v_{ik} = \frac{\partial v_i}{\partial x_k} - \frac{\partial v_k}{\partial x_i} \tag{3.39}$$

be the rate of deformation tensor in a viscous fluid [cf. (2.22)], with deviator

$$v^0_{ik} = \frac{1}{2}\left(\frac{\partial v_i}{\partial x_k} + \frac{\partial v_k}{\partial x_i}\right) - \frac{1}{3}v_{ll}\delta_{ik} \qquad \left(v_{ll} = \frac{\partial v_l}{\partial x_l}\right), \tag{3.40}$$

and let

$$p_{ik} = 2\mu v^0_{ik} + \zeta v_{ll}\delta_{ik}$$

be the viscous stress tensor, where μ and ζ are constants [cf. (2.61)], Calculate the quantity

$$D = \hat{p}_{ik}\frac{\partial v_i}{\partial x_k}$$

called the *dissipation function* of the viscous fluid.

Solution. We have

$$D = \hat{p}_{ik}\frac{\partial v_i}{\partial x_k} = \frac{1}{2}\left(\hat{p}_{ik}\frac{\partial v_i}{\partial x_k} + \hat{p}_{ik}\frac{\partial v_i}{\partial x_k}\right)$$

$$= \frac{1}{2}\left(\hat{p}_{ik}\frac{\partial v_i}{\partial x_k} + \hat{p}_{ki}\frac{\partial v_k}{\partial x_i}\right),$$

and hence, since $\hat{p}_{ik} = \hat{p}_{ki}$,

$$D = \hat{p}_{ik}v_{ik}.$$

But, according to (3.39) and (3.40),

$$v_{ik} = v_{ik}^0 + \tfrac{1}{3}v_{ll}\delta_{ik}.$$

Therefore

$$D = \hat{p}_{ik}v_{ik} = \hat{p}_{ik}v_{ik}^0 + \hat{p}_{ik}\tfrac{1}{3}v_{ll}\delta_{ik}$$

$$= 2\mu v_{ik}^0 v_{ik}^0 + \zeta v_{ll}\delta_{ik}v_{ik}^0 + \tfrac{1}{3}\hat{p}_{ii}v_{ll},$$

and hence

$$D = 2\mu v_{ik}^0 v_{ik}^0 + \zeta(v_{ll})^2,$$

since

$$\delta_{ik}v_{ik}^0 = v_{ii}^0 = 0,$$

$$\hat{p}_{ii} = 2\mu v_{ii}^0 + \zeta v_{ll}\delta_{ii} = 3\zeta v_{ll}.$$

***Problem* 2.** Prove that any axis of symmetry of a system of n particles is a principal axis of the moment of inertia tensor with respect to any point of the axis.

Solution. Let K be a system of rectangular coordinates x_1, x_2, x_3 with the x_3-axis as an axis of symmetry of the system. Then for every particle of mass m_j with coordinates $x_1^{(j)}$, $x_2^{(j)}$, $x_3^{(j)}$, there is a particle with the same mass $m_k = m_j$ and coordinates

$$x_1^{(k)} = x_1^{(j)}, \quad x_2^{(k)} = -x_2^{(j)}, \quad x_3^{(k)} = -x_3^{(j)}.$$

Hence

$$\begin{aligned} I_{12} &= \sum_{j=1}^{n} m_j x_1^{(j)}x_2^{(j)} = -\sum_{j=1}^{n/2} m_j x_1^{(j)}x_2^{(j)} - \sum_{k=(n/2)+1}^{n} m_k x_1^{(k)}x_2^{(k)} \\ &= \sum_{j=1}^{n/2} m_j x_1^{(j)}x_2^{(j)} + \sum_{j=1}^{n/2} m_j x_1^{(j)}x_2^{(j)} = 0, \end{aligned} \tag{3.41}$$

and similarly $I_{13} = 0$. Therefore the moment of inertia tensor I_{ik} in the system K is

$$\|I_{ik}\| = \left\| \begin{matrix} I_{11} & 0 & 0 \\ 0 & I_{22} & I_{23} \\ 0 & I_{32} & I_{33} \end{matrix} \right\|, \tag{3.42}$$

i.e., the x_1-axis is a principal axis of the tensor.

Problem 3. Prove that any axis l perpendicular to a plane of symmetry P of a system of n particles is a principal axis of the moment of inertia tensor with respect to the point of intersection of l and P.

Solution. Let K be a system of rectangular coordinates x_1, x_2, x_3 such that l is the x_1-axis and P is the x_2x_3-plane. Then the origin of K is the point of intersection of l and P. Since the x_2x_3-plane is a plane of symmetry of the system, for every particle of mass m_j with coordinates $x_1^{(j)}$, $x_2^{(j)}$, $x_3^{(j)}$, there is a particle with the same mass $m_k = m_j$ and coordinates

$$x_1^{(k)} = -x_1^{(j)}, \quad x_2^{(k)} = x_2^{(j)}, \quad x_3^{(k)} = x_3^{(j)}.$$

Therefore we again have $I_{12} = 0$, just as in (3.41), and similarly $I_{13} = 0$. Hence the moment of inertia tensor I_{ik} in the system K is again given by (3.42), i.e., the x_1-axis is a principal axis of the tensor, as before. Suppose the system K is rotated through the angle

$$\varphi = \tfrac{1}{2} \arctan \frac{2I_{23}}{I_{22} - I_{33}}$$

about the x_1-axis (cf. p. 113). Then in the new system K', whose x_1'-axis coincides with the x_1-axis, we have

$$I'_{ik} = \left\| \begin{matrix} I'_{11} & 0 & 0 \\ 0 & I'_{22} & 0 \\ 0 & 0 & I'_{33} \end{matrix} \right\|,$$

i.e., K' is the principal axis system.

Problem 4. Let I_{ik} be the moment of inertia tensor of a system of n particles of masses $m_1, m_2, \ldots, m_n$ with respect to the center of mass of the system of particles, and let K be the principal axis system of I_{ik}. Suppose K is given a parallel displacement along any of its coordinate axes. Prove that the resulting coordinate system K' is still a principal axis system of the moment of inertia tensor.

Solution. Since the origin of K is at the center of mass, we have

$$\sum_{j=1}^{n} m_j x_1^{(j)} = \sum_{j=1}^{n} m_j x_2^{(j)} = \sum_{j=1}^{n} m_j x_3^{(j)} = 0. \tag{3.43}$$

Moreover

$$I_{12} = I_{13} = I_{23} = 0, \tag{3.44}$$

since the axes of the system K are principal directions of I_{ik}. Suppose K is given a parallel displacement along the x_1-axis moving its origin to the point $x_1 = a$, $x_2 = 0$, $x_3 = 0$. Then

$$I_{ik} = -\sum_{j=1}^{n} m_j x_i^{(j)'} x_k^{(j)'} \qquad (i \neq k) \tag{3.45}$$

in the resulting system K', where

$$\begin{aligned} x_1^{(j)'} &= x_1^{(j)} - a, \\ x_2^{(j)'} &= x_2^{(j)}, \\ x_3^{(j)'} &= x_3^{(j)}. \end{aligned} \tag{3.46}$$

But substituting (3.46) into (3.45), and taking account of (3.43) and (3.44), we find that $I'_{ik} = 0$ $(i \neq k)$, i.e., K' is still a principal axis system of the moment of inertia tensor. The same is true if K is shifted along the x_2 or x_3-axis instead.

Problem 5. Let T_{ik} be a second-order symmetric Cartesian tensor, with positive eigenvalues, and let

$$T_{ik} x_i x_k \tag{3.47}$$

be the associated tensor ellipsoid (see p. 120). What is the geometric meaning of the invariants (3.31) in terms of this surface?

Solution. Let a, b, c be the semiaxes of the ellipsoid (3.47), regardless of its orientation relative to the system with rectangular coordinates x_1, x_2, x_3, and let $\bar{a}, \bar{b}, \bar{c}$ be the half-lengths of the segments cut off from the coordinate axes by the ellipsoid. Setting $x_2 = x_3 = 0$, we have

$$x_1 = \bar{a}, \qquad T_{11}\bar{a}^2 = 1,$$

and hence

$$\bar{a} = \frac{1}{\sqrt{T_{11}}}.$$

Similarly

$$\bar{b} = \frac{1}{\sqrt{T_{22}}}, \qquad \bar{c} = \frac{1}{\sqrt{T_{33}}},$$

and hence

$$\begin{aligned} I_1 = T_{11} + T_{22} + T_{33} &= \frac{1}{\bar{a}^2} + \frac{1}{\bar{b}^2} + \frac{1}{\bar{c}^2} \\ &= \frac{1}{a^2} + \frac{1}{b^2} + \frac{1}{c^2} = T'_{11} + T'_{22} + T'_{33}, \end{aligned}$$

where T'_{11}, T'_{22}, T'_{33} are the components of T_{ik} in the principal axis system. Thus the sum of reciprocals of the squares of the half-lengths of the segments cut off by the tensor ellipsoid from a rectangular trihedron centered at the origin is independent of the orientation of the trihedron and equals the first invariant I_1 of the tensor determining the ellipsoid.

The equation of the elliptical cross section of the ellipsoid (3.47) cut off by the plane $x_3 = 0$ is

$$T_{11}x_1^2 + 2T_{12}x_1x_2 + T_{22}x_2^2 = 1.$$

The principal semiaxes of this ellipse are $1/\sqrt{\mu_1}$ and $1/\sqrt{\mu_2}$, where μ_1 and μ_2 are the roots of the equation

$$\begin{vmatrix} T_{11} - \mu & T_{12} \\ T_{12} & T_{22} - \mu \end{vmatrix} = 0,$$

so that

$$\mu_1\mu_2 = T_{11}T_{22} - T_{12}^2.$$

Correspondingly, the area of the elliptical cross section is

$$A_3 = \frac{\pi}{\sqrt{\mu_1\mu_2}}.$$

Let A_1 and A_2 denote the areas of the cross sections cut off by the planes $x_1 = 0$ and $x_2 = 0$. Then the invariant I_2 equals

$$I_2 = \mu_1\mu_2 + \mu_1\mu_3 + \mu_2\mu_3 = \pi^2\left(\frac{1}{A_1^2} + \frac{1}{A_2^2} + \frac{1}{A_3^2}\right)$$
$$= \frac{1}{a^2b^2} + \frac{1}{a^2c^2} + \frac{1}{b^2c^2} = \lambda_1\lambda_2 + \lambda_1\lambda_3 + \lambda_2\lambda_3,$$

where λ_1, λ_2, λ_3 are the roots of the equation

$$\begin{vmatrix} T_{11} - \lambda & T_{12} & T_{13} \\ T_{12} & T_{22} - \lambda & T_{23} \\ T_{13} & T_{23} & T_{33} - \lambda \end{vmatrix} = 0.$$

Thus the sum of the reciprocals of the squares of the areas of the ellipses cut off by three perpendicular planes intersecting at the center of the ellipsoid is independent of the orientation of the planes and equals $1/\pi^2$ times the second invariant I_2 of the tensor determining the ellipsoid.

Finally we note that the volume of the tensor ellipsoid equals

$$V = \frac{4\pi}{3}abc.$$

But I_3 has the form

$$I_3 = \begin{vmatrix} \frac{1}{a^2} & 0 & 0 \\ 0 & \frac{1}{b^2} & 0 \\ 0 & 0 & \frac{1}{c^2} \end{vmatrix} = \frac{1}{a^2b^2c^2}$$

in principal axes, and hence

$$V = \frac{4\pi}{3}\frac{1}{\sqrt{I_3}}.$$

Thus the invariance of I_3 merely expresses the fact that the volume of the tensor ellipsoid is the same in all coordinate systems.

Problem 6. Represent the symmetric second-order covariant tensor T_{ik} in terms of the orthonormal basis $\mathbf{e}_1$, $\mathbf{e}_2$, $\mathbf{e}_3$ of its principal axis system.

Solution. Let $\mathbf{e}_\alpha$ have covariant components $e_{\alpha i}$ and contravariant components e_α^i. It follows from (3.20) that

$$(T_{ik} - \lambda_\alpha g_{ik})e_\alpha^k = 0 \qquad (i = 1, 2, 3), \tag{3.48}$$

where λ_α is the eigenvalue corresponding to $\mathbf{e}_\alpha$. Multiplying (3.48) by $e_{\alpha l}$ and summing over α, we obtain

$$T_{ik}\sum_{\alpha=1}^{3} e_\alpha^k e_{\alpha l} = \sum_{\alpha=1}^{3}\lambda_\alpha g_{ik} e_\alpha^k e_{\alpha l}.$$

Therefore finally

$$T_{ik} = \sum_{\alpha=1}^{3}\lambda_\alpha e_{\alpha i} e_{\alpha k}$$

after using formula (2.66).

EXERCISES

1. Form scalars by contracting the tensors with matrices

$$\begin{Vmatrix} 1 & 0 & 5 \\ 0 & 6 & 3 \\ 2 & 4 & 3 \end{Vmatrix}, \quad \begin{Vmatrix} 5 & 0 & 1 \\ 3 & 6 & 3 \\ 4 & 5 & 4 \end{Vmatrix}, \quad \begin{Vmatrix} 3 & 5 & 3 \\ 4 & 4 & 4 \\ 3 & 2 & 6 \end{Vmatrix}.$$

2. Given that

$$\|T_{ik}\| = \begin{Vmatrix} 1 & 0 & 2 \\ 3 & 4 & 1 \\ 1 & 3 & 4 \end{Vmatrix}, \quad \mathbf{A} = \mathbf{i}_1 + 2\mathbf{i}_2 + 3\mathbf{i}_3,$$

find the inner products $T_{ik}A_i$ and $T_{ik}A_k$.

3. Let T_{ik} and **A** be the same as in the preceding problem, and let

$$\mathbf{B} = 4\mathbf{i}_1 + 5\mathbf{i}_2 + 6\mathbf{i}_3.$$

Find the inner product $T_{ik}A_iB_k$.

4. Prove that if S_{ik} is a symmetric tensor and A_{ik} an antisymmetric tensor, then $S_{ik}A_{ik} = 0$.

5. Given that

$$\|T_{ik}\| = \begin{Vmatrix} 1 & 2 & 3 \\ 4 & 5 & 6 \\ 7 & 8 & 9 \end{Vmatrix}, \quad \mathbf{C} = \mathbf{i}_1 + 2\mathbf{i}_2 + 3\mathbf{i}_3,$$

find the symmetric part S_{ik} and antisymmetric part A_{ik} of the tensor T_{ik}. Calculate

a) $T_{ik}C_k$, $T_{ik}C_i$, $T_{ik}C_iC_k$;

b) $A_{ik}T_{ik}$, $A_{ik}S_{ik}$, $A_{ik}C_i$, $A_{ik}C_iC_k$;

c) $T_{ik}\delta_{ik}$, $A_{ik}\delta_{ik}$, $S_{ik}\delta_{ik}$;

d) $T_{ik} - \frac{1}{3}\delta_{ik}T_{ll}$, $(T_{ik} - \frac{1}{3}\delta_{ik}T_{ll})C_i$, $(T_{ik} - \frac{1}{3}\delta_{ik}T_{ll})C_iC_k$.

6. Find the invariants of the tensors in Exercises 1 and 5.

7. Find the characteristic vectors and principal directions of the following tensors:

$$\begin{Vmatrix} 1 & 0 & 0 \\ 0 & 2 & 3 \\ 0 & 3 & 4 \end{Vmatrix}, \quad \begin{Vmatrix} 1 & 2 & 0 \\ 2 & 2 & 0 \\ 0 & 0 & 3 \end{Vmatrix}, \quad \begin{Vmatrix} 0 & 0 & 1 \\ 0 & 2 & 1 \\ 1 & 1 & 1 \end{Vmatrix}, \quad \begin{Vmatrix} 4 & 1 & 2 \\ 1 & 5 & 0 \\ 2 & 0 & 0 \end{Vmatrix}.$$

8. Prove the following formulas expressing the invariants (3.30) in terms of the eigenvalues $\lambda_1, \lambda_2, \lambda_3$:

$$I_1 = \lambda_1 + \lambda_2 + \lambda_3,$$
$$I_2 = \lambda_1\lambda_2 + \lambda_1\lambda_3 + \lambda_2\lambda_3,$$
$$I_3 = \lambda_1\lambda_2\lambda_3.$$

Hint. The characteristic equation can be written as

$$(\lambda - \lambda_1)(\lambda - \lambda_2)(\lambda - \lambda_3) = 0.$$

9. Prove that in generalized coordinates the analogues of the invariants (3.31) are given by

$$I_1 = T_i^{\cdot i},$$
$$I_2 = \tfrac{1}{2}[(T_i^{\cdot i})^2 - T_l^{\cdot k}T_k^{\cdot l}],$$
$$I_3 = \det \|T_i^{\cdot k}\|.$$

10. Prove that the roots of the characteristic equation (3.21) are all real if the covariant tensor T_{ik} is symmetric.

11. Show that if $\lambda_1, \lambda_2, \lambda_3$ are the characteristic values of a symmetric tensor T_{ik}, then

$$\sum_{i=1}^{3} \lambda_i = g^{kl}T_{kl}, \quad \sum_{i=1}^{3} \lambda_i^2 = T^{kl}T_{kl}, \quad \sum_{i=1}^{3} \lambda_i^3 = T^{kl}T_{lm}T^{m}_{.k}.$$

12. Prove that the analogue of the deviator

$$D_{ik} = T_{ik} - \tfrac{1}{3}T_{ll}\delta_{ik}$$

in generalized coordinates has components

$$D_{ik} = T_{ik} - \tfrac{1}{3}T_{l}^{.l}g_{ik},$$
$$D^{ik} = T^{ik} - \tfrac{1}{3}T_{l}^{.l}g^{ik},$$
$$D_{i}^{.k} = T_{i}^{.k} - \tfrac{1}{3}T_{l}^{.l}g_{i}^{.k}.$$

Comment. Note that $D_{i}^{.i} = 0$.

13. Suppose T_{ik}, T^{ik} and $T_{i}^{.k}$ are such that the quantities

$$T_{ik}A^iB^k, \quad T^{ik}A_iB_k, \quad T_{i}^{.k}A^iB_k$$

are the same for all choices of the vectors **A** and **B**. Prove that T_{ik}, T^{ik} and $T_{i}^{.k}$ are the covariant, contravariant and mixed components, respectively, of a second-order tensor.

14. Show that if $T^{i}_{.kl}A^kB^lC_i$ is a scalar for arbitrary vectors **A**, **B** and **C**, then $T^{i}_{.kl}$ is a mixed third-order tensor of the indicated structure.

15. Suppose $A_{ikl}\, dx^i\, dx^k\, dx^l = 0$ for arbitrary differentials dx^i, dx^k and dx^l. Prove that

$$A_{123} + A_{231} + A_{312} + A_{132} + A_{321} + A_{213} = 0.$$

16. Let

$$u_{ik} = \frac{1}{2}\left(\frac{\partial u_i}{\partial x_k} + \frac{\partial u_k}{\partial x_i}\right)$$

be the linear part of the deformation tensor in an elastic medium (cf. p. 72) with deviator

$$u^0_{ik} = u_{ik} - \tfrac{1}{3}u_{ll}\delta_{ik} \qquad \left(u_{ll} = \frac{\partial u_l}{\partial x_l}\right),$$

and let

$$p_{ik} = 2\lambda u^0_{ik} + K u_{ll}\delta_{ik}$$

be the stress tensor, where λ and K are constants. Calculate the quantity

$$F = \tfrac{1}{2}p_{ik}u_{ik},$$

called the *free energy* of the elastic medium.

17.[3] Outline a theory of m-dimensional vectors, tensors and pseudotensors. Discuss the notions of linear dependence and linear independence, bases, coordinates, components, transformation laws, etc. in m dimensions.

---

[3] See e.g., G. E. Shilov, *op. cit.*, Sec. 39.

4

VECTOR AND TENSOR ANALYSIS: RUDIMENTS

We now make a systematic study of the differential and integral calculus of vector and tensor functions of space and time, a subject known as *vector and tensor analysis.*

4.1. The Field Concept

4.1.1. Tensor functions of a scalar argument. By a *tensor function of a scalar argument* we mean a rule assigning a unique value of a tensor $A_{i_1i_2\ldots i_n}$ to each admissible value of a scalar t (usually, but not necessarily, the time). To indicate such a function, we simply write

$$A_{i_1i_2\ldots i_n} = A_{i_1i_2\ldots i_n}(t). \tag{4.1}$$

For example, suppose the state of stress of an elastic medium varies in time. Then the stress tensor p_{ik} defined in Sec. 2.4.1 becomes a function of time:

$$p_{ik} = p_{ik}(t).$$

By the *derivative* of the function (4.1) with respect to t we mean the tensor with components

$$\frac{dA_{i_1i_2\ldots i_n}}{dt} = \lim_{\Delta t \to 0} \frac{A_{i_1i_2\ldots i_n}(t+\Delta t) - A_{i_1i_2\ldots i_n}(t)}{\Delta t}, \tag{4.2}$$

calculated in a coordinate system which does not vary in time. The derivative (4.2) is clearly of the same order as the tensor (4.1) itself.

4.1.2. Tensor fields. By a *tensor field* we mean a rule assigning a unique value of a tensor to each point of a certain volume V (V may be all of space). Let $\mathbf{r}$ be the radius vector of a variable point of V with respect to the origin of some coordinate system. Then a tensor field is indicated by writing

$$A_{i_1 i_2 \ldots i_n} = A_{i_1 i_2 \ldots i_n}(\mathbf{r}) \tag{4.3}$$

if the tensor is of order n. The scalar field

$$\varphi = \varphi(\mathbf{r})$$

and vector field

$$\mathbf{A} = \mathbf{A}(\mathbf{r})$$

are the special cases of (4.3) corresponding to $n = 0$ and $n = 1$, respectively.

Example 1. The state of the atmosphere is described by scalar fields like the pressure $p = p(\mathbf{r})$, the temperature $T = T(\mathbf{r})$, the density $\rho = \rho(\mathbf{r})$, and by vector fields like the wind velocity $\mathbf{v} = \mathbf{v}(\mathbf{r})$ and the wind acceleration $\mathbf{a} = \mathbf{a}(\mathbf{r})$.

Example 2. The state of stress of an elastic medium is described by a tensor field $p_{ik} = p_{ik}(\mathbf{r})$.

We shall also consider fields, called *nonstationary* fields, which are functions of both space and time, i.e., of both the vector $\mathbf{r}$ and the scalar t:

$$\varphi = \varphi(\mathbf{r}, t), \quad \mathbf{A} = \mathbf{A}(\mathbf{r}, t), \quad p_{ik} = p_{ik}(\mathbf{r}, t), \ldots \tag{4.4}$$

These fields can be regarded as functions of a four-dimensional vector with components x_1, x_2, x_3, t varying over a region in "space-time."

A tensor field is said to be *homogeneous* if it has no spatial dependence. In this case, (4.4) reduces to

$$\varphi = \varphi(t), \quad \mathbf{A} = \mathbf{A}(t), \quad p_{ik} = p_{ik}(t), \ldots$$

The tensor fields $A_{i_1 i_2 \ldots i_n} = A_{i_1 i_2 \ldots i_n}(\mathbf{r})$ considered in this book will always be *continuous*, i.e., such that

$$\lim_{|\Delta\mathbf{r}| \to 0} [A_{i_1 i_2 \ldots i_n}(\mathbf{r} + \Delta\mathbf{r}) - A_{i_1 i_2 \ldots i_n}(\mathbf{r})] = 0.$$

Remark. The algebraic operations considered in Chap. 3 apply equally well to tensor fields, provided the operations involve tensors associated with the same point of space. For example, the sum of two tensor fields of the same order n (and the same physical dimensions) is a new tensors field of order n.

4.1.3. Line integrals. Circulation. Let M_1M_2 by any curve in a vector field $\mathbf{A} = \mathbf{A}(\mathbf{r})$, as in Fig. 4.1. Suppose we subdivide M_1M_2 by introducing

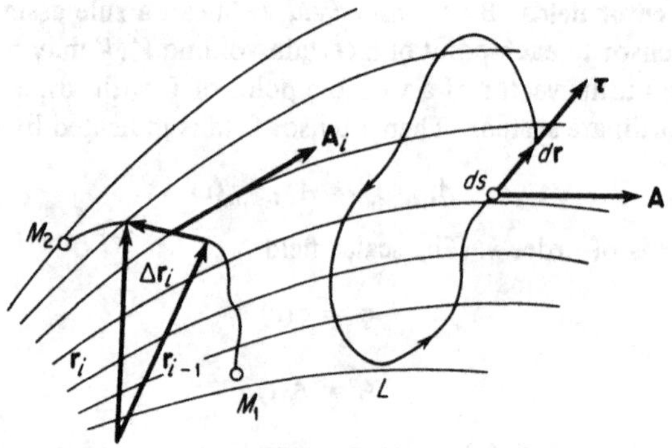

FIG. 4.1. Illustrating line integrals and circulation of a vector field.

$n - 1$ points of subdivision $\mathbf{r}_0, \mathbf{r}_1, \mathbf{r}_2, \ldots, \mathbf{r}_n$, where M_1 and M_2 have position vectors $\mathbf{r}_0$ and $\mathbf{r}_n$, respectively. Consider the sum

$$\sum_{i=1}^{n} \mathbf{A}_i \cdot \Delta\mathbf{r}_i,$$

where $\Delta\mathbf{r}_i = \mathbf{r}_i - \mathbf{r}_{i-1}$ and $\mathbf{A}_i$ is the value of the field $\mathbf{A}$ at any point of $\Delta\mathbf{r}_i$ (we ignore the difference between $\Delta\mathbf{r}_i$ and the arc of which it is the chord). The limit of this sum (provided it exists) as $n \to \infty$ and the maximum length of the $\Delta\mathbf{r}_i$ goes to zero is called the *line integral* of $\mathbf{A}$ along M_1M_2, denoted by

$$\lim_{\substack{\max|\Delta\mathbf{r}_i|\to 0\\ n\to\infty}} \sum_{i=1}^{n} \mathbf{A}_i \cdot \Delta\mathbf{r}_i = \int_{M_1M_2} \mathbf{A} \cdot d\mathbf{r} = \int_{M_1M_2} A_1\,dx_1 + A_2\,dx_2 + A_3\,dx_3.$$

Here the vector $d\mathbf{r}$ is directed along the tangent to the curve at every point, and its magnitude equals the element of arc length along the curve:

$$|d\mathbf{r}| = \sqrt{(dx_1)^2 + (dx_2)^2 + (dx_3)^2} = ds.$$

A case of particular interest is where the line integral is taken over a closed contour, like the contour L in Fig. 4.1 ($\boldsymbol{\tau}$ is the unit tangent to L and $d\mathbf{r} = \boldsymbol{\tau}\,ds$). The integral

$$\Gamma = \oint_L \mathbf{A} \cdot d\mathbf{r}$$

is then called the *circulation* of the vector $\mathbf{A}$ around the contour L. If $\mathbf{A}$ is a force field, then Γ is the work done by the force in moving a particle around L.

4.2. The Theorems of Gauss, Green and Stokes

We now prove two key theorems of mathematical analysis. One, called *Gauss' theorem*, expresses the value of a certain integral over a volume V bounded by a closed surface S in terms of a related integral over S. The other, called *Stokes' theorem*, expresses the value of a certain integral over a surface S bounded by a closed contour L in terms of a related integral along L. As a prelude to proving Stokes' theorem, we shall also prove *Green's theorem*, to which Stokes' theorem reduces if S is a plane region.

4.2.1. Gauss' theorem. First we prove

GAUSS' THEOREM. *Given a volume V bounded by a closed surface S, suppose the functions*

$$P(x_1, x_2, x_3), \quad Q(x_1, x_2, x_3), \quad R(x_1, x_2, x_3)$$

and their derivatives

$$\frac{\partial P}{\partial x_1}, \quad \frac{\partial Q}{\partial x_2}, \quad \frac{\partial R}{\partial x_3}$$

are continuous in $V \cup S$.[1] *Then*

$$\iiint_V \left(\frac{\partial P}{\partial x_1} + \frac{\partial Q}{\partial x_2} + \frac{\partial R}{\partial x_3}\right) dV = \iint_S [P \cos(\mathbf{n}, x_1) + Q \cos(\mathbf{n}, x_2) + R \cos(\mathbf{n}, x_3)]\, dS, \tag{4.5}$$

where $\mathbf{n}$ is the unit exterior normal to S (see Fig. 4.2).

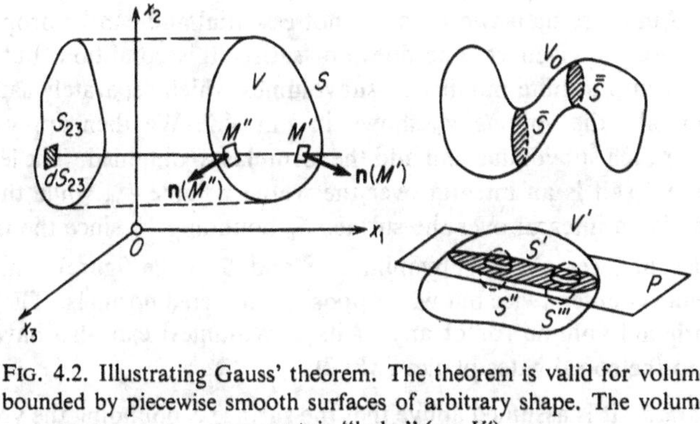

FIG. 4.2. Illustrating Gauss' theorem. The theorem is valid for volumes bounded by piecewise smooth surfaces of arbitrary shape. The volumes may contain "holes" (see V').

[1] Given two sets A and B, $A \cup B$ denotes the *union* of A and B, i.e., the set of all points belonging to A or to B (or to both).

Proof. Suppose no line parallel to the x_1-axis intersects S in more than two points M' and M'',[2] with unit exterior normals $\mathbf{n}(M')$ and $\mathbf{n}(M'')$ as shown in Fig. 4.2. Then, if S_{23} is the projection of S onto the x_2x_3-plane, we have

$$\iiint\limits_V \frac{\partial P}{\partial x_1}\,dV = \iint\limits_{S_{23}} \left(\int \frac{\partial P}{\partial x_1}\,dx_1\right) dS_{23} = \iint\limits_{S_{23}} [P(M') - P(M'')]\,dS_{23}.$$

But the element dS_{23} of the projection S_{23} can be expressed in terms of the elements of the surface S at the points M' and M'':

$$dS_{23} = dS(M')\cos\,[\mathbf{n}(M'), x_1] = -dS(M'')\cos\,[\mathbf{n}(M''), x_1].$$

Therefore

$$\iiint\limits_V \frac{\partial P}{\partial x_1}\,dV = \iint\limits_S P(M)\cos\,[\mathbf{n}(M), x_1]\,dS(M), \tag{4.6}$$

where M is a variable point of the surface S. The formulas

$$\iiint\limits_V \frac{\partial Q}{\partial x_2}\,dV = \iint\limits_S Q\cos\,(\mathbf{n}, x_2)\,dS, \tag{4.7}$$

$$\iiint\limits_V \frac{\partial R}{\partial x_3}\,dV = \iint\limits_S R\cos\,(\mathbf{n}, x_3)\,dS \tag{4.8}$$

are proved in the same way, provided no line parallel to the x_2 or x_3-axis intersects S in more than two points. Adding (4.6), (4.7) and (4.8), we obtain (4.5).

Remark 1. It is assumed that the surface S is two-sided, i.e., that S has a unique interior and exterior normal at each of its points.

Remark 2. The requirement that no line parallel to the coordinate axes intersect S in more than two points is not essential and can be dropped. In fact, suppose the given volume does not satisfy this condition, but can be partitioned into a finite number of subvolumes which separately satisfy the condition, like the volume V_0 shown in Fig. 4.2. We then apply Gauss' theorem to each subvolume and add the formulas so obtained. The left-hand side of the result is an integral over the whole volume V_0, while the right-hand side is an integral over the surface S_0 bounding V_0 since the integrals over adjacent faces of the subvolumes ($\bar{S}$ and $\bar{\bar{S}}$ in the figure) cancel each other, being counted twice but with oppositely directed normals. The surface of the original volume (or of any of its subvolumes) can also have pieces parallel to the coordinates planes (why?).

Remark 3. It is assumed above that the surface S bounding the volume V is smooth, so that the direction of the normal to S varies continuously from

[2] This restriction will be removed in Remark 2 below.

point to point. Gauss' theorem continues to hold in the case where S is *piecewise smooth*, i.e., where S is made up of a finite number of smooth pieces (e.g., prisms, pyramids, cylinders with caps, etc.). In fact, we can always partition V into a finite number of subvolumes each of which is bounded by a smooth surface. We then apply Gauss' theorem to each subvolume and add the resulting formulas, just as in Remark 2. Analytically, each smooth piece of a piecewise smooth surface has an equation of the form $x_3 = f(x_1, x_2)$ in a suitable coordinate system, where f is a function with continuous first partial derivatives.

Remark 4. Gauss' theorem continues to hold for volumes with "holes," i.e., for volumes like V' in Fig. 4.2 which are bounded by several closed surfaces (S', S'', S'''). It is only necessary to draw a surface inside V' intersecting V' in its "holes" (like the plane P in the figure) and then apply Gauss' theorem to the neighboring "hole-free" volumes which result.

Given a vector field $\mathbf{A} = \mathbf{A}(\mathbf{r})$, let

$$A_1 = P(x_1, x_2, x_3),$$
$$A_2 = Q(x_1, x_2, x_3),$$
$$A_3 = R(x_1, x_2, x_3)$$

be the components of $\mathbf{A}$ in a system of rectangular coordinates x_1, x_2, x_3. Then Gauss' theorem takes the form

$$\iiint\limits_V \left(\frac{\partial A_1}{\partial x_1} + \frac{\partial A_2}{\partial x_2} + \frac{\partial A_3}{\partial x_3}\right) dV = \iint\limits_S [A_1 \cos(\mathbf{n}, x_1) + A_2 \cos(\mathbf{n}, x_2) + A_3 \cos(\mathbf{n}, x_3)]\, dS. \tag{4.9}$$

Bearing in mind that the components of the unit exterior normal $\mathbf{n}$ are

$$n_1 = \cos(\mathbf{n}, x_1), \quad n_2 = \cos(\mathbf{n}, x_2), \quad n_3 = \cos(\mathbf{n}, x_3),$$

we can write (4.9) as

$$\iiint\limits_V \left(\frac{\partial A_1}{\partial x_1} + \frac{\partial A_2}{\partial x_2} + \frac{\partial A_3}{\partial x_3}\right) dV = \iint\limits_S \mathbf{A} \cdot \mathbf{n}\, dS. \tag{4.10}$$

The integrand of the volume integral also has a vector interpretation, which will be given in Sec. 4.4.3.

4.2.2. Green's theorem. We now prove a result which is the specialization of Stokes' theorem to the case of plane surfaces:

GREEN'S THEOREM. *Given a plane region S bounded by a closed contour L, suppose the functions*

$$P(x_1, x_2), Q(x_1, x_2)$$

and their derivatives

$$\frac{\partial P}{\partial x_2}, \quad \frac{\partial Q}{\partial x_1}$$

are continuous on $S \cup L$. Then

$$\iint_S \left(\frac{\partial Q}{\partial x_1} - \frac{\partial P}{\partial x_2}\right) dS = \oint_L P\,dx_1 + Q\,dx_2, \tag{4.11}$$

where L is traversed in the direction such that S appears to the left of an observer moving along L.

Proof. Suppose no line parallel to the x_1 or x_2-axis intersects L in more than two points,[3] as in Fig. 4.3. Then

$$\iint_S \frac{\partial P}{\partial x_2}\,dS = \iint_S \frac{\partial P}{\partial x_2}\,dx_1\,dx_2$$

$$= \int_a^b dx_1 \int_{x_2=\varphi_1(x_1)}^{x_2=\varphi_2(x_1)} \frac{\partial P}{\partial x_2}\,dx_2,$$

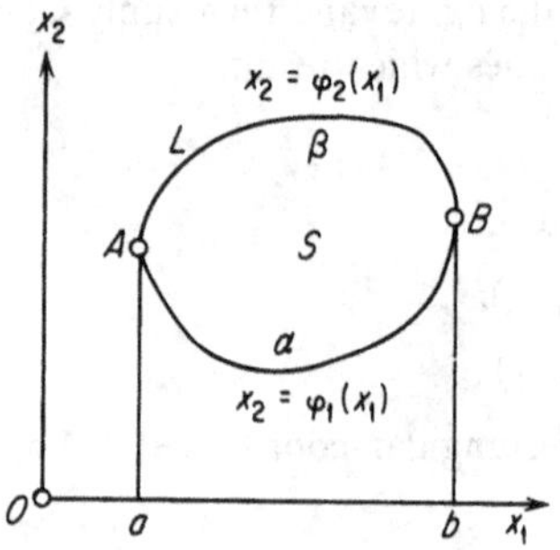

FIG. 4.3. Illustrating Green's theorem.

where the meaning of a, b, $\varphi_1(x)$ and $\varphi_2(x)$ is shown in the figure, and hence

$$\iint_S \frac{\partial P}{\partial x_2}\,dS = \int_a^b \{P[x_1, \varphi_2(x_1)] - P[x_1, \varphi_1(x_1)]\}\,dx_1 \tag{4.12}$$

$$= \int_a^b P[x_1, \varphi_2(x_1)]\,dx_1 - \int_a^b P[x_1, \varphi_1(x_1)]\,dx_1.$$

But the integrals on the right will be recognized as the line integrals of the function $P(x_1, x_2)$ along the curves $A\beta B$ and $A\alpha B$, respectively. Therefore (4.12) becomes

$$\iint_S \frac{\partial P}{\partial x_2}\,dS = -\int_{B\beta A} P(x_1, x_2)\,dx_1 - \int_{A\alpha B} P(x_1, x_2)\,dx_1$$
$$= -\oint_L P(x_1, x_2)\,dx_1. \tag{4.13}$$

The formula

$$\iint_S \frac{\partial Q}{\partial x_1}\,dS = \oint_L Q\,dx_2 \tag{4.14}$$

is proved in the same way. Subtracting (4.13) from (4.14), we obtain (4.11).

[3] It can be shown that this restriction is unnecessary, just like the analogous restriction in the proof of Gauss' theorem (cf. Remark 2, p. 138). For the details, see R. C. Buck, *Advanced Calculus*, second edition, McGraw-Hill Book Co., Inc. (1965), p. 408.

4.2.3. Stokes' theorem. We are now in a position to prove

STOKES' THEOREM. *Given a surface S bounded by a closed contour L, suppose the functions*

$$P(x_1, x_2, x_3), \quad Q(x_1, x_2, x_3), \quad R(x_1, x_2, x_3)$$

and their derivatives

$$\frac{\partial P}{\partial x_2}, \quad \frac{\partial P}{\partial x_3}, \quad \frac{\partial Q}{\partial x_1}, \quad \frac{\partial Q}{\partial x_3}, \quad \frac{\partial R}{\partial x_1}, \quad \frac{\partial R}{\partial x_2}$$

are continuous on $S \cup L$. Then

$$\begin{aligned}\iint_S \Big\{\left(\frac{\partial R}{\partial x_2} - \frac{\partial Q}{\partial x_3}\right) \cos(\mathbf{n}, x_1) &+ \left(\frac{\partial P}{\partial x_3} - \frac{\partial R}{\partial x_1}\right) \cos(\mathbf{n}, x_2) \\ + \left(\frac{\partial Q}{\partial x_1} - \frac{\partial P}{\partial x_2}\right) \cos(\mathbf{n}, x_3)\Big\}\, dS &= \oint_L P\, dx_1 + Q\, dx_2 + R\, dx_3,\end{aligned} \tag{4.15}$$

where **n** *is the unit exterior normal to S (see Fig. 4.4).*[4] *Here L is traversed in the direction such that S appears to the left of an observer moving along L with the vector* **n** *at points near L pointing from the observer's feet to his head.*

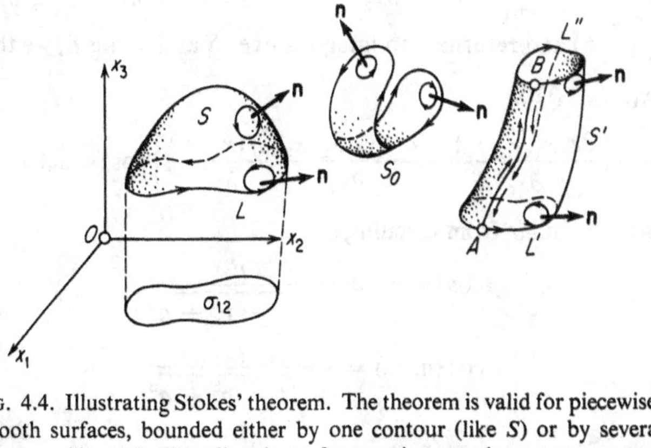

FIG. 4.4. Illustrating Stokes' theorem. The theorem is valid for piecewise-smooth surfaces, bounded either by one contour (like S) or by several contours (like S'). The direction of traversing an elementary contour about any point of the surface is consistent with the direction of traversing the boundary and would cause a right-handed screw to advance along **n**.

---

[4] It is assumed that S is a two-sided surface, with one of its sides singled out by the choice of **n**.

Proof.[5] Suppose no line parallel to the x_3-axis intersects S in more than one point, as in Fig. 4.4. Then the projection of S onto the x_1x_2-plane is a plane region σ_{12}, and the projection of L onto the x_1x_2-plane is a closed contour l which is the boundary of σ_{12}. Let l be traversed in the direction such that σ_{12} appears to the left of an observer moving along l (a right-handed screw turned in this direction would advance along the positive x_3-axis). This establishes a corresponding direction of traversing L and a corresponding direction of the exterior normal $\mathbf{n}$ to the surface S such that $\mathbf{n}$ makes an acute angle with the positive x_3-axis. The relation between the elements of area on σ_{12} and S is then

$$d\sigma_{12} = dS\cos(\mathbf{n}, x_3), \qquad \cos(\mathbf{n}, x_3) > 0. \tag{4.16}$$

We now use the fact that the contour L belongs to a surface whose equation can be written in the form $x_3 = f(x_1, x_2)$ to replace the integral along L by an integral along l:

$$\oint_L P(x_1, x_2, x_3)\,dx_1 = \oint_l P[x_1, x_2, f(x_1, x_2)]\,dx_1. \tag{4.17}$$

Applying Green's theorem to the right-hand side of (4.17) and bearing in mind that x_2 appears in the expression for P both directly and via $x_3 = f(x_1, x_2)$, we obtain

$$\oint_l P[x_1, x_2, f(x_1, x_2)]\,dx_1$$
$$= -\iint_{\sigma_{12}} \left[\frac{\partial P[x_1, x_2, f(x_1, x_2)]}{\partial x_2} + \frac{\partial P[x_1, x_2, f(x_1, x_2)]}{\partial f}\frac{\partial f}{\partial x_2}\right] d\sigma_{12}.$$

Using (4.16) and returning to integrals over S and along L, we then have

$$\oint_L P(x_1, x_2, x_3)\,dx_1$$
$$= -\iint_S \left[\frac{\partial P(x_1, x_2, x_3)}{\partial x_2} + \frac{\partial P(x_1, x_2, x_3)}{\partial x_3}\frac{\partial f(x_1, x_2)}{\partial x_2}\right]\cos(\mathbf{n}, x_3)\,dS. \tag{4.18}$$

But, as is familiar from calculus,

$$\cos(\mathbf{n}, x_1) = \frac{p}{\pm\sqrt{1 + p^2 + q^2}},$$
$$\cos(\mathbf{n}, x_2) = \frac{q}{\pm\sqrt{1 + p^2 + q^2}}, \tag{4.19}$$
$$\cos(\mathbf{n}, x_3) = \frac{1}{\mp\sqrt{1 + p^2 + q^2}},$$

where

$$p = \frac{\partial f}{\partial x_1}, \qquad q = \frac{\partial f}{\partial x_2}.$$

[5] Another, less rigorous proof of Stokes' theorem will be given in Sec. 5.2.2.

Therefore, choosing the bottom signs in (4.19) [since $\cos(\mathbf{n}, x_3) > 0$], we have

$$\frac{\partial f}{\partial x_2}\cos(\mathbf{n}, x_3) = -\cos(\mathbf{n}, x_2).$$

Then (4.18) takes the form

$$\oint_L P\,dx_1 = \iint_S \left[\frac{\partial P}{\partial x_3}\cos(\mathbf{n}, x_2) - \frac{\partial P}{\partial x_2}\cos(\mathbf{n}, x_3)\right] dS.$$

The formulas

$$\oint_L Q\,dx_2 = \iint_S \left[\frac{\partial Q}{\partial x_1}\cos(\mathbf{n}, x_3) - \frac{\partial Q}{\partial x_3}\cos(\mathbf{n}, x_1)\right] dS, \qquad (4.20)$$

$$\oint_L R\,dx_3 = \iint_S \left[\frac{\partial R}{\partial x_2}\cos(\mathbf{n}, x_1) - \frac{\partial R}{\partial x_1}\cos(\mathbf{n}, x_2)\right] dS \qquad (4.21)$$

are proved in the same way. Adding (4.18), (4.20) and (4.21) we obtain Stokes' theorem (4.15).

Remark 1. As in Gauss' theorem, it is assumed that the surface S is smooth or piecewise smooth.

Remark 2. The requirement that no line parallel to the coordinate axes intersect S in more than one point is not essential and can be dropped (none of the surfaces S, S_0 and S' shown in Fig. 4.4 satisfy this requirement). In fact, suppose the given surface does not satisfy this condition, but can be partitioned into a finite number of subsurfaces which separately satisfy the condition. We then apply Stokes' theorem to each subsurface and add the formulas so obtained. The left-hand side of the result is an integral over the whole surface, while the right-hand side is an integral over the boundary of the surface since the integrals over adjacent sides of the subsurfaces cancel each other, being counted twice but with opposite signs.

Remark 3. Stokes' theorem continues to hold for surfaces like S' in Fig. 4.4 which are bounded by several closed curves (L', L'' in this case). In fact, cutting S' along the curve AB, we get a new surface bounded by L', L'' and the two edges of AB. Applying Stokes' theorem to the cut surface, we then glue the surface back together again along AB. When this is done, the line integrals along the two edges of AB cancel each other (since they go in opposite directions), leaving just the integrals along L' and L''.

Given a vector field $\mathbf{A} = \mathbf{A}(\mathbf{r})$, let

$$A_1 = P(x_1, x_2, x_3),$$

$$A_2 = Q(x_1, x_2, x_3),$$

$$A_3 = R(x_1, x_2, x_3)$$

be the components of **A** in a system of rectangular coordinates x_1, x_2, x_3 with orthonormal basis $\mathbf{i}_1$, $\mathbf{i}_2$, $\mathbf{i}_3$. Then Stokes' theorem takes the form

$$\iint_S \left\{ \left(\frac{\partial A_3}{\partial x_2} - \frac{\partial A_2}{\partial x_3} \right) \cos(\mathbf{n}, x_1) + \left(\frac{\partial A_1}{\partial x_3} - \frac{\partial A_2}{\partial x_1} \right) \cos(\mathbf{n}, x_2) + \left(\frac{\partial A_2}{\partial x_1} - \frac{\partial A_1}{\partial x_2} \right) \cos(\mathbf{n}, x_3) \right\} dS = \oint_L A_1\, dx_1 + A_2\, dx_2 + A_3\, dx_3 = \oint_L \mathbf{A} \cdot d\mathbf{r}, \tag{4.22}$$

where in the last step we use the fact that

$$d\mathbf{r} = \mathbf{i}_1\, dx_1 + \mathbf{i}_2\, dx_2 + \mathbf{i}_3\, dx_3.$$

The integrand of the surface integral also has a vector interpretation, which will be given in Sec. 4.4.5.

4.2.4. Simply and multiply connected regions. In using Stokes' theorem to transform a contour integral into a surface integral, we must always make sure that both the closed contour L and the surface S of which L is the boundary lie entirely inside the region where the hypotheses of Stokes' theorem are satisfied. However, there are regions such that a surface S lying entirely inside the region cannot be found for certain closed contours L lying entirely inside the given region. For example, if the region is the interior of a torus, there is no surface which is bounded by the contour L_1 shown in Fig. 4.5 and lies entirely inside the torus. Similarly, if the region is the exterior of a torus, there is no surface which is bounded by the contour L_2 and lies entirely outside the torus.

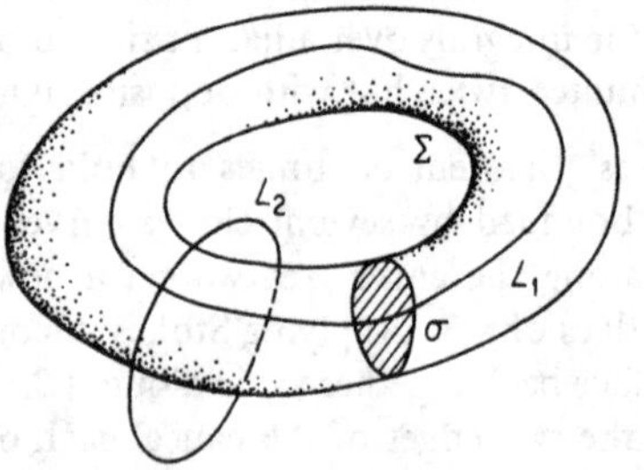

FIG. 4.5. The multiply connected region inside the torus can be made simply connected by adding a "partition" σ to its boundary. The multiply connected region outside a torus can be made simply connected by adding a "soap film" Σ to its boundary, thereby closing the hole in the torus.

A region is said to be *simply connected* if every closed contour in the region can be shrunk continuously to a point without leaving the region.

In this case, every closed contour is the boundary of some surface lying entirely in the region (why?) For example, all of three-dimensional space, the whole plane, the interior of a closed plane curve, the interior or exterior of a sphere, cube, etc. are all simply connected regions.

A region is said to be *multiply connected* if it contains a contour which cannot be shrunk to a point without leaving the region and hence a contour which is not the boundary of a surface lying entirely in the region. For example, both the interior and exterior of a torus are multiply connected, and so are the regions shown in Figs. 4.6(b) and 4.6(c). A multiply connected region can be made simply connected by enlarging its boundary in

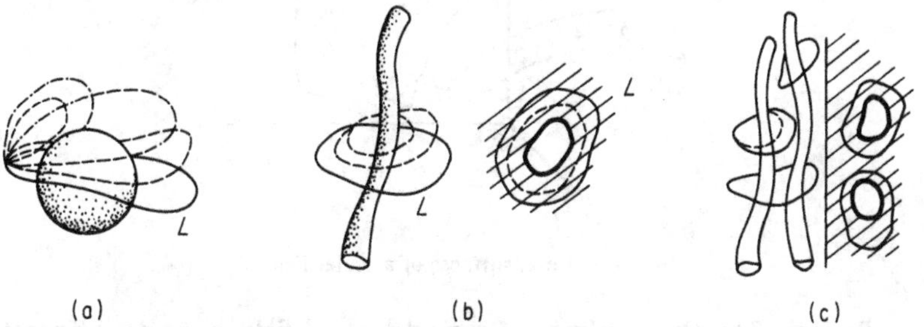

FIG. 4.6. (a) The region outside a sphere is simply connected, since any contour L can be shrunk to a point without leaving the region. (b) The three-dimensional region outside an infinite filament and the two-dimensional region outside a closed curve are multiply connected, since the indicated contours L cannot be shrunk to a point without intersecting the filament or the curve. (c) The three-dimensional region outside two infinite filaments and the two-dimensional region outside two closed curves are multiply connected in a more complicated way.

such a way as to "block off" contours which cannot be shrunk to a point without leaving the region. How this is done in the case of a torus is described in the caption to Fig. 4.5.

4.3. Scalar Fields

4.3.1. Level surfaces. Given a scalar field

$$\varphi = \varphi(\mathbf{r}) = \varphi(x_1, x_2, x_3)$$

(in a system K of rectangular coordinates x_1, x_2, x_3), those points for which φ takes a fixed value C form a surface

$$\varphi(x_1, x_2, x_3) = C,$$

called a *level surface* of the field. By giving C various values, we obtain a family of level surfaces as shown in Fig. 4.7. These surfaces serve to characterize the field geometrically. For example, in places where the level surfaces crowd together, the function φ changes rapidly in the direction perpendicular to the surfaces.

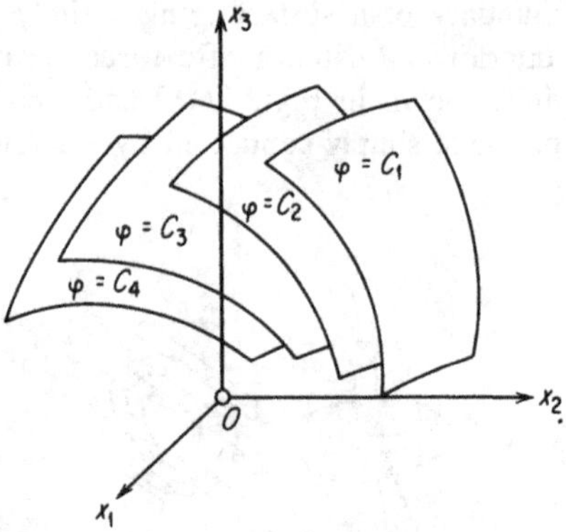

FIG. 4.7. Level surfaces of a scalar field φ.

Remark. The level surfaces of a single-valued field φ cannot intersect, since otherwise φ would take several values at the points of intersection, which is impossible.

4.3.2. The gradient and the directional derivative. The most important characteristic of a scalar field is its gradient, which is the analogue for functions $\varphi(\mathbf{r})$ of a vector argument of the notion of a derivative for functions $f(t)$ of a scalar argument.

In studying the behavior of a function $f(t)$ of a scalar argument near some point M with abscissa t_0, the derivative

$$\left(\frac{df}{dt}\right)_M = \left(\frac{df}{dt}\right)_{t=t_0}, \tag{4.23}$$

if it exists, tells us how rapidly $f(t)$ changes as t is given values exceeding t_0. Thus (4.23) serves as a measure of the rate of change of $f(t)$ *at the point M*. By analogy, in the case of a scalar field $\varphi = \varphi(\mathbf{r})$ it might be expected that the three partial derivatives

$$\left(\frac{\partial \varphi}{\partial x_1}\right)_M, \quad \left(\frac{\partial \varphi}{\partial x_2}\right)_M, \quad \left(\frac{\partial \varphi}{\partial x_3}\right)_M,$$

evaluated at some point M, would serve to describe the rate of change of $\varphi = \varphi(\mathbf{r})$ as we move away from M *in any direction*. This is in fact the case, as we now show.

First we recall from Example 1, p. 91 that the three quantities $\partial\varphi/\partial x_i$ are the components of a vector. This vector is called the *gradient* of the field φ and is denoted by grad φ. Thus

$$\operatorname{grad} \varphi = \mathbf{i}_1 \frac{\partial\varphi}{\partial x_1} + \mathbf{i}_2 \frac{\partial\varphi}{\partial x_2} + \mathbf{i}_3 \frac{\partial\varphi}{\partial x_3} = \mathbf{i}_k \frac{\partial\varphi}{\partial x_k}, \tag{4.24}$$

in terms of the orthonormal basis vectors $\mathbf{i}_1, \mathbf{i}_2, \mathbf{i}_3$ of the rectangular coordinate system K.

Now, given any two points M and M' in the field, let $\boldsymbol{l}$ be a unit vector in the direction from M to M'. Then the average rate of change of φ in the direction $\boldsymbol{l}$ equals

$$\frac{\varphi(M') - \varphi(M)}{M'M}, \tag{4.25}$$

where $MM' = |\overrightarrow{MM'}|$ (see Fig. 4.8). As shown in the figure, let M'' be another point on the level surface of M'. Then, although $\varphi(M') = \varphi(M'')$ it is clear that the ratio

$$\frac{\varphi(M'') - \varphi(M)}{M''M},$$

giving the average rate of change of φ in the direction $\boldsymbol{l}_1$ (from M to M''), is different from (4.25), since $M'M \neq M''M$. In other words, the field φ changes more rapidly in some directions than in others.

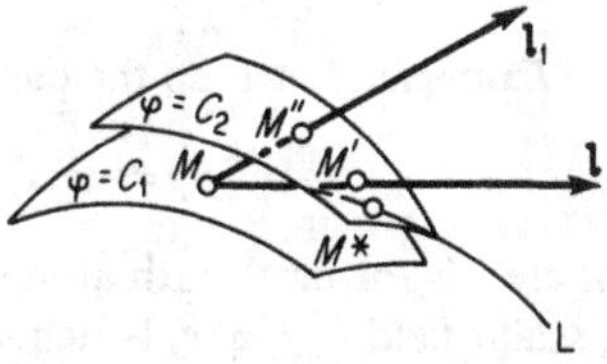

FIG. 4.8. The directional derivative and the derivative along a curve. The field changes more rapidly in the direction $\boldsymbol{l}_1$ than in the direction $\boldsymbol{l}$.

The limit of the ratio (4.25), if it exists, as M' approaches M along $M'M$ is called the *directional derivative* of φ at M in the direction $\boldsymbol{l}$ and is denoted by $d\varphi/dl$:

$$\frac{d\varphi}{dl} = \lim_{M' \to M} \frac{\varphi(M') - \varphi(M)}{M'M}.$$

This derivative describes the rate of change of the field φ at the point M in the direction $\boldsymbol{l}$. If $\varphi(M) = \varphi(x_1, x_2, x_3)$, then

$$\varphi(M') = \varphi[x_1 + M'M\cos(\boldsymbol{l}, x_1), x_2 + M'M\cos(\boldsymbol{l}, x_2), x_3 + M'M\cos(\boldsymbol{l}, x_3)].$$

Hence, expanding $\varphi(M')$ in a Taylor series, we have

$$\varphi(M') = \varphi(M) + \left[\frac{\partial\varphi}{\partial x_1}\cos(\boldsymbol{l}, x_1) + \frac{\partial\varphi}{\partial x_2}\cos(\boldsymbol{l}, x_2) + \frac{\partial\varphi}{\partial x_3}\cos(\boldsymbol{l}, x_3)\right] M'M + O\{(M'M)^2\},$$

where $O\{(M'M)^2\}$ is a quantity of the second order of smallness in the displacement $M'M$. Thus, calculating $d\varphi/dl$, we find that

$$\frac{d\varphi}{dl} = \frac{\partial\varphi}{\partial x_1}\cos(\boldsymbol{l}, x_1) + \frac{\partial\varphi}{\partial x_2}\cos(\boldsymbol{l}, x_2) + \frac{\partial\varphi}{\partial x_3}\cos(\boldsymbol{l}, x_3) = l_k\frac{\partial\varphi}{\partial x_k}. \tag{4.26}$$

Together (4.26) and (4.24) imply

$$\frac{d\varphi}{dl} = \boldsymbol{l}\cdot \operatorname{grad}\varphi, \tag{4.27}$$

i.e., *the directional derivative of* φ *in the direction characterized by the unit vector* $\boldsymbol{l}$ *equals the scalar product of* $\boldsymbol{l}$ *and* grad φ. In the right-hand side of (4.27), grad φ characterizes the field φ while the vector $\boldsymbol{l}$ is independent of φ and characterizes the direction in which the derivative is evaluated. Thus, if grad φ is defined at a point M in the field φ, we can always find the rate of change of φ at M along any direction at all. In this sense, grad φ describes the inhomogeneity of the field φ.

Example. Let L be the curve specified by the parametric equations

$$x_1 = x_1(s), \quad x_2 = x_2(s), \quad x_3 = x_3(s),$$

where s is the arc length along L measured from a fixed point of L. Suppose a scalar field $\varphi = \varphi(\mathbf{r})$ is defined at every point of L. Then by the *derivative of* φ *along the curve* L at the point M is meant the limit

$$\frac{d\varphi}{ds} = \lim_{M^*\to M}\frac{\varphi(M^*) - \varphi(M)}{\Delta s}$$

(provided it exists), where M^* is a variable point of L (see Fig. 4.8), and Δs is the length of the arc MM^* of the curve L. If $\boldsymbol{l}$ is the unit tangent vector to L at the point M, as in the figure, then

$$\frac{d\varphi}{ds} = \frac{d\varphi}{dl}. \tag{4.28}$$

In fact, on the curve L we have

$$\varphi = \varphi[x_1(s), x_2(s), x_3(s)],$$

and hence, by the chain rule of partial differentiation,

$$\frac{d\varphi}{ds} = \frac{\partial\varphi}{\partial x_1}\frac{dx_1}{ds} + \frac{\partial\varphi}{\partial x_2}\frac{dx_2}{ds} + \frac{\partial\varphi}{\partial x_3}\frac{dx_3}{ds}$$

$$= \frac{\partial\varphi}{\partial x_1}\cos(\boldsymbol{l}, x_1) + \frac{\partial\varphi}{\partial x_2}\cos(\boldsymbol{l}, x_2) + \frac{\partial\varphi}{\partial x_3}\cos(\boldsymbol{l}, x_3) = \frac{d\varphi}{dl}.$$

4.3.3. Properties of the gradient. The operator ∇. Writing (4.27) in the form

$$\frac{d\varphi}{dl} = |\text{grad } \varphi| \cos (\boldsymbol{l}, \text{grad } \varphi),$$

we arrive at the following important conclusions:

1) The rate of change of the field φ is greatest in the direction of grad φ [since then $\cos (\boldsymbol{l}, \text{grad } \varphi) = 1$], and equals

$$\left(\frac{d\varphi}{dl}\right)_{\max} = |\text{grad } \varphi| = \sqrt{\left(\frac{\partial \varphi}{\partial x_1}\right)^2 + \left(\frac{\partial \varphi}{\partial x_2}\right)^2 + \left(\frac{\partial \varphi}{\partial x_3}\right)^2}.$$

2) The vector grad φ at the point M points in the direction of increasing φ along the normal to the level surface $\varphi = C$ (say) containing M. To see this, let $\boldsymbol{l}$ lie in the plane tangent to the level surface $\varphi = C$. Then, by the definition of a level surface, $d\varphi/ds = 0$ along any curve lying in the surface $\varphi = C$, and hence by (4.28),

$$\frac{d\varphi}{dl} = 0.$$

But then grad φ is directed along the normal to the level surface $\varphi = C$ ($|\text{grad } \varphi| \neq 0$), in fact in the direction of increasing φ since

$$\left(\frac{d\varphi}{dl}\right)_{\max} = |\text{grad } \varphi| > 0$$

(see Fig. 4.9). In particular, if $\mathbf{n}$ is the unit vector normal to the level surface $\varphi = C$, we have

$$\frac{d\varphi}{dn} = |\text{grad } \varphi|, \qquad \text{grad } \varphi = \mathbf{n} \frac{d\varphi}{dn}.$$

The vector field obtained by taking the gradient of a scalar field φ has a number of special features to be discussed in Sec. 5.4. At this point, we merely note that taking the gradient of φ entails the following sequence of operations [cf. (4.24)]:

1) Form the partial derivatives of φ;
2) Multiply them by the corresponding unit basis vectors of the underlying rectangular coordinate system;
3) Add the resulting expressions.

The effect of these three operations can be described by a single differential operator, denoted by the symbol ∇ and

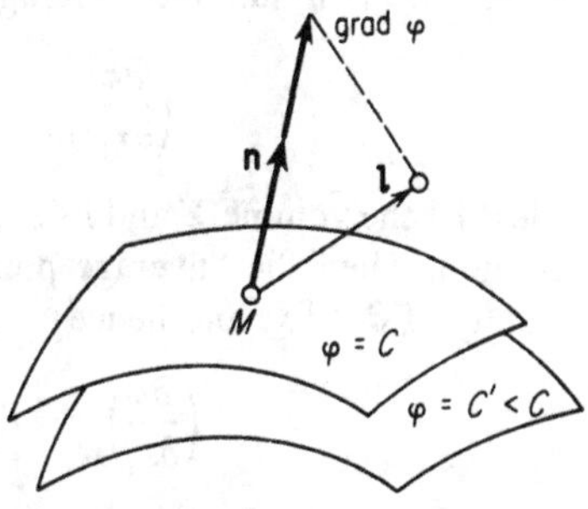

Fig. 4.9. The gradient of the scalar function φ is directed along the normal to the level surface $\varphi = C$ in the direction of increasing φ.

read "del" or "nabla," which takes the form

$$\nabla \equiv \mathbf{i}_1 \frac{\partial}{\partial x_1} + \mathbf{i}_2 \frac{\partial}{\partial x_2} + \mathbf{i}_3 \frac{\partial}{\partial x_3} \tag{4.29}$$

in rectangular coordinates. When applied to a scalar φ, this operator produces the vector field grad φ, i.e.,

$$\nabla\varphi \equiv \text{grad } \varphi = \mathbf{i}_k \frac{\partial \varphi}{\partial x_k}.$$

4.3.4. Another definition of grad φ. We now give a way of defining the vector grad φ which is independent of any coordinate system. Applying Gauss' theorem (4.10) to a vector field of the special type

$$\mathbf{A} = \mathbf{c}\varphi(x_1, x_2, x_3),$$

where the vector $\mathbf{c}$ is fixed but arbitrary, we find that

$$\mathbf{c} \cdot \left(\iiint\limits_V \text{grad } \varphi \, dV - \iint\limits_S \varphi\mathbf{n} \, dS \right) = 0.$$

Since $\mathbf{c}$ is arbitrary, the fact that the scalar product of $\mathbf{c}$ with another vector vanishes implies that the other vector vanishes, i.e., that

$$\iiint\limits_V \text{grad } \varphi \, dV = \iint\limits_S \varphi\mathbf{n} \, dS. \tag{4.30}$$

Now let V be a small volume surrounding some point M in the field φ, and consider any component of grad φ, say $\partial\varphi/\partial x_1$. Then, by the mean-value theorem for integrals,

$$\iiint\limits_V \frac{\partial \varphi}{\partial x_1} \, dV = \left(\frac{\partial \varphi}{\partial x_1} \right)_{M'} V,$$

where M' is a suitable "average point" in the volume V. It follows that

$$\left(\frac{\partial \varphi}{\partial x_1} \right)_{M'} = \frac{1}{V} \iint\limits_S \varphi \cos(\mathbf{n}, x_1) \, dS.$$

Next let the volume V and its surface S shrink to the point M in an arbitrary fashion. Then the "average point" M' approaches M because of the continuity of $\partial\varphi/\partial x_1$, and hence

$$\left(\frac{\partial \varphi}{\partial x_1} \right)_{M} = \lim_{V \to 0} \frac{1}{V} \iint\limits_S \varphi \cos(\mathbf{n}, x_1) \, dS,$$

and similarly for the other two components of grad φ. Thus we have

$$\text{grad } \varphi = \lim_{V \to 0} \frac{1}{V} \iint\limits_S \varphi\mathbf{n} \, dS \tag{4.31}$$

(at the point M). This formula can serve as a definition of grad φ, provided the limit on the right exists. The advantage of (4.31) is that it is independent of the choice of coordinate system and hence can be used to define the components of grad φ in any coordinate system at all (oblique, curvilinear, etc.). This will be done in Sec. 4.7.

4.4. Vector Fields

4.4.1. Trajectories of a vector field. A curve whose tangent at every point has the same direction as a vector field $\mathbf{A} = \mathbf{A}(\mathbf{r})$ is called a *trajectory* of the field (see Fig. 4.10).

Example 1. The trajectories of the field $\mathbf{A} = \text{grad } \varphi$ are the curves orthogonal to the level surfaces $\varphi = \text{const}$ at every point, i.e., the lines of most rapid change of the function $\varphi = \varphi(\mathbf{r})$.

Example 2. The trajectories of the velocity field of a rigid body rotating about an axis are concentric circles with centers on the axis, while the trajectories of the velocity field of a rigid body moving in a straight line are themselves straight lines.

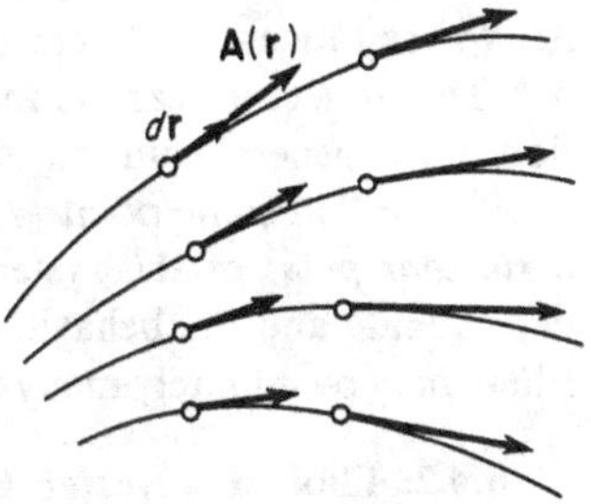

FIG. 4.10. Trajectories of a vector field $\mathbf{A} = \mathbf{A}(\mathbf{r})$.

Example 3. The trajectories of the velocity field $\mathbf{v}$ of a moving fluid are called *streamlines*. In general, the streamlines change with time $[\mathbf{v} = \mathbf{v}(\mathbf{r}, t)]$ and do not coincide with the paths of the fluid particles. However, if the velocity field is stationary $[\mathbf{v} = \mathbf{v}(\mathbf{r})]$, the streamlines do not change in time and represent the actual paths of the fluid particles.

Let $\mathbf{r} = \mathbf{r}(s)$ be a trajectory of a vector field $\mathbf{A} = \mathbf{A}(\mathbf{r})$, in terms of some parameter s (usually the arc length). Then the condition for the tangent to the trajectory to be collinear with $\mathbf{A}$ can be written concisely as

$$d\mathbf{r} \times \mathbf{A}(\mathbf{r}) = 0. \tag{4.32}$$

This is the vector form of the differential equation of the trajectories of the field $\mathbf{A}(\mathbf{r})$. Since the components of collinear vectors must be proportional, the trajectories of the field $\mathbf{A}(\mathbf{r})$ also satisfy the system of scalar differential equations

$$\frac{dx_1}{A_1(x_1, x_2, x_3)} = \frac{dx_2}{A_2(x_1, x_2, x_3)} = \frac{dx_3}{A_3(x_1, x_2, x_3)} \tag{4.33}$$

in a system of rectangular coordinates x_1, x_2, x_3 [note that (4.33) is an immediate consequence of (4.32)].

Integration of (4.32) or (4.33) gives the family of trajectories of the field **A**(**r**). If the field is nonstationary, (4.32) and (4.33) are replaced by

$$d\mathbf{r} \times \mathbf{A}(\mathbf{r}, t) = 0 \tag{4.32'}$$

and

$$\frac{dx_1}{A_1(x_1, x_2, x_3, t)} = \frac{dx_2}{A_2(x_1, x_2, x_3, t)} = \frac{dx_3}{A_3(x_1, x_2, x_3, t)}. \tag{4.33'}$$

In other words, in the case of a nonstationary field **A**(**r**, t), the differential equations of the trajectories have the same form except that t appears as a parameter determining a family of trajectories at every given instant of time.

If $\mathbf{A} \neq 0$ at some point M, there is a unique trajectory passing through M, whose tangent at M has a well-defined direction coinciding with that of **A**.[6] This trajectory can be found by choosing suitable constants of integration in the general solution of (4.33).

If $\mathbf{A} = 0$ at some point M, all the denominators in (4.33) vanish. At such a *singular point* of the system (4.33), the direction of the trajectory is indeterminate and the behavior of the trajectories becomes more complicated (there may be infinitely many trajectories through M or even none at all).

4.4.2. Flux of a vector field. Let S be a two-sided piecewise-smooth surface (which may or may not be closed) immersed in a vector field $\mathbf{A} = \mathbf{A}(\mathbf{r})$. Let dS be an element of S, and let **n** be the unit normal to dS at one of its points. Then by the *flux* of the field **A**(**r**) through the element dS we mean the quantity

$$\mathbf{A} \cdot \mathbf{n} \, dS = A_n \, dS,$$

where **A** is taken at the same point as **n**. Similarly, by the flux of **A**(**r**) through the whole surface S we mean the integral

$$\iint\limits_S \mathbf{A} \cdot \mathbf{n} \, dS = \iint\limits_S A_n \, dS.$$

This surface integral is independent of the choice of coordinate system. In a system of rectangular coordinates x_1, x_2, x_3, it takes the form

$$\iint\limits_S \mathbf{A} \cdot \mathbf{n} \, dS = \iint\limits_S [A_1 \cos(\mathbf{n}, x_1) + A_2 \cos(\mathbf{n}, x_2) + A_3 \cos(\mathbf{n}, x_3)] \, dS,$$

where A_1, A_2, A_3 are the components of the vector **A**.

The following physical example clarifies the meaning of the concept of the flux of a vector field:

---

[6] It is assumed that **A** has the smoothness required to invoke the appropriate existence and uniqueness theorems for the system (4.33).

Example. Consider the stationary flow of an incompressible fluid totally occupying a certain region of space. Such a flow is characterized by a continuous velocity field $\mathbf{v} = \mathbf{v}(\mathbf{r})$. Let S be a smooth surface immersed in the flow. Then, according to (4.33), the flux of the field $\mathbf{v}(\mathbf{r})$ through S is given by the integral

$$Q = \iint_S \mathbf{v} \cdot \mathbf{n}\, dS. \tag{4.34}$$

As we now show, this is just the amount (i.e., volume) of fluid flowing through dS per unit time.

In fact, let dS be an element of S, with unit normal $\mathbf{n}$ (see Fig. 4.11). The boundary of dS defines an elementary *tube of flow AB*, i.e., the surface formed

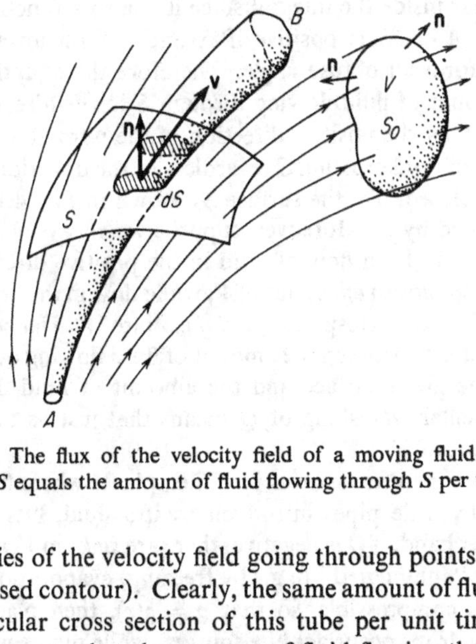

FIG. 4.11. The flux of the velocity field of a moving fluid through a surface S equals the amount of fluid flowing through S per unit time.

by the trajectories of the velocity field going through points of the boundary of S (a small closed contour). Clearly, the same amount of fluid flows through every perpendicular cross section of this tube per unit time. To calculate this quantity, we note that the amount of fluid flowing through dS in time dt equals the volume of the cylinder with base dS and generator $|\mathbf{v}|\, dt$. The altitude h of this cylinder obviously equals the projection of $\mathbf{v}dt$ onto the normal to the base, i.e.,

$$h = |\mathbf{v}|\, dt \cos(\mathbf{v}, \mathbf{n}) = |\mathbf{v} \cdot \mathbf{n}|\, dt.$$

Therefore the amount of fluid flowing through dS in time dt equals

$$dQ = |\mathbf{v}| \cos(\mathbf{v}, \mathbf{n})\, dS = \mathbf{v} \cdot \mathbf{n}\, dS. \tag{4.35}$$

Integrating (4.35) over the whole surface S, we obtain (4.34), as required.

Thus the flux of the velocity field $\mathbf{v}(\mathbf{r})$ through S equals the amount of fluid flowing through S per unit time.

If the fluid is incompressible, then the mass m of fluid flowing through the surface S per unit time is obtained by multiplying (4.34) by the density ρ of the fluid:

$$m = \rho Q = \rho \iint_S \mathbf{v} \cdot \mathbf{n}\, dS. \tag{4.36}$$

On the other hand, if the fluid is compressible with density $\rho = \rho(\mathbf{r})$, itself a (scalar) field, equation (4.36) must be replaced by

$$m = \iint_S \rho \mathbf{v} \cdot \mathbf{n}\, dS, \tag{4.37}$$

i.e., ρ must appear inside the integral since it is also a function of $\mathbf{r}$.

The quantity $\mathbf{A} \cdot \mathbf{n}\, dS$ is positive if $\mathbf{A}$ and $\mathbf{n}$ form an acute angle and negative if they form an obtuse angle. Therefore the quantity (4.34) represents the *net* amount of fluid flowing through S in the direction determined at each point of S by the positive direction of the normal $\mathbf{n}$, rather than the absolute amount of fluid crossing S regardless of the direction of flow.

Suppose S is closed, like the surface S_0 shown in Fig. 4.11, and let V be the volume enclosed by S. Moreover, suppose we always choose $\mathbf{n}$ to be the unit *exterior* normal. Then flow of fluid in the positive direction (the direction of $+\mathbf{n}$) corresponds to *efflux* out of V, while flow in the negative direction (the direction of $-\mathbf{n}$) corresponds to *influx* into V. Hence the (net) flux Q equals the difference between the amount of fluid flowing out of the volume V enclosed by the given surface and the amount of fluid flowing into the volume. In particular, vanishing of Q means that just as much fluid flows into V as flows out of V.

If Q is positive, there are *sources* in V, i.e., places where fluid is somehow "created" (e.g., by little pipes introducing extra fluid, bits of melting ice, etc.). On the other hand, if Q is negative, there are *sinks* in V, i.e., places where fluid is somehow "annihilated" (e.g., by freezing, evaporation, etc.).

If the fluid is compressible, so that $\rho = \rho(\mathbf{r})$, then places where rarefaction or density drops occur act like sources, while places where condensation or density rises occur act like sinks. For example, a density drop means that the same mass of fluid occupies a larger volume near the point of rarefaction. But this causes a greater mass of fluid to appear in the rest of the volume occupied by the fluid.

Thus the flux of a vector field through a closed surface allows us to form some idea of the behavior of the field in the volume V bounded by the surface. However, since V is finite, such estimates can be very crude. For example, the fact that the flux of the velocity field of a moving fluid through a closed surface vanishes can mean the absence of sources and sinks inside the volume

V bounded by the surface. But it can also mean that V contains sources and sinks of equal strength,[7] or a distribution of sources and sinks whose total strength is zero. This suggests introducing a quantity characterizing the "local" or "pointwise" distribution of sources and sinks. In this way, we are led naturally to the concept of the *divergence* of a vector field, which is the subject of Sec. 4.4.3.

4.4.3. Divergence of a vector field. Given any point M in a vector field $\mathbf{A} = \mathbf{A}(\mathbf{r})$, let S be an arbitrary closed surface surrounding M and enclosing a volume V. Calculating the flux of $\mathbf{A}$ through S and dividing by V, we obtain

$$\frac{1}{V}\iint\limits_S \mathbf{A} \cdot \mathbf{n}\, dS. \tag{4.38}$$

Interpreted geometrically, the quantity (4.38) is the average strength of the sources and sinks inside V. The limit of (4.38) as the volume V and its surface S shrink to the point M in an arbitrary fashion (if the limit exists) is called the *divergence* of the field $\mathbf{A}$ (at the point M), denoted by div $\mathbf{A}$. Thus, by definition,

$$\operatorname{div} \mathbf{A} = \lim_{V \to 0} \frac{1}{V}\iint\limits_S \mathbf{A} \cdot \mathbf{n}\, dS. \tag{4.39}$$

Note that the divergence of a vector field $\mathbf{A}(\mathbf{r})$ is a scalar function of $\mathbf{r}$ and hence a scalar field.

Remark. Just like the analogous definition (4.31) of grad φ, the definition (4.39) of div $\mathbf{A}$ is independent of the choice of coordinate system and hence can be used to define div $\mathbf{A}$ in any coordinate system at all (see Sec. 4.7).

The field div $\mathbf{A}$ does not exist for every field $\mathbf{A}$. However, div $\mathbf{A}$ exists at every point where the components A_1, A_2, A_3 of $\mathbf{A}$ (in a system of rectangular coordinates x_1, x_2, x_3) and their derivatives

$$\frac{\partial A_1}{\partial x_1}, \quad \frac{\partial A_2}{\partial x_2}, \quad \frac{\partial A_3}{\partial x_3}$$

are continuous. In fact, it follows from Gauss' theorem in the form (4.10) that

$$\frac{1}{V}\iint\limits_S \mathbf{A} \cdot \mathbf{n}\, dS = \frac{1}{V}\iiint\limits_V \left(\frac{\partial A_1}{\partial x_1} + \frac{\partial A_2}{\partial x_2} + \frac{\partial A_3}{\partial x_3}\right) dV.$$

As V shrinks to some interior point M, the right-hand side obviously has a limit equal to the value of

$$\frac{\partial A_1}{\partial x_1} + \frac{\partial A_2}{\partial x_2} + \frac{\partial A_3}{\partial x_3}$$

[7] By the *strength* of a source (or sink) we mean the amount of fluid emitted (or absorbed) by the source (or sink) per unit time, as on p. 160.

at the point M. Therefore the left-hand side also has a limit equal by definition to the divergence of **A** at M. Hence

$$\operatorname{div} \mathbf{A} = \frac{\partial A_1}{\partial x_1} + \frac{\partial A_2}{\partial x_2} + \frac{\partial A_3}{\partial x_3} \tag{4.40}$$

in rectangular coordinates x_1, x_2, x_3.

Formula (4.40) can also be derived directly by choosing the arbitrary volume V in (4.39) to be an infinitely small parallelepiped with faces perpendicular to the coordinate axes (see Fig. 4.12). The unit normals of the

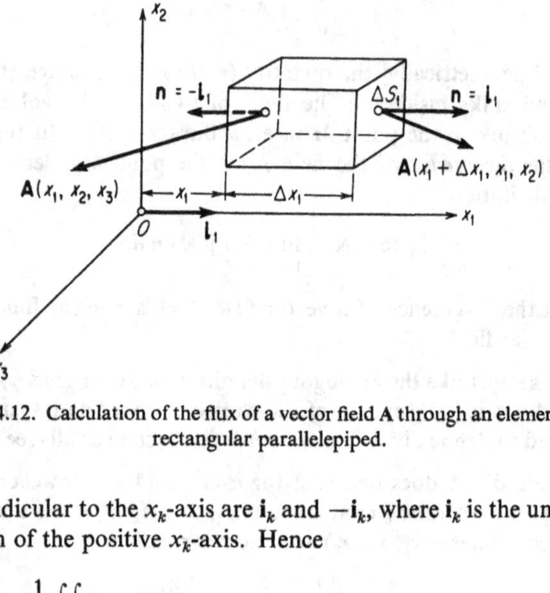

FIG. 4.12. Calculation of the flux of a vector field **A** through an elementary rectangular parallelepiped.

faces perpendicular to the x_k-axis are $\mathbf{i}_k$ and $-\mathbf{i}_k$, where $\mathbf{i}_k$ is the unit vector in the direction of the positive x_k-axis. Hence

$$\begin{aligned}\operatorname{div} \mathbf{A} &= \lim_{V\to 0} \frac{1}{V} \iint_S \mathbf{A} \cdot \mathbf{n}\, dS \\ &= \lim_{V\to 0} \frac{1}{V} \{[\mathbf{i}_1 \cdot \mathbf{A}(x_1 + \Delta x_1, x_2, x_3) - \mathbf{i}_1 \cdot \mathbf{A}(x_1, x_2, x_3)]\, \Delta S_1 + \cdots\} \\ &= \lim_{V\to 0} \frac{1}{V} \{[A_1(x_1 + \Delta x_1, x_2, x_3) - A_1(x_1, x_2, x_3)]\, \Delta S_1 + \cdots\} \\ &= \lim_{V\to 0} \frac{1}{V} \sum_{k=1}^{3} \frac{\partial A_k}{\partial x_k} \Delta x_k\, \Delta S_k,\end{aligned} \tag{4.41}$$

where ΔS_k is the area of the two faces perpendicular to the x_k-axis and Δx_k is their distance apart. But then (4.41) implies (4.40), since

$$\Delta x_1\, \Delta S_1 = \Delta x_2\, \Delta S_2 = \Delta x_3\, \Delta S_3 = V.$$

Using the concept of divergence, we can write Gauss' theorem (4.10) as

$$\iiint\limits_V \operatorname{div} \mathbf{A}\, dV = \iint\limits_S \mathbf{A} \cdot \mathbf{n}\, dS. \tag{4.42}$$

This form of Gauss' theorem, often called the *divergence theorem*, has widespread physical applications. Geometrically, (4.42) means that the integral of the divergence of a vector field over a volume V equals the flux of the field through the surface S bounding V (provided the field is suitably smooth inside V and on S).

The expression (4.29) for the operator ∇ implies the following representation of the divergence of $\mathbf{A}$:

$$\operatorname{div} \mathbf{A} = \frac{\partial A_k}{\partial x_k} = \mathbf{i}_k \frac{\partial}{\partial x_k} \cdot \mathbf{A} = \nabla \cdot \mathbf{A}.$$

In other words, div $\mathbf{A}$ is just the scalar product of ∇ and the vector $\mathbf{A}$.

A coordinate-free symbolic representation of the operator ∇ is

$$\nabla(\cdots) = \lim_{V \to 0} \frac{1}{V} \iint\limits_S \mathbf{n}(\cdots)\, dS, \tag{4.43}$$

where $(\cdots)$ is some expression (possibly preceded by a dot or a cross) on which the given operator acts. In fact, according to (4.31) and (4.39),

$$\nabla \varphi = \lim_{V \to 0} \frac{1}{V} \iint\limits_S \mathbf{n} \varphi\, dS, \tag{4.44}$$

$$\nabla \cdot \mathbf{A} = \lim_{V \to 0} \frac{1}{V} \iint\limits_S \mathbf{n} \cdot \mathbf{A}\, dS. \tag{4.45}$$

4.4.4. Physical examples. We now give some examples clarifying the physical meaning of the concept of divergence.

***Example 1* (*Divergence of the velocity field of a fluid*).** Let $\mathbf{v}(\mathbf{r})$ be the stationary velocity field of a moving fluid. Choosing any point M in the field, we surround M with a surface S enclosing a volume V. If the flux of the velocity field through S is positive, i.e., if

$$\frac{1}{V} \iint\limits_S \mathbf{v} \cdot \mathbf{n}\, dS > 0,$$

then a larger volume of fluid flows out of V (through S) than into V. Suppose V contains neither sources nor sinks. Then the fluid inside V must expand, i.e., its density must decrease. The quantity

$$\frac{1}{V} \iint\limits_S \mathbf{v} \cdot \mathbf{n}\, dS \tag{4.46}$$

characterizes the average expansion of the fluid inside V per unit time, or equivalently, the average rate of volume expansion [contraction if (4.46) is negative] of the fluid inside V. Let V and its surface S shrink to the point M. Then the limit

$$\operatorname{div} \mathbf{v} = \lim_{V \to 0} \frac{1}{V} \iint_S \mathbf{v} \cdot \mathbf{n}\, dS,$$

if it exists, characterizes the rate of change of the volume of the fluid at the point M. Thus a fluid element at the point M which originally had volume ΔV has volume

$$\Delta V' = \Delta V(1 + \operatorname{div} \mathbf{v})$$

one unit of time later. Naturally, if the fluid is incompressible and contains no sources or sinks, then

$$\operatorname{div} \mathbf{v} = 0 \tag{4.47}$$

at every point of the velocity field.

***Example 2* (*Equation of continuity*).** Let S be any closed surface immersed in the stationary velocity field of a moving fluid, and suppose S encloses a volume V. Then the amount of fluid flowing into V per unit time equals the amount of fluid flowing out of V, provided V contains no sources or sinks. Hence, taking account of (4.37), we find that

$$\iint_S \rho \mathbf{v} \cdot \mathbf{n}\, dS = 0$$

for any closed surface S immersed in the fluid.

If the density and velocity of the fluid can vary in time, so that

$$\rho = \rho(\mathbf{r}, t), \qquad \mathbf{v} = \mathbf{v}(\mathbf{r}, t),$$

then the change in mass of the fluid inside V per unit time equals

$$\frac{\partial}{\partial t} \iiint_V \rho\, dV.$$

Since the position of V does not change in time,

$$\frac{\partial}{dt} \iiint_V \rho\, dV = \iiint_V \frac{\partial \rho}{\partial t}\, dV.$$

This change in the mass of fluid inside the fixed volume V must equal the mass flowing into V through its fixed surface S, i.e.,[8]

$$\iiint_V \frac{\partial \rho}{\partial t}\, dV = -\iint_S \rho \mathbf{v} \cdot \mathbf{n}\, dS. \tag{4.48}$$

---

[8] We must put a minus sign before the surface integral since

$$+\iint_S \rho \mathbf{v} \cdot \mathbf{n}\, dS$$

is the mass flowing out of V ($\mathbf{n}$ is the unit *exterior* normal to S).

Using Gauss' theorem to transform the right-hand side of (4.48), we obtain

$$\iiint_V \left(\frac{\partial \rho}{\partial t} + \operatorname{div}(\rho \mathbf{v})\right) dV = 0. \tag{4.49}$$

But (4.49) must hold for an arbitrary volume V, and hence

$$\frac{\partial \rho}{\partial t} + \operatorname{div}(\rho \mathbf{v}) = 0, \tag{4.50}$$

provided the integrand of (4.49) is continuous. Equation (4.50) is the familiar hydrodynamical *equation of continuity*.

Example 3 (Fields due to sources and sinks). Consider a vector field of the form

$$\mathbf{A}(\mathbf{r}) = q\frac{\mathbf{r}}{r^3} \tag{4.51}$$

where $q = \text{const}$ and $\mathbf{r} = \mathbf{i}_1x_1 + \mathbf{i}_2x_2 + \mathbf{i}_3x_3$ is the radius vector. Calculating the divergence of this field, we obtain

$$\frac{\partial A_1}{\partial x_1} = \frac{q(r^2 - 3x_1^2)}{r^5},$$

$$\frac{\partial A_2}{\partial x_2} = \frac{q(r^2 - 3x_2^2)}{r^5},$$

$$\frac{\partial A_3}{\partial x_3} = \frac{q(r^2 - 3x_3^2)}{r^5}.$$

Therefore

$$\operatorname{div} \mathbf{A} = \frac{\partial A_1}{\partial x_1} + \frac{\partial A_2}{\partial x_2} + \frac{\partial A_3}{\partial x_3} = 3\frac{q}{r^3} - \frac{3qr^3}{r^5} = 0$$

everywhere except at the origin of coordinates ($r = 0$). The origin does not belong to the field, since div **A** (like **A** itself) is not defined at the origin.

If S is any closed surface which does not surround the origin, then the flux of the field $\mathbf{A}(\mathbf{r})$ through S vanishes. This follows at once from Gauss' theorem (4.42), since div **A** vanishes throughout the volume V enclosed by S.

On the other hand, if S is a closed surface surrounding the origin, then the volume enclosed by S contains a "singular point" where both **A** and $\partial A_k/\partial x_k$ are undefined, so that Gauss' theorem is not applicable to S. To calculate the flux of the field $\mathbf{A} = \mathbf{A}(\mathbf{r})$ through S, we first surround the origin with a little sphere ε of radius ρ lying entirely inside S, as shown in Fig. 4.13. We can then apply Gauss' theorem to the volume V

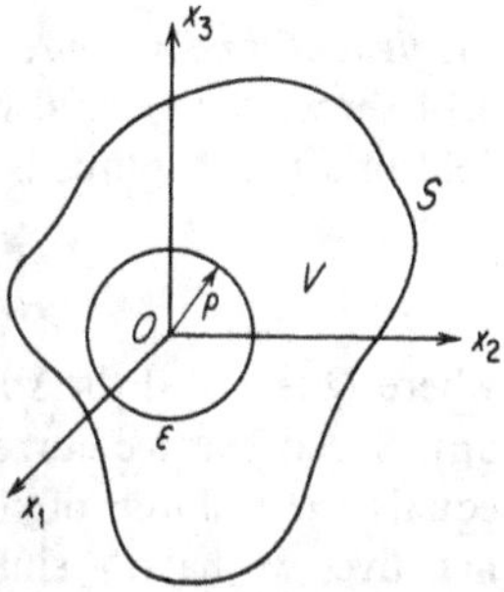

FIG. 4.13. Isolation of a singular point at the origin.

between S and ε. Since div $\mathbf{A} = 0$ everywhere in V, we have

$$\iint_S \mathbf{A} \cdot \mathbf{n}\, dS + \iint_\varepsilon \mathbf{A} \cdot \mathbf{n}\, d\varepsilon = 0. \tag{4.52}$$

But on the sphere ε

$$\mathbf{A}\big|_\varepsilon = \mathbf{A}\big|_{\mathbf{r}=\boldsymbol{\rho}} = q\frac{\boldsymbol{\rho}}{\rho^3},$$

$$\mathbf{n}\big|_\varepsilon = -\frac{\boldsymbol{\rho}}{\rho},$$

where $\boldsymbol{\rho}$ is the radius vector of a variable point on ε. It follows from (4.52) that

$$\iint_S \mathbf{A} \cdot \mathbf{n}\, dS = -\iint_\varepsilon \mathbf{A} \cdot \mathbf{n}\, d\varepsilon = \iint_\varepsilon \frac{q\boldsymbol{\rho}}{\rho^3} \cdot \frac{\boldsymbol{\rho}}{\rho}\, d\varepsilon = \frac{q}{\rho^2}\iint_\varepsilon d\varepsilon = 4\pi q. \tag{4.53}$$

According to (4.53), the flux of the field (4.51) through a surface S surrounding the origin is nonzero and equals $4\pi q$. The field (4.51) is called

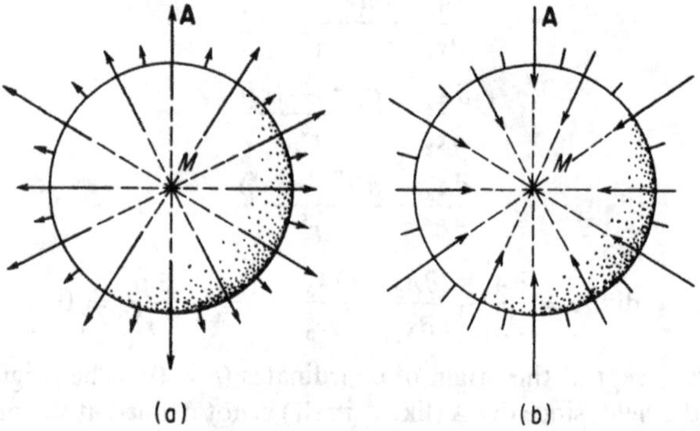

FIG. 4.14. (a) A source; (b) A sink.

the *field of a point source* if $q > 0$ or the *field of a point sink* if $q < 0$. Figure 4.14 shows the general nature of the trajectories of fields of this kind. The field of a point source is usually written in the form

$$\mathbf{A} = \frac{Q}{4\pi}\frac{\mathbf{r}}{r^3},$$

where Q is called the *strength* of the source, equal to the flux of $\mathbf{A}$ through any closed surface surrounding the source. Thus the strength of a source equals the volume of fluid emitted by the source per unit time. If Q is negative, we have a sink of strength $|Q|$, where $|Q|$ is now the volume of fluid absorbed by the sink per unit time.

It is easy to see that the field due to n points sources of strengths Q_1,

$Q_2, \ldots, Q_n$ at the points with radius vectors $\mathbf{r}_1, \mathbf{r}_2, \ldots, \mathbf{r}_n$ equals

$$\mathbf{A} = \frac{1}{4\pi}\left(Q_1 \frac{\mathbf{r}-\mathbf{r}_1}{|\mathbf{r}-\mathbf{r}_1|^3} + Q_2 \frac{\mathbf{r}-\mathbf{r}_2}{|\mathbf{r}-\mathbf{r}_2|^3} + \cdots + Q_n \frac{\mathbf{r}-\mathbf{r}_n}{|\mathbf{r}-\mathbf{r}_n|^3}\right)$$

$$= \frac{1}{4\pi}\sum_{k=1}^{n} Q_k \frac{\mathbf{r}-\mathbf{r}_k}{|\mathbf{r}-\mathbf{r}_k|^3}.$$

4.4.5. Curl of a vector field. Besides the divergence of a vector field $\mathbf{A} = \mathbf{A}(\mathbf{r})$, there is another important differential characteristic of $\mathbf{A}$, namely the curl of $\mathbf{A}$, denoted by curl $\mathbf{A}$. Given a point M in the field $\mathbf{A}$, let S be a closed surface surrounding M with unit exterior normal $\mathbf{n}$, and suppose S encloses a volume V. Then by the value of curl $\mathbf{A}$ at M is meant the limit

$$\operatorname{curl} \mathbf{A} = \lim_{V\to 0} \frac{1}{V} \iint_S \mathbf{n} \times \mathbf{A}\, dS, \tag{4.54}$$

(provided it exists) where both V and its surface S shrink to the point M in an arbitrary fashion. Since $\mathbf{n} \times \mathbf{A}$ is an axial vector (pseudovector), so is curl $\mathbf{A}$. Note the similarity between (4.54) and the definition (4.31) of the gradient and the definition (4.39) of the divergence. In each case, the definition is independent of the choice of coordinate system. Moreover, comparing (4.54) and (4.43), we find that

$$\nabla \times \mathbf{A} = \lim_{V\to 0} \frac{1}{V} \iint_S \mathbf{n} \times \mathbf{A}\, dS \tag{4.55}$$

[cf. (4.44) and (4.45)], and hence

$$\operatorname{curl} \mathbf{A} = \nabla \times \mathbf{A}.$$

In other words, curl $\mathbf{A}$ is just the vector product of the operator ∇ and the vector $\mathbf{A}$.

To explain the geometric meaning of curl $\mathbf{A}$, we choose the surface S to be a right cylinder of infinitesimal cross section, whose generator is of length h and points in the direction specified by the unit vector $\boldsymbol{l}$ (see Fig. 4.15).

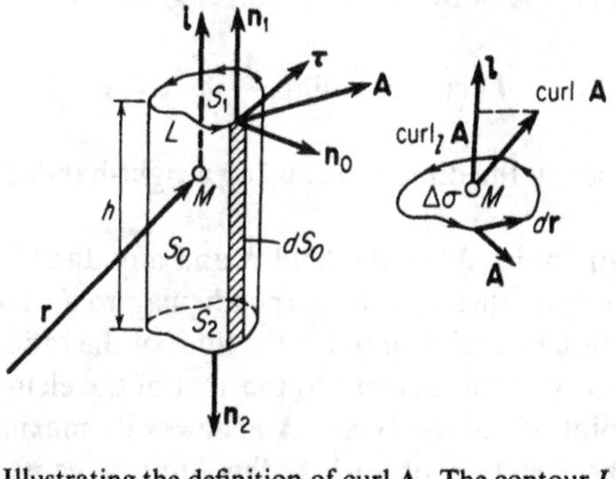

FIG. 4.15. Illustrating the definition of curl A. The contour L is traversed in the direction causing a right-handed screw to advance along $\boldsymbol{l}$.

Let S_0 be the lateral surface of the cylinder, and let S_1 and S_2 be its end faces of common area $\Delta\sigma$. Moreover, let the lateral surface and the faces have unit exterior normals $\mathbf{n}_0$ and $\mathbf{n}_1$, $\mathbf{n}_2$, respectively, where $\boldsymbol{l} = \mathbf{n}_1 = -\mathbf{n}_2$.

We now calculate the projection of curl **A** onto the direction $\boldsymbol{l}$ of the generator of the cylinder, obtaining

$$\begin{aligned}\boldsymbol{l}\cdot\text{curl}\,\mathbf{A} = \text{curl}_l\,\mathbf{A} = \lim_{V\to 0}\frac{1}{h\Delta\sigma}\Bigg\{&\iint_{S_1}\boldsymbol{l}\cdot(\mathbf{n}_1\times\mathbf{A})\,dS_1 \\ &+\iint_{S_2}\boldsymbol{l}\cdot(\mathbf{n}_2\times\mathbf{A})\,dS_2+\iint_{S_0}\boldsymbol{l}\cdot(\mathbf{n}_0\times\mathbf{A})\,dS_0\Bigg\},\end{aligned} \tag{4.56}$$

where V is the volume enclosed by S. Since $\mathbf{n}_1 = -\mathbf{n}_2$, the integrals over the faces S_1 and S_2 cancel each other. As for the integral over the lateral surface S_0, we first note that

$$\boldsymbol{l}\cdot(\mathbf{n}_0\times\mathbf{A}) = \mathbf{A}\cdot(\boldsymbol{l}\times\mathbf{n}_0) = \mathbf{A}\cdot(\mathbf{n}_1\times\mathbf{n}_0)$$

[cf. (1.27)] and then that

$$\mathbf{n}_1\times\mathbf{n}_0 = \boldsymbol{\tau},$$

where $\boldsymbol{\tau}$ is the unit tangent to the contour L bounding any perpendicular cross section of the cylinder. But

$$dS_0 = h\,ds$$

(see Fig. 4.15), where ds is the element of arc length along L. Therefore the right-hand side of (4.56) reduces to

$$\lim_{\substack{h\to 0\\ \Delta\sigma\to 0}}\frac{1}{h\,\Delta\sigma}\iint_{S_0}\boldsymbol{l}\cdot(\mathbf{n}_0\times\mathbf{A})\,dS_0 = \lim_{\Delta\sigma\to 0}\frac{1}{\Delta\sigma}\oint_L\mathbf{A}\cdot d\mathbf{r},$$

where $d\mathbf{r} = \boldsymbol{\tau}\,ds$ as on p. 136 and we can choose L to be the boundary of the perpendicular cross section of the cylinder containing M. Hence, finally, the projection of curl **A** onto any direction is given by

$$\boldsymbol{l}\cdot\text{curl}\,\mathbf{A} = \lim_{\Delta\sigma\to 0}\frac{1}{\Delta\sigma}\oint_L\mathbf{A}\cdot d\mathbf{r}, \tag{4.57}$$

where L is traversed in the direction causing a right-handed screw to advance along $\boldsymbol{l}$.

Thus, given any point M of the field **A** and any direction $\boldsymbol{l}$, consider any element of plane area through M perpendicular to $\boldsymbol{l}$. Then, according to (4.57), the projection of curl **A** onto $\boldsymbol{l}$ is the limit of the ratio of the circulation around the boundary of the element to the area of the element as the element shrinks to the point M. Since $\boldsymbol{l}\cdot$curl **A** achieves its maximum value when $\boldsymbol{l}$ coincides with the direction of curl **A**, this limit takes its maximum value, equal to |curl **A**|, when the area is perpendicular to curl **A**.

The components of curl **A** in a rectangular coordinate system can be found in two ways, One way consists in choosing the surface S to be a rectangular parallelepiped with faces perpendicular to the coordinate axes, and then explicitly calculating the right-hand side of (4.54), just as was done on p. 156 for the case of the divergence (the details are left as an exercise). Another way is to use the fact that

$$\nabla = \mathbf{i}_k \frac{\partial}{\partial x_k}$$

in rectangular coordinates [recall (4.29)] to deduce the representation

$$\operatorname{curl} \mathbf{A} = \nabla \times \mathbf{A} = \begin{vmatrix} \mathbf{i}_1 & \mathbf{i}_2 & \mathbf{i}_3 \\ \dfrac{\partial}{\partial x_1} & \dfrac{\partial}{\partial x_2} & \dfrac{\partial}{\partial x_3} \\ A_1 & A_2 & A_3 \end{vmatrix} \tag{4.58}$$

It follows from (4.58) that

$$\begin{aligned} \operatorname{curl}_1 \mathbf{A} &= \frac{\partial A_3}{\partial x_2} - \frac{\partial A_2}{\partial x_3}, \\ \operatorname{curl}_2 \mathbf{A} &= \frac{\partial A_1}{\partial x_3} - \frac{\partial A_3}{\partial x_1}, \\ \operatorname{curl}_3 \mathbf{A} &= \frac{\partial A_2}{\partial x_1} - \frac{\partial A_1}{\partial x_2}, \end{aligned} \tag{4.59}$$

or more concisely,

$$\operatorname{curl}_i \mathbf{A} = \frac{\partial A_k}{\partial x_j} - \frac{\partial A_j}{\partial x_k} \qquad (i = 1, 2, 3),$$

where the indices i, j, k are a cyclic permutation of the numbers 1, 2, 3.

Example. Consider a rigid body rotating about a fixed point O with angular velocity $\boldsymbol{\omega}$. Then the point with radius vector r has velocity

$$\mathbf{v} = \boldsymbol{\omega} \times \mathbf{r}$$

(see Prob. 11, p. 45), and hence

$$\operatorname{curl} \mathbf{v} = \operatorname{curl} (\boldsymbol{\omega} \times \mathbf{r}).$$

Therefore

$$\operatorname{curl}_1 \mathbf{v} = \frac{\partial v_3}{\partial x_2} - \frac{\partial v_2}{\partial x_3} = \frac{\partial}{\partial x_2}(\omega_1 x_2 - \omega_2 x_1) - \frac{\partial}{\partial x_3}(\omega_3 x_1 - \omega_1 x_3) = 2\omega_1$$

($\boldsymbol{\omega}$ is independent of **r**), and similarly

$$\begin{aligned} \operatorname{curl}_2 \mathbf{v} &= 2\omega_2, \\ \operatorname{curl}_3 \mathbf{v} &= 2\omega_3. \end{aligned}$$

It follows that

$$\operatorname{curl} \mathbf{v} = 2\boldsymbol{\omega},$$

i.e., the curl of the velocity field of a rotating body equals twice the angular velocity of the body.

Using the concept of curl, we can write Stokes' theorem (4.22) in the form

$$\iint_S \mathbf{n} \cdot \operatorname{curl} \mathbf{A} \, dS = \oint_L \mathbf{A} \cdot d\mathbf{r}. \tag{4.60}$$

This form of Stokes' theorem has widespread physical applications. Geometrically, (4.60) means that the flux of the curl of a vector field through a surface S bounded by a contour L equals the circulation of the field around L (provided the field is sufficiently smooth on S and L).

4.4.6. Directional derivative of a vector field. Given a vector field $\mathbf{A} = \mathbf{A}(\mathbf{r})$, let M and M' be two points in the field and let $\boldsymbol{l}$ be a unit vector in the direction from M to M'. Then by the *directional derivative* of $\mathbf{A}$ at M in the direction $\boldsymbol{l}$, denoted by $d\mathbf{A}/dl$, we mean the limit

$$\frac{d\mathbf{A}}{dl} = \lim_{M' \to M} \frac{\mathbf{A}(M') - \mathbf{A}(M)}{M'M},$$

provided it exists. Suppose $\mathbf{A}$ has components A_1, A_2, A_3 in a system of rectangular coordinates. Then, recalling Sec. 4.3.2, we find that $d\mathbf{A}/dl$ is the vector with components[9]

$$\frac{dA_1}{dl}, \quad \frac{dA_2}{dl}, \quad \frac{dA_3}{dl}.$$

Just as formula (4.27) expresses the directional derivative of a scalar field φ in any direction $\boldsymbol{l}$ in terms of the scalar product

$$\frac{d\varphi}{dl} = \boldsymbol{l} \cdot \operatorname{grad} \varphi$$

we can express the components dA_i/dl of the directional derivative $d\mathbf{A}/dl$ in terms of the scalar product

$$\frac{dA_i}{dl} = \boldsymbol{l} \cdot \operatorname{grad} A_i. \tag{4.61}$$

We can also write (4.61) as an inner product

$$\frac{dA_i}{dl} = l_k \frac{\partial A_i}{\partial x_k}, \tag{4.62}$$

[9] Note that

$$\left(\frac{d\mathbf{A}}{dl}\right)_i = \left(\lim_{M' \to M} \frac{\mathbf{A}(M') - \mathbf{A}(M)}{M'M}\right)_i = \lim_{M' \to M} \frac{A_i(M') - A_i(M)}{M'M} = \frac{dA_i}{dl}.$$

where l_1, l_2, l_3 are the components of $\boldsymbol{l}$ and $\partial A_i/\partial x_k$ is a second-order tensor since

$$\frac{\partial A'_i}{\partial x'_k} = \alpha_{i'l}\frac{\partial A_l}{\partial x'_k} = \alpha_{i'l}\frac{\partial A_l}{\partial x_m}\frac{\partial x_m}{\partial x'_k} = \alpha_{i'l}\alpha_{k'm}\frac{\partial A_l}{\partial x_m}.$$

Another way of writing (4.61) is

$$\frac{d\mathbf{A}}{dl} = (\boldsymbol{l}\cdot\nabla)\mathbf{A}, \tag{4.63}$$

in terms of the differential operator

$$\boldsymbol{l}\cdot\nabla = (l_1\mathbf{i}_1 + l_2\mathbf{i}_2 + l_3\mathbf{i}_3)\cdot\left(\mathbf{i}_1\frac{\partial}{\partial x_1} + \mathbf{i}_2\frac{\partial}{\partial x_2} + \mathbf{i}_3\frac{\partial}{\partial x_3}\right)$$
$$= l_1\frac{\partial}{\partial x_1} + l_2\frac{\partial}{\partial x_2} + l_3\frac{\partial}{\partial x_3}.$$

Example (Acceleration field of a moving fluid). Let $\mathbf{v} = \mathbf{v}(\mathbf{r}, t)$ be the velocity field (in general, inhomogeneous and nonstationary) of a moving fluid. Suppose that in time dt a fluid particle moves from M to M', thereby undergoing a velocity change $d\mathbf{v}$. Then there are two contributions to the increment $d\mathbf{v}$. One is a "local" increment

$$d\mathbf{v}_{\text{loc}} = \frac{\partial\mathbf{v}}{\partial t}dt \tag{4.64}$$

(stemming from the nonstationarity of the velocity field) equal to the change in the velocity at M in the time dt it takes the particle to go from M to M' [see Fig. 4.16(a)]. The other contribution is a "convective" increment

$$d\mathbf{v}_{\text{conv}} = \frac{d\mathbf{v}}{dl}dl$$

(stemming from the inhomogeneity of the velocity field) equal to the difference in velocities at the points M and M' at the same time t [see Fig. 4.16(b)].

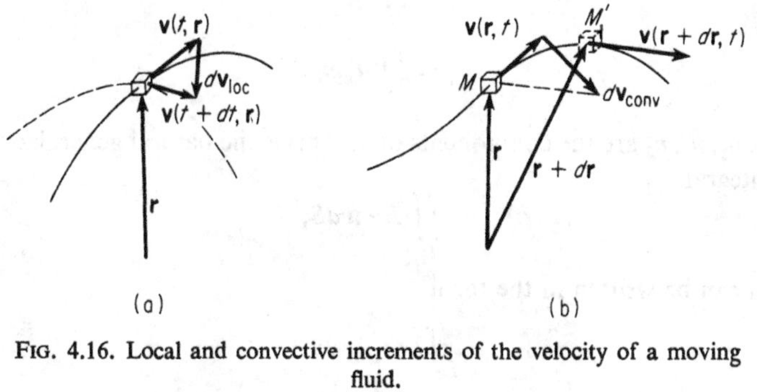

FIG. 4.16. Local and convective increments of the velocity of a moving fluid.

Here $d\mathbf{v}/dl$ is the directional derivative of the velocity field in the direction from M to M' (characterized by the unit vector $\boldsymbol{l}$), and $dl = |d\mathbf{r}|$ is the distance between M and M'. But clearly

$$\boldsymbol{l} = \frac{\mathbf{v}}{|\mathbf{v}|},$$

since the particle moves in the direction given by its velocity. Therefore, using (4.63), we have

$$\frac{d\mathbf{v}}{dl} = \left(\frac{\mathbf{v}}{|\mathbf{v}|} \cdot \nabla\right)\mathbf{v} = \frac{1}{|\mathbf{v}|}(\mathbf{v} \cdot \nabla)\mathbf{v},$$

and hence

$$d\mathbf{v}_{\text{conv}} = (\mathbf{v} \cdot \nabla)\mathbf{v}\, dt \tag{4.65}$$

since $dl = |\mathbf{v}|\, dt$. Adding (4.64) and (4.65), we find that the total velocity increment is

$$d\mathbf{v} = \frac{\partial \mathbf{v}}{\partial t} dt + (\mathbf{v} \cdot \nabla)\mathbf{v}\, dt.$$

Therefore, finally, the acceleration field of the moving fluid is

$$\frac{d\mathbf{v}}{dt} = \frac{\partial \mathbf{v}}{\partial t} + (\mathbf{v} \cdot \nabla)\mathbf{v}. \tag{4.66}$$

In component form, (4.66) becomes

$$\frac{dv_i}{dt} = \frac{\partial v_i}{\partial t} + v_k \frac{\partial v_i}{\partial x_k}. \tag{4.67}$$

4.5. Second-Order Tensor Fields

Let $T_{ik} = T_{ik}(\mathbf{r})$ be a second-order tensor field (see Sec. 4.1.2), and let S be a two-sided piecewise smooth surface in the field of T_{ik} with variable unit normal $\mathbf{n}$. Then by the *flux* of T_{ik} through S we mean either of the vectors $\mathbf{W}$ with components

$$W_i = \iint_S T_{ik} n_k\, dS \tag{4.68}$$

or

$$W_i = \iint_S T_{ki} n_k\, dS, \tag{4.69}$$

where n_1, n_2, n_3 are the components of $\mathbf{n}$. This is the natural generalization of the integral

$$\iint_S \mathbf{A} \cdot \mathbf{n}\, dS,$$

which can be written in the form

$$\iint_S A_k n_k\, dS.$$

Note that whereas the flux of a vector field is a scalar, the flux of a second-order tensor field is a vector.

Example 1. Let $T_{ik} = p_{ik}$ be the stress tensor of an elastic medium. Choosing a surface S in the medium (S may or may not be closed), we calculate the resultant $\mathbf{P}$ of all stresses acting on S. If $\mathbf{p}_n$ is the stress on the element dS with unit normal $\mathbf{n}$, then $\mathbf{P}$ is the vector

$$\mathbf{P} = \iint_S \mathbf{p}_n \, dS,$$

with components

$$P_k = \iint_S p_{nk} \, dS.$$

But according to (2.14), $p_{nk} = p_{ik} n_i$ and hence

$$P_k = \iint_S p_{ik} n_i \, dS,$$

i.e., the total stress on S is the flux [as defined by (4.68)] of the stress tensor p_{ik} through S.

Example 2. Given a closed surface S, the flux of the unit tensor δ_{ik} through S, this time defined by (4.69), is the vector $\mathbf{W}$ with components

$$W_i = \iint_S \delta_{ki} n_k \, dS = \iint_S n_i \, dS.$$

It follows that

$$\mathbf{W} = \iint_S \mathbf{n} \, dS.$$

But

$$\iint_S \mathbf{n} \, dS = 0,$$

as can be seen at once by setting $\varphi = \text{const}$ in (4.30). Therefore the flux of the unit tensor through any closed surface vanishes.

Next let S be a closed surface surrounding a point M in a tensor field $T_{ik}(\mathbf{r})$ and enclosing a volume V. Then, by analogy with (4.39), we define the divergence of $T_{ik}(\mathbf{r})$ at M as either of the limits[10]

$$(\operatorname{div} \mathbf{T})_i = \lim_{V \to 0} \frac{1}{V} \iint_S T_{ik} n_k \, dS \tag{4.70}$$

or

$$(\operatorname{div} \mathbf{T})_i = \lim_{V \to 0} \frac{1}{V} \iint_S T_{ki} n_k \, dS \tag{4.71}$$

---

[10] The definition (4.70) corresponds to (4.68), while (4.71) corresponds to (4.69). Concerning the meaning of $\mathbf{T}$, see the remark on p. 93.

(provided they exist) as the volume V and its surface S shrink to the point M in an arbitrary fashion. Note that whereas the divergence of a vector field is a scalar field, the divergence of a second-order tensor field is a vector field. In a system of rectangular coordinates x_1, x_2, x_3, we have

$$(\text{div } T)_i = \frac{\partial T_{ik}}{\partial x_k}$$

with the definition (4.70) and

$$(\text{div } T)_i = \frac{\partial T_{ki}}{\partial x_k}$$

with the definition (4.71), by the same arguments as on pp. 155–156.

Finally, by the natural generalization of (4.26) and (4.62), the directional derivative of a tensor field $T_{ik}(\mathbf{r})$ along the direction $\boldsymbol{l}$ is defined as

$$\frac{dT_{ik}}{dl} = l_j \frac{\partial T_{ik}}{\partial x_j},$$

where it is easily verified that $\partial T_{ik}/\partial x_j$ is a third-order tensor.

Remark. The operation of taking the curl of a vector field has no analogue for the case of higher-order tensor fields.

4.6. The Operator ∇ and Related Differential Operators

The first-order differential operator ∇ has already been encountered in the expressions

$$\begin{aligned} \nabla\varphi &= \text{grad } \varphi, \\ \nabla \cdot \mathbf{A} &= \text{div } \mathbf{A}, \\ \nabla \times \mathbf{A} &= \text{curl } \mathbf{A}. \end{aligned} \tag{4.72}$$

Applying ∇ once again to (4.72), we obtain the following expressions involving second-order differential operators:[11]

$$\begin{aligned} \nabla \cdot \nabla\varphi &= \text{div grad } \varphi \equiv \nabla^2\varphi \equiv \Delta\varphi, \\ \nabla \times \nabla\varphi &= \text{curl grad } \varphi \\ \nabla(\nabla \cdot \mathbf{A}) &= \text{grad div } \mathbf{A}, \\ \nabla \cdot (\nabla \times \mathbf{A}) &= \text{div curl } \mathbf{A} \\ \nabla \times (\nabla \times \mathbf{A}) &= \text{curl curl } \mathbf{A}. \end{aligned}$$

[11] As we will see in a moment, two of these expressions vanish identically.

It will be recalled that the operator ∇ has the coordinate-free representation

$$\nabla(\cdots) = \lim_{V\to 0} \frac{1}{V} \iint_S \mathbf{n}(\cdots)\, dS$$

[cf. (4.43)] and the particularly simple representation

$$\nabla = \mathbf{i}_k \frac{\partial}{\partial x_k} \tag{4.73}$$

in rectangular coordinates [cf. (4.29)]. Using (4.73), we find that the expressions curl grad φ and div rot $\mathbf{A}$ both vanish in rectangular coordinates, and hence in any coordinate system (why?). The operator $\nabla \cdot \nabla \equiv \nabla^2 \equiv \Delta$ is called the *Laplacian*. In retangular coordinates, it takes the form

$$\Delta \equiv \nabla \cdot \nabla = \mathbf{i}_k \frac{\partial}{\partial x_k} \cdot \mathbf{i}_l \frac{\partial}{\partial x_l} = (\mathbf{i}_k \cdot \mathbf{i}_l) \frac{\partial^2}{\partial x_k\, \partial x_l}$$

$$= \frac{\partial^2}{\partial x_k\, \partial x_k} \equiv \frac{\partial^2}{\partial x_1^2} + \frac{\partial^2}{\partial x_2^2} + \frac{\partial^2}{\partial x_3^2},$$

so that

$$\Delta\varphi = \frac{\partial^2 \varphi}{\partial x_k\, \partial x_k} = \frac{\partial^2 \varphi}{\partial x_1^2} + \frac{\partial^2 \varphi}{\partial x_2^2} + \frac{\partial^2 \varphi}{\partial x_3^2}.$$

Clearly the result of applying the operator ∇ to a sum of two or more terms is the sum of the results of applying ∇ to each term separately. Thus we have

$$\text{grad}\,(\varphi + \chi) = \text{grad}\,\varphi + \text{grad}\,\chi,$$

$$\text{div}\,(\mathbf{A} + \mathbf{B}) = \text{div}\,\mathbf{A} + \text{div}\,\mathbf{B},$$

$$\text{curl}\,(\mathbf{A} + \mathbf{B}) = \text{curl}\,\mathbf{A} + \text{curl}\,\mathbf{B},$$

$$\text{grad div}\,(\mathbf{A} + \mathbf{B}) = \text{grad div}\,\mathbf{A} + \text{grad div}\,\mathbf{B},$$

$$\Delta(\varphi + \chi) = \Delta\varphi + \Delta\chi,$$

$$\text{curl curl}\,(\mathbf{A} + \mathbf{B}) = \text{curl curl}\,\mathbf{A} + \text{curl curl}\,\mathbf{B}.$$

On the other hand, when applied to the expressions

$$\varphi\chi, \quad \varphi\mathbf{A}, \quad \mathbf{A} \cdot \mathbf{B}, \quad \mathbf{A} \times \mathbf{B},$$

the operator ∇ acts on each factor separately with the other held fixed. Thus ∇ should be written after any factor regarded as a constant in a given term and before any factor regarded as variable. In this way, we obtain the following formulas (the subscript c denotes that the quantity to which it is attached is momentarily being held fixed):[12]

$$\operatorname{grad}(\varphi\chi) = \nabla(\varphi\chi) = \nabla\varphi_c\chi + \nabla\varphi\chi_c = \varphi_c\,\nabla\chi + \chi_c\,\nabla\varphi$$
$$= \varphi\,\nabla\chi + \chi\,\nabla\varphi = \varphi\operatorname{grad}\chi + \chi\operatorname{grad}\varphi,$$

$$\operatorname{div}(\varphi\mathbf{A}) = \nabla\cdot(\varphi\mathbf{A}) = \nabla\cdot\varphi_c\mathbf{A} + \nabla\cdot\varphi\mathbf{A}_c = \varphi_c\,\nabla\cdot\mathbf{A} + \mathbf{A}_c\cdot\nabla\varphi$$
$$= \varphi\,\nabla\cdot\mathbf{A} + \mathbf{A}\cdot\nabla\varphi = \varphi\operatorname{div}\mathbf{A} + \mathbf{A}\cdot\operatorname{grad}\varphi,$$

$$\operatorname{div}(\mathbf{A}\times\mathbf{B}) = \nabla\cdot(\mathbf{A}\times\mathbf{B}) = \nabla\cdot(\mathbf{A}_c\times\mathbf{B}) + \nabla\cdot(\mathbf{A}\times\mathbf{B}_c)$$
$$= -\mathbf{A}_c\cdot(\nabla\times\mathbf{B}) + \mathbf{B}_c\cdot(\nabla\times\mathbf{A})$$
$$= -\mathbf{A}\cdot(\nabla\times\mathbf{B}) + \mathbf{B}\cdot(\nabla\times\mathbf{A})$$
$$= \mathbf{B}\cdot\operatorname{curl}\mathbf{A} - \mathbf{A}\cdot\operatorname{curl}\mathbf{B},$$

$$\operatorname{curl}(\varphi\mathbf{A}) = \nabla\times(\varphi\mathbf{A}) = \nabla\times(\varphi_c\mathbf{A}) + \nabla\times(\varphi\mathbf{A}_c)$$
$$= \varphi_c(\nabla\times\mathbf{A}) + \nabla\varphi\times\mathbf{A}_c = \varphi(\nabla\times\mathbf{A}) + \nabla\varphi\times\mathbf{A}$$
$$= \varphi\operatorname{curl}\mathbf{A} - \mathbf{A}\times\operatorname{grad}\varphi.$$

A case of great practical importance is where φ and $\mathbf{A}$ are composite functions, i.e.,

$$\varphi = \varphi[f(\mathbf{r})], \qquad \mathbf{A} = \mathbf{A}[f(\mathbf{r})],$$

where $f(\mathbf{r})$ is a scalar function of the radius vector $\mathbf{r}$. Then we have

$$\operatorname{grad}\varphi[f(r)] = \frac{d\varphi}{df}\operatorname{grad}f,$$
$$\operatorname{div}\mathbf{A}[f(\mathbf{r})] = \operatorname{grad}f\cdot\frac{d\mathbf{A}}{df}, \tag{4.74}$$
$$\operatorname{curl}\mathbf{A}[f(\mathbf{r})] = \operatorname{grad}f\times\frac{d\mathbf{A}}{df}.$$

To prove (4.74), say, we start from the definition

$$\operatorname{div}\mathbf{A}\big|_M = \lim_{V\to 0}\frac{1}{V}\iint_S \mathbf{A}\cdot\mathbf{n}\,dS \tag{4.75}$$

[12] The reader should bear in mind the explicit nature of the operator ∇, involving first partial differentiation, then multiplication by the basis vectors and finally addition (cf. pp. 149–150).

and expand the integrand in a neighborhood of a point M, obtaining

$$\mathbf{A} = \mathbf{A}(M) + \left.\frac{d\mathbf{A}}{df}\right|_M [f - f(M)] + \cdots. \tag{4.76}$$

Substituting (4.76) into (4.75) and bearing in mind that

$$\iint_S \mathbf{n}\, dS = 0$$

(see Example 2, p. 167), we find that

$$\operatorname{div} \mathbf{A}\big|_M = \lim_{V\to 0} \frac{1}{V} \left.\frac{d\mathbf{A}}{df}\right|_M \cdot \iint_S f\mathbf{n}\, dS = \left.\frac{d\mathbf{A}}{df}\right|_M \cdot \lim_{V\to 0} \frac{1}{V} \iint_S f\mathbf{n}\, dS.$$

Using the definition (4.31) and dropping the explicit dependence on M, we finally obtain

$$\operatorname{div} \mathbf{A} = \frac{d\mathbf{A}}{df} \cdot \operatorname{grad} f,$$

as required.

4.6.1. Differential operators in orthogonal curvilinear coordinates. We now derive expressions for the quantities[13]

$$\operatorname{grad} f, \quad \operatorname{div} \mathbf{A}, \quad \operatorname{curl} \mathbf{A}, \quad \Delta f$$

in a system of orthogonal curvilinear coordinates q^1, q^2, q^3 with *orthogonal* local basis $\mathbf{e}_1$, $\mathbf{e}_2$, $\mathbf{e}_3$. By the natural generalization of the argument given in Sec. 4.3.2, we find that

$$\operatorname{grad} f = \frac{\partial f}{\partial s_1} \mathbf{e}_1^0 + \frac{\partial f}{\partial s_2} \mathbf{e}_2^0 + \frac{\partial f}{\partial s_3} \mathbf{e}_3^0, \tag{4.77}$$

where $ds_i = |\mathbf{e}_i|\, dq^i$ is the element of arc length along the coordinate curve (q^i) [recall Sec. 2.8.4] and $\mathbf{e}_i^0$ is the unit vector $\mathbf{e}_i/|\mathbf{e}_i|$. In fact, with this definition of grad f, the directional derivative of f along any direction $\boldsymbol{l}$ is still given by the formula

$$\frac{df}{dl} = \boldsymbol{l} \cdot \operatorname{grad} f = l^1 \frac{\partial f}{\partial s_1} + l^2 \frac{\partial f}{\partial s_2} + l^3 \frac{\partial f}{d s_3}$$

($\boldsymbol{l} = l^1\mathbf{e}_1^0 + l^2\mathbf{e}_2^0 + l^3\mathbf{e}_3^0$), and is a maximum in the direction of grad f. But on p. 87 we found that

$$ds_i = h_i dq^i \quad \text{(no summation over } i\text{)},$$

in orthogonal curvilinear coordinates, where

$$h_i = \sqrt{g_{ii}}.$$

[13] Here we prefer the symbol f for a scalar field (to avoid confusion with the angle φ in cylindrical and spherical coordinates).

Therefore (4.77) can be written in the form

$$\operatorname{grad} f = \frac{1}{h_1}\frac{\partial f}{\partial q^1}\mathbf{e}_1^0 + \frac{1}{h_2}\frac{\partial f}{\partial q^2}\mathbf{e}_2^0 + \frac{1}{h_3}\frac{\partial f}{\partial q^3}\mathbf{e}_3^0. \tag{4.78}$$

To calculate div **A** in orthogonal curvilinear coordinates, we use the definition[14]

$$\operatorname{div}\mathbf{A} = \lim_{V\to 0}\frac{1}{V}\iint\limits_S \mathbf{A}\cdot\mathbf{n}\,dS,$$

choosing V to be an elementary "curvilinear parallelepiped" of volume

$$ds_1\,ds_2\,ds_3 = h_1h_2h_3\,dq^1\,dq^2\,dq^3, \tag{4.79}$$

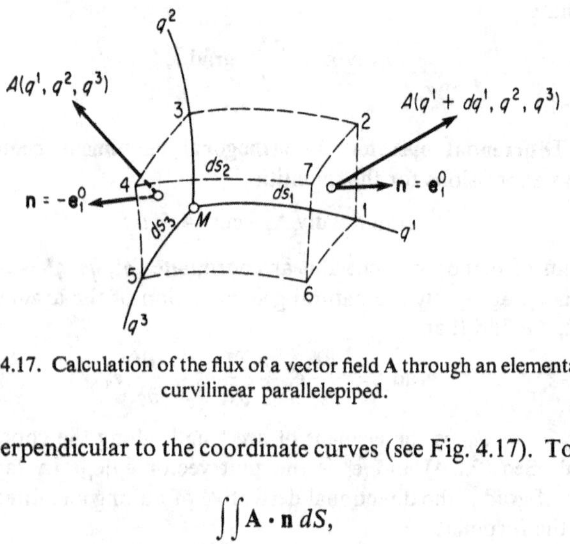

FIG. 4.17. Calculation of the flux of a vector field **A** through an elementary curvilinear parallelepiped.

with faces perpendicular to the coordinate curves (see Fig. 4.17). To calculate the flux

$$\iint\limits_S \mathbf{A}\cdot\mathbf{n}\,dS,$$

we first consider the face M345, with exterior normal $\mathbf{n} = -\mathbf{e}_1^0$. The flux through this face is clearly

$$-\mathbf{A}\cdot\mathbf{e}_1^0\,ds_2\,ds_3 = -A_1h_2h_3\,dq^2\,dq^3, \tag{4.80}$$

where $A_1 = \mathbf{A}\cdot\mathbf{e}_i^0$. On the face 1276, with exterior normal $\mathbf{n} = \mathbf{e}_i^0$, q^1 increases by dq^1 and correspondingly (4.80) is replaced by

$$\left(A_1h_2h_3 + \frac{\partial(A_1h_2h_3)}{\partial q^1}\,dq^1\right)dq^2\,dq^3.$$

[14] As usual, V and its surface S shrink to some point M in an arbitrary fashion.

Calculating the flux through the other two pairs of faces in the same way, adding the resulting expressions and dividing by (4.79), we finally find that

$$\operatorname{div} \mathbf{A} = \frac{1}{h_1h_2h_3}\left[\frac{\partial}{\partial q^1}(A_1h_2h_3) + \frac{\partial}{\partial q^2}(A_2h_1h_3) + \frac{\partial}{\partial q^3}(A_3h_1h_2)\right] \tag{4.81}$$

$$(A_2 = \mathbf{A}\cdot\mathbf{e}_2^0,\quad A_3 = \mathbf{A}\cdot\mathbf{e}_3^0).$$

Similarly, we can calculate curl **A**, starting from the formula

$$\operatorname{curl} \mathbf{A} = \lim_{V\to 0}\frac{1}{V}\iint_S \mathbf{n}\times\mathbf{A}\,dS$$

(the details are left as an exercise). The result is

$$\begin{aligned}\operatorname{curl} \mathbf{A} &= \frac{1}{h_2h_3}\left[\frac{\partial}{\partial q^2}(A_3h_3) - \frac{\partial}{\partial q^3}(A_2h_2)\right]\mathbf{e}_1^0\\ &+ \frac{1}{h_1h_3}\left[\frac{\partial}{\partial q^3}(A_1h_1) - \frac{\partial}{\partial q^1}(A_3h_3)\right]\mathbf{e}_2^0\\ &+ \frac{1}{h_1h_2}\left[\frac{\partial}{\partial q^1}(A_2h_2) - \frac{\partial}{\partial q^2}(A_1h_1)\right]\mathbf{e}_3^0.\end{aligned} \tag{4.82}$$

Finally, to find an expression for

$$\Delta f = \nabla\cdot\nabla f = \operatorname{div}\operatorname{grad} f$$

in curvilinear coordinates, we use (4.78) and (4.81), obtaining

$$\Delta f = \frac{1}{h_1h_2h_3}\left[\frac{\partial}{\partial q^1}\left(\frac{h_2h_3}{h_1}\frac{\partial f}{\partial q^1}\right) + \frac{\partial}{\partial q^2}\left(\frac{h_1h_3}{h_2}\frac{\partial f}{\partial q^2}\right) + \frac{\partial}{\partial q^3}\left(\frac{h_1h_2}{h_3}\frac{\partial f}{\partial q^3}\right)\right]. \tag{4.83}$$

Example 1. In cylindrical coordinates

$$q^1 = R,\quad q^2 = \varphi,\quad q^3 = z,$$
$$h_1 = 1,\quad h_2 = R,\quad h_3 = 1$$

(see Sec. 2.8), and hence in this case (4.78) and (4.81)–(4.83) give

$$\operatorname{grad} f = \frac{\partial f}{\partial R}\mathbf{e}_R + \frac{1}{R}\frac{\partial f}{\partial\varphi}\mathbf{e}_\varphi + \frac{\partial f}{\partial z}\mathbf{e}_z,$$

$$\operatorname{div} \mathbf{A} = \frac{1}{R}\frac{\partial}{\partial R}(RA_R) + \frac{1}{R}\frac{\partial A_\varphi}{\partial\varphi} + \frac{\partial A_z}{\partial z},$$

$$\begin{aligned}\operatorname{curl} \mathbf{A} &= \left(\frac{1}{R}\frac{\partial A_z}{\partial\varphi} - \frac{\partial A_\varphi}{\partial z}\right)\mathbf{e}_R + \left(\frac{\partial A_R}{\partial z} - \frac{\partial A_z}{\partial R}\right)\mathbf{e}_\varphi\\ &+ \frac{1}{R}\left(\frac{\partial}{\partial R}(RA_\varphi) - \frac{\partial A_R}{\partial\varphi}\right)\mathbf{e}_z,\end{aligned}$$

$$\Delta f = \frac{1}{R}\frac{\partial}{\partial R}\left(R\frac{\partial f}{\partial R}\right) + \frac{1}{R^2}\frac{\partial^2 f}{\partial\varphi^2} + \frac{\partial^2 f}{\partial z^2},$$

where $\mathbf{e}_1^0 = \mathbf{e}_R$, $\mathbf{e}_2^0 = e_\varphi$, $\mathbf{e}_3^0 = \mathbf{e}_z$ is the local *orthonormal* basis, and $\mathbf{A}$ has components $A_1 = A_R$, $A_2 = A_\varphi$, $A_3 = A_z$ with respect to this basis.

Example 2. In spherical coordinates

$$q^1 = R, \quad q^2 = \theta, \quad q^3 = \varphi,$$

$$h_1 = 1, \quad h_2 = R, \quad h_3 = R\sin\theta,$$

so that

$$\operatorname{grad} f = \frac{\partial f}{\partial R}\mathbf{e}_R + \frac{1}{R}\frac{\partial f}{\partial\theta}\mathbf{e}_\theta + \frac{1}{R\sin\theta}\frac{\partial f}{\partial\varphi}\mathbf{e}_\varphi,$$

$$\operatorname{div}\mathbf{A} = \frac{1}{R^2}\frac{\partial}{\partial R}(R^2 A_R) + \frac{1}{R\sin\theta}\frac{\partial}{\partial\theta}(A_\theta\sin\theta) + \frac{1}{R\sin\theta}\frac{\partial A_\varphi}{\partial\varphi},$$

$$\begin{aligned}\operatorname{curl}\mathbf{A} = {} & \frac{1}{R\sin\theta}\left(\frac{\partial}{\partial\theta}(A_\varphi\sin\theta) - \frac{\partial A_\theta}{\partial\varphi}\right)\mathbf{e}_R \\ & + \left(\frac{1}{R\sin\theta}\frac{\partial A_R}{\partial\varphi} - \frac{1}{R}\frac{\partial}{\partial R}(RA_\varphi)\right)\mathbf{e}_\theta \\ & + \frac{1}{R}\left(\frac{\partial}{\partial R}(RA_\theta) - \frac{\partial A_R}{\partial\theta}\right)\mathbf{e}_\varphi,\end{aligned}$$

$$\Delta f = \frac{1}{R^2\sin\theta}\left[\frac{\partial}{\partial R}\left(R^2\sin\theta\frac{\partial f}{\partial R}\right) + \frac{\partial}{\partial\theta}\left(\sin\theta\frac{\partial f}{\partial\theta}\right) + \frac{\partial}{\partial\varphi}\left(\frac{1}{\sin\theta}\frac{\partial f}{\partial\varphi}\right)\right],$$

where $\mathbf{e}_1^0 = \mathbf{e}_R$, $\mathbf{e}_2^0 = \mathbf{e}_\theta$, $\mathbf{e}_3^0 = \mathbf{e}_\varphi$ is the local orthonormal basis and $\mathbf{A}$ has components $A_1 = A_R$, $A_2 = A_\theta$, $A_3 = A_\varphi$ with respect to this basis.

SOLVED PROBLEMS

Problem 1 (The Frenet-Serret formulas). Let

$$\mathbf{r} = \mathbf{r}(s) \tag{4.84}$$

be the equation of a (directed) space curve, where $\mathbf{r}$ is the radius vector of a variable point M on the curve and s is the arc length measured from some fixed point on the curve (s increases in the direction chosen to be positive). Suppose that with each point M there is associated a unique trihedral, consisting of a unit tangent $\boldsymbol{\tau}$, a unit normal $\mathbf{n}$ and a unit binormal $\mathbf{b}$, as shown in Fig. 4.18. Then

$$\boldsymbol{\tau} = \frac{d\mathbf{r}}{ds},$$

since $\boldsymbol{\tau}$ and $d\mathbf{r}/ds$ have the same direction (recall Sec. 1.7.2) and

$$\left|\frac{d\mathbf{r}}{ds}\right| = \lim_{\Delta s\to 0}\left|\frac{\Delta\mathbf{r}}{\Delta s}\right| = 1.$$

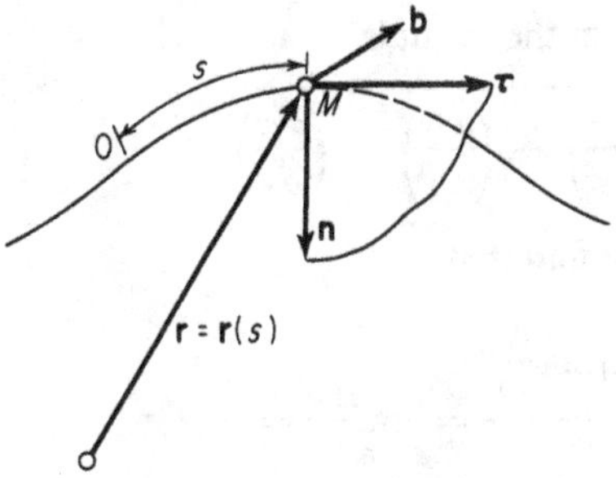

FIG. 4.18. The tangent **τ**, normal **n** and binormal **b** to a space curve.

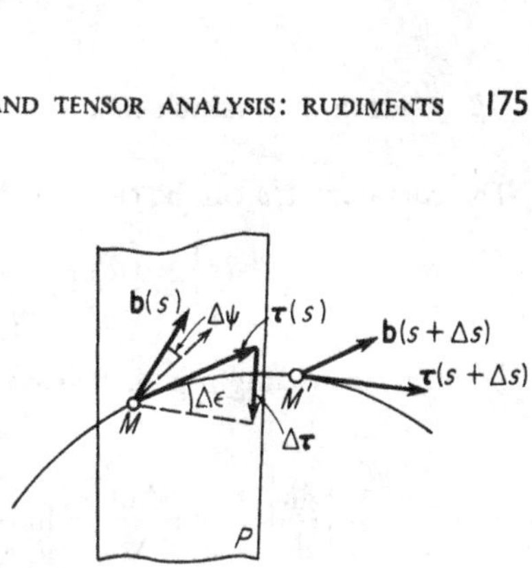

FIG. 4.19. The moving trihedral of a space curve. As $M' \to M$, the plane P approaches the osculating plane at M.

The vector **n** is orthogonal to **τ** and lies in the osculating plane[15] at M pointing in the direction of concavity of the curve, while the binormal **b** is defined as the vector product

$$\mathbf{b} = \boldsymbol{\tau} \times \mathbf{n}.$$

We now pose the problem of finding the derivatives[16]

$$\frac{d\boldsymbol{\tau}}{ds}, \quad \frac{d\mathbf{n}}{ds}, \quad \frac{d\mathbf{b}}{ds}.$$

Solution. According to Fig. 4.19,

$$\left|\frac{d\boldsymbol{\tau}}{ds}\right| = \lim_{\Delta s \to 0} \frac{|\Delta\boldsymbol{\tau}|}{\Delta s} = \lim_{\Delta s \to 0} \frac{2\,|\boldsymbol{\tau}| \sin \dfrac{\Delta\varepsilon}{2}}{\Delta s} = \lim_{\Delta s \to 0} \frac{\Delta\varepsilon}{\Delta s} = \frac{1}{\rho}, \tag{4.85}$$

where $\Delta\varepsilon$ is the angle between two neighboring tangents $\boldsymbol{\tau}(s)$ and $\boldsymbol{\tau}(s + \Delta s)$, and the quantity $1/\rho$ is called the *curvature* of the curve (ρ itself is called the *radius of curvature*). The vector $\Delta\boldsymbol{\tau}/\Delta s$ lies in the plane P whose limiting position is the osculating plane. Since $|\boldsymbol{\tau}| = \text{const}$, $d\boldsymbol{\tau}/ds$ is orthogonal to **τ** and points in the direction of concavity of the curve, as shown in the figure. In other words, $d\boldsymbol{\tau}/ds$ points in the direction of **n**. Together with (4.85), this implies

$$\frac{d\boldsymbol{\tau}}{ds} = \frac{\mathbf{n}}{\rho}. \tag{4.86}$$

[15] Let M and M' be two points of the curve (4.84). Then the *osculating plane* of the curve at M is the limiting position as M' approaches M of the plane P through the tangent at M parallel to the tangent at M' (see Fig. 4.19), provided this limit exists. A plane curve lies in its osculating plane.

[16] Clearly, each of the vectors **τ**, **n** and **b** is a vector function of the scalar argument s (the arc length). The three functions $\boldsymbol{\tau}(s)$, $\mathbf{n}(s)$ and $\mathbf{b}(s)$ are called the *moving trihedral* of the space curve.

The curvature $1/\rho$ can be calculated from the formula

$$\frac{1}{\rho} = \left|\frac{d\boldsymbol{\tau}}{ds}\right| = \left|\frac{d^2\mathbf{r}}{ds^2}\right| = \sqrt{\left(\frac{d^2x}{ds^2}\right)^2 + \left(\frac{d^2y}{ds^2}\right)^2 + \left(\frac{d^2z}{ds^2}\right)^2}.$$

Next, inspecting Fig. 4.19 again, we find that

$$\left|\frac{d\mathbf{b}}{ds}\right| = \lim_{\Delta s\to 0}\frac{|\Delta\mathbf{b}|}{\Delta s} = \lim_{\Delta s\to 0}\frac{2\,|\mathbf{b}|\sin\dfrac{\Delta\psi}{2}}{2} = \lim_{\Delta s\to 0}\frac{\Delta\psi}{\Delta s} = \frac{1}{T},$$

where $\Delta\psi$ is the angle between two neighboring binormals $\mathbf{b}(s)$ and $\mathbf{b}(s + \Delta s)$, and the quantity $1/T$ is called the *torsion of the curve* (T itself is called the *radius of torsion*). Moreover

$$\frac{d\mathbf{b}}{ds} = \frac{d}{ds}(\boldsymbol{\tau}\times\mathbf{n}) = \frac{d\boldsymbol{\tau}}{ds}\times\mathbf{n} + \boldsymbol{\tau}\times\frac{d\mathbf{n}}{ds} = \boldsymbol{\tau}\times\frac{d\mathbf{n}}{ds},$$

and hence the vector $d\mathbf{b}/ds$ must be orthogonal to both $\boldsymbol{\tau}$ and $\mathbf{b}$ (since $|\mathbf{b}| = 1$), i.e., $d\mathbf{b}/ds$ and $\mathbf{n}$ have parallel directions. Therefore the osculating plane of the curve rotates about the tangent to the curve as the point M moves along the curve. The torsion will be considered positive if the osculating plane rotates in the direction from $\mathbf{n}$ to $\mathbf{b}$ as s increases, and negative if it rotates in the direction from $\mathbf{b}$ to $\mathbf{n}$. It follows that

$$\frac{d\mathbf{b}}{ds} = -\frac{\mathbf{n}}{T}. \tag{4.87}$$

Finally, the quantity $d\mathbf{n}/ds$ is found by the following simple calculation:

$$\frac{d\mathbf{n}}{ds} = \frac{d}{ds}(\mathbf{b}\times\boldsymbol{\tau}) = \frac{d\mathbf{b}}{ds}\times\boldsymbol{\tau} + \mathbf{b}\times\frac{d\boldsymbol{\tau}}{ds} = -\frac{1}{T}(\mathbf{n}\times\boldsymbol{\tau}) + \frac{1}{\rho}(\mathbf{b}\times\mathbf{n}),$$

i.e.,

$$\frac{d\mathbf{n}}{ds} = \frac{\mathbf{b}}{T} - \frac{\boldsymbol{\tau}}{\rho}. \tag{4.88}$$

Together, formulas (4.86)–(4.88) are called the *Frenet-Serret formulas.*

***Problem* 2.** Find the torsion of the curve $\mathbf{r} = \mathbf{r}(s)$.

Solution. It follows from (4.87) that

$$\frac{1}{T} = -\mathbf{n}\cdot\frac{d\mathbf{b}}{ds}. \tag{4.89}$$

But

$$\mathbf{b} = \boldsymbol{\tau}\times\mathbf{n}, \quad \boldsymbol{\tau} = \frac{d\mathbf{r}}{ds}, \quad \mathbf{n} = \rho\frac{d^2\mathbf{r}}{ds^2},$$

and hence

$$\mathbf{b} = \frac{\dfrac{d\mathbf{r}}{ds} \times \dfrac{d^2\mathbf{r}}{ds^2}}{\left|\dfrac{d\mathbf{r}}{ds} \times \dfrac{d^2\mathbf{r}}{ds^2}\right|} = \rho\left(\frac{d\mathbf{r}}{ds} \times \frac{d^2\mathbf{r}}{ds^2}\right),$$

so that

$$\frac{1}{T} = -\rho^2 \frac{d^2\mathbf{r}}{ds^2} \cdot \frac{d}{ds}\left(\frac{d\mathbf{r}}{ds} \times \frac{d^2\mathbf{r}}{ds^2}\right) = \rho^2 \frac{d\mathbf{r}}{ds} \cdot \left(\frac{d^2\mathbf{r}}{ds^2} \times \frac{d^3\mathbf{r}}{ds^3}\right).$$

Problem 3. Find the components of the acceleration of a particle with respect to the moving trihedral of its trajectory.

Solution. Let

$$\mathbf{r} = \mathbf{r}(t) \tag{4.90}$$

be the equation of the particle's trajectory, where the parameter t is the time. Then the velocity and acceleration of the particle are

$$\mathbf{v} = \frac{d\mathbf{r}}{dt}, \qquad \mathbf{a} = \frac{d^2\mathbf{r}}{dt^2} = \frac{d\mathbf{v}}{dt}.$$

Starting from (4.90), we can always find the function $s = s(t)$ relating the arc length s (measured from some fixed point) to the value of the parameter t. Then, writing

$$\mathbf{r} = \mathbf{r}(t) = \mathbf{r}[s(t)],$$

we have

$$\mathbf{v} = \frac{d\mathbf{r}}{dt} = \frac{d\mathbf{r}}{ds}\frac{ds}{dt} = \boldsymbol{\tau} v,$$

where $v = ds/dt$ is the speed. Hence

$$\mathbf{a} = \frac{d\mathbf{v}}{dt} = \frac{d\boldsymbol{\tau}}{dt} v + \boldsymbol{\tau}\frac{dv}{dt} = \frac{d\boldsymbol{\tau}}{ds}\frac{ds}{dt} v + \boldsymbol{\tau}\frac{dv}{dt} = \frac{v^2}{\rho}\mathbf{n} + \frac{dv}{dt}\boldsymbol{\tau},$$

where (4.86) is used in the last step. Thus the acceleration lies entirely in the osculating plane, and its components with respect to the moving trihedral are

$$a_\tau = \frac{dv}{dt}, \quad a_n = \frac{v^2}{\rho}, \quad a_b = 0.$$

Problem 4. Let $\mathbf{g}_1(\varphi)$ and $\mathbf{g}_2(\varphi)$ be two unit vector functions ($|\mathbf{g}_1| = |\mathbf{g}_2| = 1$) which lie in the x_1x_2-plane and make angles φ and $\varphi + \frac{1}{2}\pi$ with

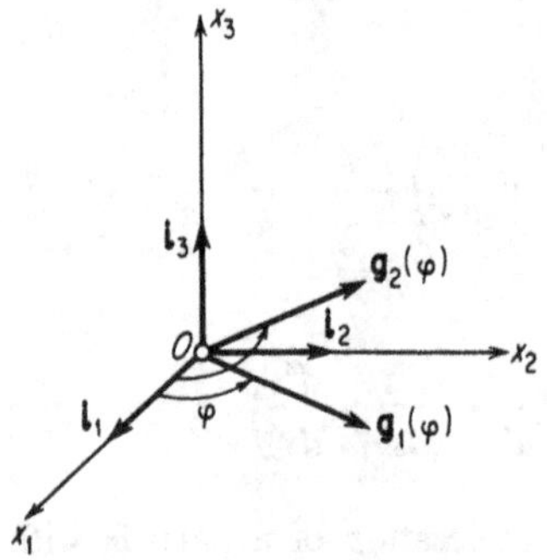

FIG. 4.20. Illustrating the vector functions $\mathbf{g}_1(\varphi)$ and $\mathbf{g}_2(\varphi)$.

respect to the x_1-axis (see Fig. 4.20).. Expand $\mathbf{g}_1(\varphi)$, $\mathbf{g}_2(\varphi)$ with respect to the fixed orthonormal basis $\mathbf{i}_1$, $\mathbf{i}_2$, $\mathbf{i}_3$, and express $d\mathbf{g}_1/d\varphi$, $d\mathbf{g}_2/d\varphi$ in terms of $\mathbf{g}_1$ and $\mathbf{g}_2$.

Solution. By definition,

$$\mathbf{g}_1(\varphi) = \mathbf{i}_1 \cos \varphi + \mathbf{i}_2 \sin \varphi,$$
$$\mathbf{g}_2(\varphi) = -\mathbf{i}_1 \sin \varphi + \mathbf{i}_2 \cos \varphi.$$

Differentiating with respect to φ, we have

$$\frac{d\mathbf{g}_1}{d\varphi} = \mathbf{g}_2, \qquad \frac{d\mathbf{g}_2}{d\varphi} = -\mathbf{g}_1. \tag{4.91}$$

Problem 5. Find the radial and tangential components of the velocity and acceleration of a particle moving in the plane,[17] given that the particle's trajectory has a parametric equations

$$r = r(t), \qquad \varphi = \varphi(t),$$

where $\mathbf{r}$ is the radius vector of the particle, t is the time, $r = |\mathbf{r}|$ and φ is the angle between $\mathbf{r}$ and the x_1-axis.

Solution. Let $\mathbf{g}_1(\varphi)$ and $\mathbf{g}_2(\varphi)$ be the same as in the preceding problem, so that $\mathbf{r} = r\mathbf{g}_1(\varphi)$. Then, using (4.91), we have

$$\mathbf{v} = \frac{d\mathbf{r}}{dt} = \dot{r}\mathbf{g}_1 + r\frac{d\mathbf{g}_1}{d\varphi}\dot{\varphi} = \dot{r}\mathbf{g}_1 + r\dot{\varphi}\mathbf{g}_2,$$

$$\mathbf{a} = \frac{d\mathbf{v}}{dt} = \ddot{r}\mathbf{g}_1 + \dot{r}\frac{d\mathbf{g}_1}{d\varphi}\dot{\varphi} + (\dot{r}\dot{\varphi} + r\ddot{\varphi})\mathbf{g}_2 + r\dot{\varphi}^2\frac{d\mathbf{g}_2}{d\varphi}$$
$$= (\ddot{r} - r\dot{\varphi}^2)\mathbf{g}_1 + (2\dot{r}\dot{\varphi} + r\ddot{\varphi})\mathbf{g}_2,$$

where the overdot denotes differentiation with respect to t. Hence the radial and tangential components of $\mathbf{v}$ and $\mathbf{a}$ are

$$v_r = r, \qquad v_\varphi = r\dot{\varphi}$$

and

$$a_r = \ddot{r} - r\dot{\varphi}^2, \qquad a_\varphi = 2\dot{r}\dot{\varphi} + r\ddot{\varphi} = \frac{1}{r}\frac{d}{dt}(r^2\dot{\varphi}).$$

Problem 6. A point simultaneously undergoes uniform rotation about the x_3-axis and uniform translation along the x_3-axis. Find the equation of

[17] Given a vector **A** in the plane, the *radial component* of **A** is the projection of **A** onto the direction of the radius vector **r**, while the *tangential component* of **A** is the projection of **A** onto the direction of the vector obtained by rotating **r** through 90° counterclockwise.

the resulting helical trajectory. Find the arc length of the helix and show that the unit tangent to the helix makes a constant angle with the x_1x_2-plane.

Solution. As in Prob. 4, let $\mathbf{g}_1(\varphi)$ be the unit vector making angle φ with the x_1-axis, and let $\mathbf{r}$ be the radius vector of a variable point M of the helix (see Fig. 4.21). Then

$$\mathbf{r} = R\mathbf{g}_1(\varphi) + c(\varphi)\mathbf{i}_3,$$

where R is the radius of the circle representing the projection of the helix onto the x_1x_2-plane and $c(\varphi)$ is the projection of $\mathbf{r}$ onto the x_3-axis. By definition of the helix,

$$\varphi = \omega t, \qquad c = vt,$$

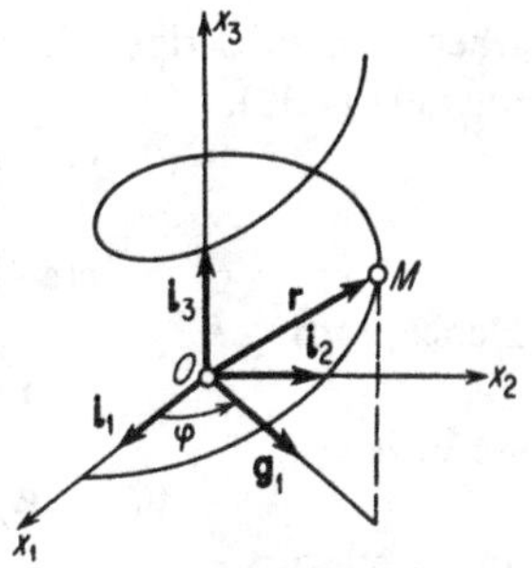

FIG. 4.21. A helix.

where M rotates about the x_3-axis with constant angular velocity ω and simultaneously moves along the x_3-axis with constant speed v. Therefore the equation of the helix is

$$\mathbf{r} = \mathbf{r}(\varphi) = R\mathbf{g}_1(\varphi) + vt\mathbf{i}_3 = R\mathbf{g}_1(\varphi) + \frac{v}{\omega}\varphi\mathbf{i}_3. \tag{4.92}$$

If $v = 0$, (4.92) reduces to

$$\mathbf{r} = R\mathbf{g}_1(\varphi),$$

which is the equation of a circle of radius R in the x_1x_2-plane with its center at the origin.

To find the arc lengths of the helix, we note that

$$s = \int_0^\varphi \left|\frac{d\mathbf{r}}{d\varphi}\right| d\varphi = \int_0^\varphi \sqrt{R^2 + \left(\frac{v}{\omega}\right)^2}\, d\varphi = \sqrt{R^2 + \left(\frac{v}{\omega}\right)^2}\,\varphi.$$

The unit tangent $\boldsymbol{\tau}$ equals

$$\boldsymbol{\tau} = \frac{d\mathbf{r}}{ds} = \frac{d\mathbf{r}}{d\varphi}\frac{d\varphi}{ds} = \frac{R\mathbf{g}_2 + \dfrac{v}{\omega}\mathbf{i}_3}{\sqrt{R^2 + \left(\dfrac{v}{\omega}\right)^2}},$$

which implies

$$\boldsymbol{\tau}\cdot\mathbf{i}_3 = \frac{\dfrac{v}{\omega}}{\sqrt{R^2 + \left(\dfrac{v}{\omega}\right)^2}} = \text{const},$$

i.e., $\boldsymbol{\tau}$ makes a constant angle with the x_3-axis and hence with the x_1x_2-plane.

Problem 7. Find $\nabla(\mathbf{A}\cdot\mathbf{B})$.

Solution. Clearly

$$\nabla(\mathbf{A}\cdot\mathbf{B}) = \nabla(\mathbf{A}_c\cdot\mathbf{B}) + \nabla(\mathbf{A}\cdot\mathbf{B}_c),$$

where the subscript c has the same meaning as on p. 170. According to formula (1.30),

$$\mathbf{c}(\mathbf{a}\cdot\mathbf{b}) = (\mathbf{a}\cdot\mathbf{c})\mathbf{b} - \mathbf{a}\times(\mathbf{b}\times\mathbf{c})$$

or

$$\mathbf{c}(\mathbf{a}\cdot\mathbf{b}) = (\mathbf{a}\cdot\mathbf{c})\mathbf{b} + \mathbf{a}\times(\mathbf{c}\times\mathbf{b}).$$

Hence, setting

$$\mathbf{a} = \mathbf{A}_c, \quad \mathbf{b} = \mathbf{B}, \quad \mathbf{c} = \nabla,$$

we have

$$\nabla(\mathbf{A}_c\cdot\mathbf{B}) = (\mathbf{A}_c\cdot\nabla)\mathbf{B} + \mathbf{A}_c\times(\nabla\times\mathbf{B}),$$

and similarly

$$\nabla(\mathbf{A}\cdot\mathbf{B}_c) \times \nabla(\mathbf{B}_c\cdot\mathbf{A}) = (\mathbf{B}_c\cdot\nabla)\mathbf{A} + \mathbf{B}_c\times(\nabla\times\mathbf{A}).$$

Thus, finally,

$$\nabla(\mathbf{A}\cdot\mathbf{B}) = (\mathbf{A}\cdot\nabla)\mathbf{B} + (\mathbf{B}\cdot\nabla)\mathbf{A} + \mathbf{A}\times\operatorname{curl}\mathbf{B} + \mathbf{B}\times\operatorname{curl}\mathbf{A}. \quad (4.93)$$

Problem 8. Find curl $(\mathbf{A}\times\mathbf{B})$.

Solution. Clearly

$$\operatorname{curl}(\mathbf{A}\times\mathbf{B}) = \nabla\times(\mathbf{A}\times\mathbf{B}) = \nabla\times(\mathbf{A}_c\times\mathbf{B}) + \nabla\times(\mathbf{A}\times\mathbf{B}_c),$$

and moreover

$$\nabla\times(\mathbf{A}_c\times\mathbf{B}) = \mathbf{A}_c(\nabla\cdot\mathbf{B}) - (\mathbf{A}_c\cdot\nabla)\mathbf{B},$$
$$\nabla\times(\mathbf{A}\times\mathbf{B}_c) = (\mathbf{B}_c\cdot\nabla)\mathbf{A} - \mathbf{B}_c(\nabla\cdot\mathbf{A}),$$

where we write each vector triple product in a form such that the operator ∇ acts only on the vector regarded as variable. It follows that

$$\operatorname{curl}(\mathbf{A}\times\mathbf{B}) = (\mathbf{B}\cdot\nabla)\mathbf{A} - (\mathbf{A}\cdot\nabla)\mathbf{B} + \mathbf{A}\operatorname{div}\mathbf{B} - \mathbf{B}\operatorname{div}\mathbf{A}.$$

Problem 9. Show that the rays drawn from the foci F_1 and F_2 of an ellipse to an arbitrary point M of the ellipse make equal angles with the tangent at M.

Solution. First we note that the gradient of the magnitude of the vector $\mathbf{r}_{AM}$ drawn from a fixed point $A = (x_{10}, x_{20}, x_{30})$ to a variable point $M = (x_1, x_2, x_3)$ is just the unit vector in the direction from A to M, since

$$\nabla r_{AM} = \nabla\sqrt{(x_1 - x_{10})^2 + (x_2 - x_{20})^2 + (x_3 - x_{30})^2}$$
$$= \frac{(x_1 - x_{10})\mathbf{i}_1 + (x_2 - x_{20})\mathbf{i}_2 + (x_3 - x_{30})\mathbf{i}_3}{\sqrt{(x_1 - x_{10})^2 + (x_2 - x_{20})^2 + (x_3 - x_{30})^2}} = \frac{\mathbf{r}_{AM}}{r_{AM}}.$$

The ellipse is the locus of all points the sum of whose distances from the foci is a constant, and hence

$$r_{F_1M} + r_{F_2M} = \text{const.}$$

Since $\nabla(r_{F_1M} + r_{F_2M})$ is directed along the normal to the ellipse at M, we have

$$\nabla(r_{F_1M}) \cdot \boldsymbol{\tau}_M = -\nabla(r_{F_2M}) \cdot \boldsymbol{\tau}_M,$$

where $\boldsymbol{\tau}_M$ is the tangent at M. Therefore

$$\frac{\mathbf{r}_{F_1M} \cdot \boldsymbol{\tau}_M}{r_{F_1M}} = -\frac{\mathbf{r}_{F_2M} \cdot \boldsymbol{\tau}_M}{r_{F_2M}},$$

i.e., the angle between $\mathbf{r}_{F_1M}$ and $\boldsymbol{\tau}_M$ equals the angle between $\mathbf{r}_{F_2M}$ and $-\boldsymbol{\tau}_M$. Thus, if the ellipse is made of reflecting material, a ray of light leaving one focus is reflected back through the other focus.

Problem 10. Find the components of the velocity $\mathbf{v}$ and acceleration $\mathbf{a}$ of a particle with respect to the local basis $\mathbf{e}_1, \mathbf{e}_2, \mathbf{e}_3$ of a system of orthogonal curvilinear coordinates q^1, q^2, q^3.

Solution. The total differential of the radius vector $\mathbf{r} = \mathbf{r}(q^1, q^2, q^3)$ can be written in the form

$$d\mathbf{r} = \frac{\partial \mathbf{r}}{\partial q^1}\,dq^1 + \frac{\partial \mathbf{r}}{\partial q^2}\,dq^2 + \frac{\partial \mathbf{r}}{\partial q^3}\,dq^3 = h_1\mathbf{e}_1\,dq^1 + h_2\mathbf{e}_2\,dq^2 + h_3\mathbf{e}_3\,dq^3$$

(recall p. 88), and hence

$$\mathbf{v} = \frac{d\mathbf{r}}{dt} = h_1\dot{q}_1\mathbf{e}_1 + h_2\dot{q}_2\mathbf{e}_2 + h_3\dot{q}_3\mathbf{e}_3. \tag{4.94}$$

It follows, for example, that the components of $\mathbf{v}$ in a cylindrical coordinate system are

$$v_R = h_1\dot{R} = \dot{R}, \quad v_\varphi = h_2\dot{\varphi} = R\dot{\varphi}, \quad v_z = h_3\dot{z} = \dot{z}.$$

To find the components of the acceleration $\mathbf{a}$, we write

$$a_i = \mathbf{a} \cdot \mathbf{e}_i = \frac{d\mathbf{v}}{dt} \cdot \frac{1}{h_i}\frac{\partial \mathbf{r}}{\partial q^i} \qquad \text{(no summation over } i\text{)}.$$

Therefore

$$h_i a_i = \frac{d\mathbf{v}}{dt} \cdot \frac{\partial \mathbf{r}}{\partial q^i} = \frac{d}{dt}\left(\mathbf{v} \cdot \frac{\partial \mathbf{r}}{\partial q^i}\right) - \mathbf{v} \cdot \frac{d}{dt}\left(\frac{\partial \mathbf{r}}{\partial q^i}\right). \tag{4.95}$$

But differentiation of (4.94) with respect to $\dot{q}^i$ gives

$$\frac{\partial \mathbf{v}}{\partial \dot{q}^i} = h_i\mathbf{e}_i = \frac{\partial \mathbf{r}}{\partial q^i},$$

and moreover

$$\frac{d}{dt}\frac{\partial \mathbf{r}}{\partial q^i} = \sum_{k=1}^{3} \frac{\partial}{\partial q^k}\left(\frac{\partial \mathbf{r}}{\partial q^i}\right)\dot{q}^k = \sum_{k=1}^{3} \frac{\partial}{\partial q^i}\left(\frac{\partial \mathbf{r}}{\partial q^k}\right)\dot{q}^k$$

$$= \frac{\partial}{\partial q^i}\sum_{k=1}^{3} \frac{\partial \mathbf{r}}{\partial q^k}\dot{q}^k = \frac{\partial}{\partial q^i}\frac{d\mathbf{r}}{dt} = \frac{\partial \mathbf{v}}{\partial q^i}.$$

Therefore, in (4.95) we can replace $\partial\mathbf{r}/\partial q^i$ by $\partial\mathbf{v}/\partial\dot{q}^i$ and $(d/dt)(\partial\mathbf{r}/\partial q^i)$ by $\partial\mathbf{v}/\partial q^i$, obtaining

$$h_i a_i = \frac{d}{dt}\left(\mathbf{v}\cdot\frac{\partial \mathbf{v}}{\partial \dot{q}^i}\right) - \mathbf{v}\cdot\frac{\partial \mathbf{v}}{\partial q^i} = \frac{d}{dt}\frac{\partial}{\partial \dot{q}^i}\left(\frac{v^2}{2}\right) - \frac{\partial}{\partial q^i}\left(\frac{v^2}{2}\right)$$

or finally

$$a_i = \frac{1}{h_i}\left(\frac{d}{dt}\frac{\partial T}{\partial \dot{q}^i} - \frac{\partial T}{\partial q^i}\right) \qquad \text{(no summation over } i\text{)},$$

where

$$T = \frac{v^2}{2} = \tfrac{1}{2}(h_1^2\dot{q}_1^2 + h_2^2\dot{q}_2^2 + h_3^2\dot{q}_3^2)$$

is the kinetic energy of the particle (assumed to have unit mass).

EXERCISES

1. A particle with radius vector $\mathbf{r}$ has the law of motion

$$\mathbf{r} = \mathbf{a}\cos\omega t + \mathbf{b}\sin\omega t,$$

where $\mathbf{a}$, $\mathbf{b}$ and ω are constants. Show that the force acting on the particle is *central*, i.e., is always directed toward the origin.

2. Prove that

a) $\operatorname{grad} r = \dfrac{\mathbf{r}}{r}$; b) $\operatorname{grad} r^n = nr^{n-2}\mathbf{r}$; c) $\operatorname{grad}\dfrac{k}{r} = -\dfrac{k}{r^3}\mathbf{r}$;

d) $\operatorname{grad}\ln r = \dfrac{\mathbf{r}}{r^2}$; e) $\operatorname{grad}\varphi(r) = \varphi'(r)\dfrac{\mathbf{r}}{r}$;

f) $\operatorname{grad}(\mathbf{a}\cdot\mathbf{r}) = \mathbf{a}$; g) $\operatorname{grad}[\mathbf{a}\cdot\mathbf{r}\varphi(r)] = \mathbf{a}\varphi(r) + (\mathbf{a}\cdot\mathbf{r})\dfrac{\mathbf{r}\varphi'(r)}{r}$,

where n, k and $\mathbf{a}$ are constants.

3. Prove that

a) $\operatorname{div}\mathbf{r} = 3$; b) $\operatorname{div}(k\mathbf{r}) = 3k$; c) $\operatorname{div}\dfrac{\mathbf{r}}{r^3} = 0$;

d) $\operatorname{div}(r^n\mathbf{r}) = (n+3)r^n$; e) $\operatorname{div}[\varphi(r)\mathbf{r}] = 3\varphi(r) + r\varphi'(r)$;

f) $\operatorname{div}(r^n\mathbf{c}) = nr^{n-2}(\mathbf{r}\cdot\mathbf{c})$; g) $\operatorname{div}(\mathbf{c}\times\mathbf{r}) = 0$;

h) $\operatorname{div}[\mathbf{r}(\mathbf{c}\cdot\mathbf{r})] = 4(\mathbf{c}\cdot\mathbf{r})$; i) $\operatorname{div}[\mathbf{a}(\mathbf{c}\cdot\mathbf{r})] = \operatorname{div}[\mathbf{c}(\mathbf{a}\cdot\mathbf{r})] = \mathbf{a}\cdot\mathbf{c}$;

j) $\operatorname{div}[(\mathbf{r} \times \mathbf{a}) \times \mathbf{c}] = -2(\mathbf{a} \cdot \mathbf{c})$; k) $\operatorname{div}[(\mathbf{r} \times \mathbf{a}) \times \mathbf{r}] = -2(\mathbf{a} \cdot \mathbf{r})$,

where n, k, $\mathbf{a}$ and $\mathbf{c}$ are constants.

4. Prove that

a) $\operatorname{curl} \mathbf{r} = 0$; b) $\operatorname{curl}[\mathbf{r}\varphi(r)] = 0$; c) $\operatorname{curl}[\mathbf{r}(\mathbf{c} \cdot \mathbf{r})] = \mathbf{c} \times \mathbf{r}$;

d) $\operatorname{curl}(\mathbf{c} \times \mathbf{r}) = 2\mathbf{c}$; e) $\operatorname{curl}[\mathbf{c}(\mathbf{a} \cdot \mathbf{r})] = \mathbf{a} \times \mathbf{c}$;

f) $\operatorname{curl}[(\mathbf{c} \times \mathbf{r}) \times \mathbf{a}] = \mathbf{a} \times \mathbf{c}$; g) $\operatorname{curl}[(\mathbf{c} \times \mathbf{r}) \times \mathbf{r}] = 3\mathbf{c} \times \mathbf{r}$,

where $\mathbf{a}$ and $\mathbf{c}$ are constants.

5. Find the divergence and curl of the velocity field $\mathbf{v}$ and acceleration field $\mathbf{a}$ of a rigid body rotating about a fixed point, given that

$$\mathbf{v} = \boldsymbol{\omega} \times \mathbf{r}, \quad \mathbf{a} = \boldsymbol{\epsilon} \times \mathbf{r} + \boldsymbol{\omega} \times (\boldsymbol{\omega} \times \mathbf{r}),$$

where $\boldsymbol{\omega}$ (the angular velocity) and $\boldsymbol{\epsilon}$ are constant vectors.

6. Show that the flux of the radius vector $\mathbf{r}$ through any closed surface S bounding a volume V equals $3V$.

7. Find the circulation of each of the following fields $\mathbf{A} = \mathbf{A}(\mathbf{r})$ around a circle of radius R with its center at the origin:

a) $\mathbf{A} = \frac{1}{2}(-x_2\mathbf{i}_1 + x_1\mathbf{i}_2)$;

b) $\mathbf{A} = (x_1x_2 + 1)\mathbf{i}_2 + (\frac{1}{2}x_1^2 + x_1 + 2)\mathbf{i}_2$.

8. Find the flux of the field

$$A = 4x_1x_3\mathbf{i}_1 - x_2^2\mathbf{i}_2 + x_2x_3\mathbf{i}_3$$

through the surface of the unit cube bounded by the planes

$$x_1 = 0, \quad x_1 = 1, \quad x_2 = 0, \quad x_2 = 1, \quad x_3 = 0, \quad x_3 = 1.$$

Ans. $\frac{3}{2}$.

9. Find the flux of the curl of the field

$$\mathbf{A} = (x_1^2 + x_2 - 4)\mathbf{i}_1 + 3x_1x_2\mathbf{i}_2 + (2x_1x_3 + x_3^2)\mathbf{i}_3$$

through the hemisphere $x_1^2 + x_2^2 + x_3^2 = 16$, $x_3 > 0$.

Ans. -16π.

10. Prove that

$$\operatorname{curl}\operatorname{curl}\mathbf{A} = \operatorname{grad}\operatorname{div}\mathbf{A} - \Delta\mathbf{A}. \tag{4.96}$$

11. Starting from the formula

$$\Delta\mathbf{A} = \operatorname{grad}\operatorname{div}\mathbf{A} - \operatorname{curl}\operatorname{curl}\mathbf{A}$$

implied by (4.96), find the components of $\Delta\mathbf{A}$ in cylindrical coordinates.

$$Ans.\ (\Delta\mathbf{A})_R = \Delta A_R - \frac{A_R}{R^2} - \frac{2}{R^2}\frac{\partial A_\varphi}{\partial \varphi},$$

$$(\Delta\mathbf{A})_\varphi = \Delta A_\varphi - \frac{A_\varphi}{R^2} + \frac{2}{R^2}\frac{\partial A_R}{\partial \varphi},$$

$$(\Delta\mathbf{A})_z = \Delta A_z.$$

12. Find the components of $\Delta\mathbf{A}$ in spherical coordinates.

$$Ans.\ (\Delta\mathbf{A})_R = \Delta A_R - \frac{2A_R}{R^2} - \frac{2}{R^2}\frac{\partial A_\theta}{\partial \theta} - \frac{2A_\theta}{R^2}\cot\theta - \frac{2}{R^2 \sin\theta}\frac{\partial A_\varphi}{\partial \varphi},$$

$$(\Delta\mathbf{A})_\theta = \Delta A_\theta + \frac{2}{R^2}\frac{\partial A_R}{\partial \theta} - \frac{A_\theta}{R^2\sin^2\theta} - \frac{2\cos\theta}{R^2\sin^2\theta}\frac{\partial A_\varphi}{\partial \varphi},$$

$$(\Delta\mathbf{A})_\varphi = \Delta A - \frac{A_\varphi}{R^2\sin^2\theta} + \frac{2}{R^2\sin\theta}\frac{\partial A_R}{\partial \varphi} + \frac{2\cos\theta}{R^2\sin^2\theta}\frac{\partial A_\theta}{\partial \varphi}.$$

13. What is the relation between the unit vector functions $\mathbf{g}_1(\varphi)$, $\mathbf{g}_2(\varphi)$ of Prob. 4, p. 177 and the local orthonormal basis $\mathbf{e}_R$, $\mathbf{e}_\varphi$ (cf. p. 11) of a system of polar coordinates R, φ?

14. Find the curvature and torsion of the helix (4.92).

$$Ans.\ \frac{1}{\rho} = \frac{R}{R^2 + \left(\dfrac{v}{\omega}\right)^2}, \qquad \frac{1}{T} = \frac{\dfrac{v}{\omega}}{R^2 + \left(\dfrac{v}{\omega}\right)^2}.$$

5

VECTOR AND TENSOR ANALYSIS: RAMIFICATIONS

5.1. Covariant Differentiation

5.1.1. Covariant differentiation of vectors. Suppose a vector field $\mathbf{A} = \mathbf{A}(\mathbf{r})$ has components A_1, A_2, A_3 in a system of rectangular coordinates with orthonormal basis $\mathbf{i}_1, \mathbf{i}_2, \mathbf{i}_3$. Then the differential of $\mathbf{A}$ equals

$$d\mathbf{A} = d(A_j\mathbf{i}_j) = \mathbf{i}_j \, dA_j$$

in terms of the differentials of the components of $\mathbf{A}$, where $d\mathbf{i}_j = 0$ since the basis $\mathbf{i}_1, \mathbf{i}_2, \mathbf{i}_3$ does not vary from point to point in a rectangular coordinate system. Similarly, suppose a vector $\mathbf{A}$ has covariant components A_1, A_2, A_3 and contravariant components A^1, A^2, A^3 in a system of oblique coordinates with basis $\mathbf{e}_1, \mathbf{e}_2, \mathbf{e}_3$. Then the differential of $\mathbf{A}$ equals

$$d\mathbf{A} = d(A_j\mathbf{e}^j) = \mathbf{e}^j \, dA_j,$$

$$d\mathbf{A} = d(A^j\mathbf{e}_j) = \mathbf{e}_j \, dA^j,$$

where now $d\mathbf{e}_j = d\mathbf{e}^j = 0$ since the bases $\mathbf{e}_1, \mathbf{e}_2, \mathbf{e}_3$ and $\mathbf{e}^1, \mathbf{e}^2, \mathbf{e}^3$ do not vary from point to point in an oblique coordinate system.[1]

On the other hand, in a system of generalized coordinates x^1, x^2, x^3, the basis $\mathbf{e}_1, \mathbf{e}_2, \mathbf{e}_3$ is local, i.e., in general each basis vector is a vector function of x^1, x^2, x^3:

$$\mathbf{e}_j = \mathbf{e}_j(x^1, x^2, x^3), \qquad \mathbf{e}^j = \mathbf{e}^j(x^1, x^2, x^3).$$

---

[1] The basis $\mathbf{e}_1, \mathbf{e}_2, \mathbf{e}_3$ does vary from point to point unless the coordinate system is rectangular or oblique (see Fig. 2.12).

It follows that

$$\begin{aligned} d\mathbf{A} &= d(A_j\mathbf{e}^j) = \mathbf{e}^j\,dA_j + A_j\,d\mathbf{e}^j, \\ d\mathbf{A} &= d(A^j\mathbf{e}_j) = \mathbf{e}_j\,dA^j + A^j\,d\mathbf{e}_j. \end{aligned} \tag{5.1}$$

Thus, besides a term expressing the change of the components of the vector in going from point to point, the differential $d\mathbf{A}$ contains a term $A_j\,d\mathbf{e}^j$ or $A^j\,d\mathbf{e}_j$ stemming from the fact that the basis of the generalized coordinate system also varies from point to point.

Since

$$d\mathbf{A} = \frac{\partial \mathbf{A}}{\partial x^k}\,dx^k,$$

(5.1) implies the following formula for the partial derivatives of the vector $\mathbf{A}$ with respect to the generalized coordinate x^k:

$$\frac{\partial \mathbf{A}}{\partial x^k} = \frac{\partial A_j}{\partial x^k}\mathbf{e}^j + A_j\frac{\partial \mathbf{e}^j}{\partial x^k} = \frac{\partial A^j}{\partial x^k}\mathbf{e}_j + A^j\frac{\partial \mathbf{e}_j}{\partial x^k}. \tag{5.2}$$

It will be shown presently that the covariant or contravariant components of the vector $\partial\mathbf{A}/\partial x^k$ $(k = 1, 2, 3)$ are themselves the components of a second-order tensor called the *covariant derivative* of the given (covariant or contravariant) vector. The covariant derivative of the covariant vector has components

$$\frac{\partial \mathbf{A}}{\partial x^k}\cdot\mathbf{e}_i \equiv A_{i,k}, \tag{5.3}$$

while the covariant derivative of the contravariant vector has components

$$\frac{\partial \mathbf{A}}{\partial x^k}\cdot\mathbf{e}^i \equiv A^i_{\ ,k} \tag{5.4}$$

Example. A vector field $\mathbf{A} = \mathbf{A}(\mathbf{r})$ is said to be *homogeneous* if it does not vary from point to point, i.e., if its magnitude and direction are constant. Suppose $\mathbf{A}$ is homogeneous. Then

$$\mathbf{A}(x^1, x^2, x^3) = \mathbf{A}(x^1 + dx^1, x^2 + dx^2, x^3 + dx^3)$$

at any two neighboring points x^i and $x^i + dx^i$, even though the components of $\mathbf{A}$ and the local basis vary from point to point. Therefore

$$\mathbf{A} = A^i\mathbf{e}_i = (A^i + dA^i)(\mathbf{e}_i + d\mathbf{e}_i),$$

and hence

$$\mathbf{e}_i\,dA^i + A^i\,d\mathbf{e}_i + dA^i\,d\mathbf{e}_i = 0.$$

Retaining only first-order terms, we have

$$\mathbf{e}_i\,dA^i + A^i\,d\mathbf{e}_i = d\mathbf{A} = 0,$$

where $d\mathbf{A}$ is the total differential of the vector $\mathbf{A}$. Hence

$$\frac{\partial \mathbf{A}}{\partial x^k}\,dx^k = 0$$

or

$$A^i_{,k}\,dx^k = 0.$$

But then

$$A^i_{,k} = 0,$$

since the increments dx^k are arbitrary. In other words, the covariant derivative of a homogeneous vector field vanishes.

A homogeneous vector field can be regarded as the result of displacing the vector $\mathbf{A}$ parallel to itself at every point of the field. With this interpretation, $A^i_{,k} = 0$ becomes the *parallel displacement condition.*

5.1.2. Christoffel symbols. It follows from (5.2) and the definitions (5.3) and (5.4) that

$$\begin{aligned} A_{i,k} &= \frac{\partial \mathbf{A}}{\partial x^k}\cdot \mathbf{e}_i = \frac{\partial A_i}{\partial x^k} + A_j\frac{\partial \mathbf{e}^j}{\partial x^k}\cdot \mathbf{e}_i, \\ A^i_{,k} &= \frac{\partial \mathbf{A}}{\partial x^k}\cdot \mathbf{e}^i = \frac{\partial A^i}{\partial x^k} + A^j\frac{\partial \mathbf{e}_j}{\partial x^k}\cdot \mathbf{e}^i. \end{aligned} \tag{5.5}$$

Bearing in mind that the components of $g_i^{\cdot j} = \mathbf{e}_i \cdot \mathbf{e}^j$ are either zero or one, we have

$$\frac{\partial}{\partial x^k}(\mathbf{e}_i \cdot \mathbf{e}^j) = 0$$

and hence

$$\mathbf{e}_i \cdot \frac{\partial \mathbf{e}^j}{\partial x^k} = -\mathbf{e}^j \cdot \frac{\partial \mathbf{e}_i}{\partial x^k}. \tag{5.6}$$

Introducing the notation

$$\begin{Bmatrix} i \\ j \quad k \end{Bmatrix} \equiv \mathbf{e}^i \cdot \frac{\partial \mathbf{e}_j}{\partial x^k}, \tag{5.7}$$

called the *Christoffel symbol of the second kind* (with 27 components) and using (5.5) and (5.6), we find that the formulas (5.5) can be written in the form

$$\begin{aligned} A_{i,k} &= \frac{\partial A_i}{\partial x^k} - \begin{Bmatrix} j \\ i \quad k \end{Bmatrix} A_j, \\ A^i_{,k} &= \frac{\partial A^i}{\partial x^k} + \begin{Bmatrix} i \\ j \quad k \end{Bmatrix} A^j. \end{aligned} \tag{5.8}$$

According to (5.8), the covariant derivative of a vector field involves not only the rate of change of the field itself, as we move along the coordinate curves (the terms $\partial A_i/\partial x^k$, $\partial A^i/\partial x_k$), but also the rate of change of the local basis $\left(\text{the terms } -\left\{\begin{matrix} j \\ i \quad k \end{matrix}\right\}A_j, \left\{\begin{matrix} i \\ j \quad k \end{matrix}\right\}A^j\right)$. If the basis does not vary from point to point (as in rectangular or oblique coordinates), it follows from (5.7) that all the Christoffel symbols of the second kind vanish. Then the covariant derivatives $A_{i,k}$ and $A^i_{,k}$ reduce simply to $\partial A_i/\partial x^k$ and $\partial A^i/\partial x^k$, respectively.

Thus the terms $-\left\{\begin{matrix} j \\ i \quad k \end{matrix}\right\}A_j$ and $\left\{\begin{matrix} i \\ j \quad k \end{matrix}\right\}A^j$ are due entirely to introducing a local basis which varies from point to point. Therefore, as we now show, it must be possible to express the Christoffel symbols (5.7) in terms of derivatives of components of the metric tensor.

First we note that (5.7) implies

$$\frac{\partial \mathbf{e}_j}{\partial x^k} = \left\{\begin{matrix} i \\ j \quad k \end{matrix}\right\}\mathbf{e}_i, \tag{5.9}$$

so that the quantities $\left\{\begin{matrix} i \\ j \quad k \end{matrix}\right\}$ are the expansion coefficients of the vector $\partial \mathbf{e}_j/\partial x_k$ with respect to the basis $\mathbf{e}_1$, $\mathbf{e}_2$, $\mathbf{e}_3$. Let the quantities $[i, jk]$, called the *Christoffel symbols of the first kind*, be the expansion coefficients of $\partial \mathbf{e}_j/\partial x_k$ with respect to the reciprocal basis $\mathbf{e}^1$, $\mathbf{e}^2$, $\mathbf{e}^3$:

$$\frac{\partial \mathbf{e}_j}{\partial x_k} = [i, jk]\mathbf{e}^i. \tag{5.10}$$

Then (5.9) and (5.10) imply

$$[i, jk] = \mathbf{e}_i \cdot \frac{\partial \mathbf{e}_j}{\partial x^k}, \tag{5.11}$$

$$[i, jk] = g_{il}\left\{\begin{matrix} l \\ j \quad k \end{matrix}\right\}, \qquad \left\{\begin{matrix} i \\ j \quad k \end{matrix}\right\} = g^{il}[l, jk]. \tag{5.12}$$

Since

$$\frac{\partial \mathbf{e}_j}{\partial x^k} = \frac{\partial}{\partial x^k}\frac{\partial \mathbf{r}}{\partial x^j} = \frac{\partial}{\partial x^j}\frac{\partial \mathbf{r}}{\partial x^k} = \frac{\partial \mathbf{e}_k}{\partial x^j} \tag{5.13}$$

($\mathbf{r} = x^1\mathbf{e}_1 + x^2\mathbf{e}_2 + x^3\mathbf{e}_3$), we see from (5.7) and (5.11) that the Christoffel symbols $\left\{\begin{matrix} i \\ j \quad k \end{matrix}\right\}$ and $[i, jk]$ are symmetric in the indices j and k, i.e.,

$$[i, jk] = [i, kj], \qquad \left\{\begin{matrix} i \\ j \quad k \end{matrix}\right\} = \left\{\begin{matrix} i \\ k \quad j \end{matrix}\right\}.$$

Using the symmetry of $[i, jk]$ and (5.13), we find that

$$[i, jk] = \mathbf{e}_i \cdot \frac{\partial \mathbf{e}_j}{\partial x^k} = \frac{1}{2}\left(\mathbf{e}_i \cdot \frac{\partial \mathbf{e}_j}{\partial x^k} + \mathbf{e}_i \cdot \frac{\partial \mathbf{e}_k}{\partial x^j}\right)$$

$$= \frac{1}{2}\left[\frac{\partial}{\partial x^k}(\mathbf{e}_i \cdot \mathbf{e}_j) + \frac{\partial}{\partial x^j}(\mathbf{e}_i \cdot \mathbf{e}_k) - \mathbf{e}_j \cdot \frac{\partial \mathbf{e}_i}{\partial x^k} - \mathbf{e}_k \cdot \frac{\partial \mathbf{e}_i}{\partial x^j}\right]$$

$$= \frac{1}{2}\left[\frac{\partial g_{ij}}{\partial x^k} + \frac{\partial g_{ik}}{\partial x^j} - \mathbf{e}_j \cdot \frac{\partial \mathbf{e}_k}{\partial x^i} - \mathbf{e}_k \cdot \frac{\partial \mathbf{e}_j}{\partial x^i}\right]$$

$$= \frac{1}{2}\left[\frac{\partial g_{ij}}{\partial x^k} + \frac{\partial g_{ik}}{\partial x^j} - \frac{\partial}{\partial x^i}(\mathbf{e}_j \cdot \mathbf{e}_k)\right].$$

Therefore

$$[i, jk] = \frac{1}{2}\left(\frac{\partial g_{ij}}{\partial x^k} + \frac{\partial g_{ik}}{\partial x^j} - \frac{\partial g_{jk}}{\partial x^i}\right) = [i, kj], \tag{5.14}$$

and hence, by (5.12),

$$\begin{Bmatrix} i \\ j \quad k \end{Bmatrix} = \frac{1}{2} g^{il}\left(\frac{\partial g_{lj}}{\partial x^k} + \frac{\partial g_{lk}}{\partial x^j} - \frac{\partial g_{jk}}{\partial x^l}\right) = \begin{Bmatrix} i \\ k \quad j \end{Bmatrix}. \tag{5.15}$$

Formulas (5.14) and (5.15) express the Christoffel symbols of the first and second kinds in terms of the components of the metric tensor of the underlying curvilinear coordinate system.

Under changes of coordinate system, the Christoffel symbol of the first kind transforms as follows:

$$[i, jk]' = \mathbf{e}'_i \cdot \frac{\partial \mathbf{e}'_j}{\partial x'^k} = \alpha^l_{i'}\mathbf{e}_l \cdot \frac{\partial}{\partial x^m}(\alpha^n_{j'}\mathbf{e}_n)\frac{\partial x^m}{\partial x'^k}$$

$$= \alpha^l_{i'}\alpha^m_{k'}\alpha^n_{j'}\mathbf{e}_l \cdot \frac{\partial \mathbf{e}_n}{\partial x^m} + \alpha^l_{i'}\alpha^m_{k'}(\mathbf{e}_l \cdot \mathbf{e}_n)\frac{\partial \alpha^n_{j'}}{\partial x^m}$$

$$= \alpha^l_{i'}\alpha^m_{k'}\alpha^n_{j'}[l, nm] + \alpha^l_{i'}\alpha^m_{k'}\frac{\partial \alpha^n_{j'}}{\partial x^m} g_{ln}.$$

Similarly, the transformation law of the Christoffel symbol of the second kind is

$$\begin{Bmatrix} i \\ j \quad k \end{Bmatrix}' = \alpha^{i'}_l \alpha^m_{k'}\alpha^n_{j'}\begin{Bmatrix} l \\ n \quad m \end{Bmatrix} + \alpha^{i'}_n \alpha^m_{k'}\frac{\partial \alpha^n_{j'}}{\partial x^m}. \tag{5.16}$$

Thus the Christoffel symbols are *not* tensors.

As already noted, the covariant derivatives $A_{i,k}$ and $A^i_{,k}$ are second-order tensors. In fact, using (5.16), we have

$$A'_{i,k} = \frac{\partial A'_i}{\partial x'^k} - \begin{Bmatrix} j \\ i \quad k \end{Bmatrix}' A'_j$$

$$= \frac{\partial}{\partial x^m}(\alpha^l_{i'} A_l)\frac{\partial x^m}{\partial x'^k} - \alpha^r_{j'} A_r\left(\alpha^{j'}_n \alpha^m_{k'} \alpha^l_{i'} \begin{Bmatrix} n \\ l \quad m \end{Bmatrix} + \alpha^{j'}_n \alpha^m_{k'} \frac{\partial \alpha^n_{i'}}{\partial x^m}\right)$$

$$= \alpha^l_{i'} \alpha^m_{k'}\left(\frac{\partial A_l}{\partial x^m} - \begin{Bmatrix} n \\ l \quad m \end{Bmatrix} A_n\right) = \alpha^l_{i'} \alpha^m_{k'} A_{l,m}$$

and

$$A'^i_{,k} = \frac{\partial A'^i}{\partial x'^k} + \begin{Bmatrix} i \\ j \quad k \end{Bmatrix}' A'^j$$

$$= \frac{\partial}{\partial x^m}(\alpha^{i'}_l A^l)\frac{\partial x^m}{\partial x'^k} + \alpha^{j'}_r A^r\left(\alpha^{i'}_l \alpha^m_{k'} \alpha^n_{j'} \begin{Bmatrix} l \\ n \quad m \end{Bmatrix} + \alpha^{i'}_n \alpha^m_{k'} \frac{\partial \alpha^n_{j'}}{\partial x^m}\right)$$

$$= \alpha^{i'}_l \alpha^m_{k'}\left(\frac{\partial A^l}{\partial x^m} + \begin{Bmatrix} l \\ n \quad m \end{Bmatrix} A^n\right) = \alpha^{i'}_l \alpha^m_{k'} A^l_{,m}, \tag{5.17}$$

where in (5.17) we have used the relation

$$\alpha^{j'}_r \frac{\partial \alpha^n_{j'}}{\partial x^m} = -\alpha^n_{j'} \frac{\partial \alpha^{j'}_r}{\partial x^m}$$

obtained by differentiating the identity $\alpha^{j'}_r \alpha^n_{j'} = g_r^{\cdot n}$.[2] Thus the quantities $A_{i,k}$ transform like the covariant components of a second-order tensor, and the quantities $A^i_{,k}$ like the mixed components of a second-order tensor. Moreover, it follows from (5.3) and (5.4), together with

$$\mathbf{e}^i = g^{il}\mathbf{e}_l, \qquad \mathbf{e}_i = g_{il}\mathbf{e}^l,$$

that

$$A_{i,k} = g_{il}A^l_{,k}, \qquad A^i_{,k} = g^{il}A_{l,k}.$$

Hence $A_{i,k}$ and $A^i_{,k}$ are the (covariant and mixed) components *of the same tensor*, called the covariant derivative of the vector **A**.

5.1.3. Covariant differentiation of tensors. The following formulas for the components of the covariant derivative of a second-order tensor are a natural generalization of the corresponding formulas (5.5) for covariant

[2] Itself another way of writing the transformation law

$$g_r^{\cdot n} = \alpha^{i'}_r \alpha^n_{j'} g_{i'}^{\cdot j'}.$$

differentiation of a vector (first-order tensor):

$$T_{ik,l} = \frac{\partial T_{ik}}{\partial x^l} - \begin{Bmatrix} m \\ i \quad l \end{Bmatrix} T_{mk} - \begin{Bmatrix} m \\ k \quad l \end{Bmatrix} T_{im},$$

$$T^{ik}_{,l} = \frac{\partial T^{ik}}{\partial x^l} + \begin{Bmatrix} i \\ m \quad l \end{Bmatrix} T^{mk} + \begin{Bmatrix} k \\ m \quad l \end{Bmatrix} T^{im}, \qquad (5.18)$$

$$T^{i}_{.k,l} = \frac{\partial T^{i}_{.k}}{\partial x^l} + \begin{Bmatrix} i \\ m \quad l \end{Bmatrix} T^{m}_{.k} - \begin{Bmatrix} m \\ k \quad l \end{Bmatrix} T^{i}_{.m}.$$

It can easily be shown that these quantities transform under coordinate changes like the components of a third-order tensor ($T_{ik,l}$ like covariant components, $T^{ik}_{,l}$ like mixed components with two contravariant indices and one covariant index, etc.). The covariant derivative of a tensor of arbitrary order n is defined similarly: The first term is a partial derivative of the components of the tensor with respect to the coordinate x^l (say) and the remaining terms, n in all, are sums of components of the tensor multiplied by Christoffel symbols of the second kind with each index of the tensor and the "opposite" index of the Christoffel symbol being in turn an index of summation (if this dummy index is a subscript of the tensor it is a superscript of the Christoffel symbol, and vice versa). Moreover, a given sum involving tensor components and Christoffel symbols appears with a minus sign if the dummy index is a covariant index (subscript) and with a plus sign if it is a contravariant index (superscript). It can be shown that the covariant derivative of a tensor of order n is a tensor of order $n + 1$.

Example 1. The covariant derivative

$$\lambda^{..l}_{ik,m} = \frac{\partial \lambda^{..l}_{ik}}{\partial x^m} - \begin{Bmatrix} n \\ i \quad m \end{Bmatrix} \lambda^{..l}_{nk} - \begin{Bmatrix} n \\ k \quad m \end{Bmatrix} \lambda^{..l}_{in} + \begin{Bmatrix} l \\ n \quad m \end{Bmatrix} \lambda^{..n}_{ik}$$

is a mixed fourth-order tensor, with three covariant indices and one contravariant index.

Example 2. In the case of a zeroth-order tensor, i.e., a scalar, the covariant derivative reduces to the partial derivative with respect to the coordinates

$$f_{,i} = \frac{\partial f}{\partial x^i}.$$

Thus the covariant derivative of a scalar f is a covariant vector, with components equal to the covariant components of grad f (recall Example 1, p. 91).

5.1.4. Ricci's theorem. The covariant derivative of the metric tensor vanishes. This result, known as *Ricci's theorem*, is proved by the following simple calculation, based on (5.18), (5.12) and (5.14):

$$
\begin{aligned}
g_{ik,l} &= \frac{\partial g_{ik}}{\partial x^l} - \begin{Bmatrix} m \\ k \quad l \end{Bmatrix} g_{im} - \begin{Bmatrix} m \\ i \quad l \end{Bmatrix} g_{mk} \\
&= \frac{\partial g_{ik}}{\partial x^l} - [i, kl] - [k, il] \\
&= \frac{\partial g_{ik}}{\partial x^l} - \frac{1}{2}\left(\frac{\partial g_{ik}}{\partial x^l} + \frac{\partial g_{il}}{\partial x^k} - \frac{\partial g_{kl}}{\partial x^i}\right) \\
&\quad - \frac{1}{2}\left(\frac{\partial g_{ik}}{\partial x^l} + \frac{\partial g_{kl}}{\partial x^i} - \frac{\partial g_{il}}{\partial x^k}\right) = 0.
\end{aligned} \tag{5.19}
$$

Moreover, we have

$$g_{ik,l} = g_{ir}g_{ks}g^{rs}_{,l} = 0.$$

In particular, (5.19) implies the useful formula

$$\frac{\partial g_{ik}}{\partial x^l} = [i, kl] + [k, il]. \tag{5.20}$$

Because of Ricci's theorem, the components of the metric tensor can be regarded as constants under covariant differentiation. Thus, for example,

$$
\begin{aligned}
g_{il}A^l_{,k} &= (g_{il}A^l)_{,k} = A_{i,k}, \\
g_{il}T^{lm}_{,k} &= (g_{il}T^{lm})_{,k} = T_{i,k}^{\cdot m}, \\
T_{ik,l}g^{im}g^{kn} &= (T_{ik}g^{im}g^{kn})_{,l} = T^{mn}_{,l},
\end{aligned}
$$

and so on.

5.1.5. Differential operators in generalized coordinates. Next we define the quantities grad φ, div $\mathbf{A}$, $\Delta\varphi$ and curl $\mathbf{A}$ in a system of generalized coordinates x^1, x^2, x^3:

1) By the *gradient* of a scalar field $\varphi = \varphi(x^1, x^2, x^3)$ we mean the vector with covariant components

$$\frac{\partial \varphi}{\partial x^i}.$$

Thus, introducing the "del" operator

$$\nabla = \mathbf{e}^i \frac{\partial}{\partial x^i},$$

we have

$$\operatorname{grad} \varphi = \nabla\varphi = \mathbf{e}^i \frac{\partial \varphi}{\partial x^i}. \tag{5.21}$$

According to Sec. 1.6.4, the physical components of $\nabla\varphi$ are

$$(\nabla\varphi)_i^* = \frac{1}{\sqrt{g_{ii}}}\frac{\partial\varphi}{\partial x^i} \qquad \text{(no summation over } i\text{).} \tag{5.22}$$

In the case of orthogonal coordinates, (5.22) becomes

$$(\nabla\varphi)_i^* = \frac{1}{h_i}\frac{\partial\varphi}{\partial x^i}$$

in terms of the metric coefficients h_i. All the properties of the gradient established in Sec. 4.3 continue to hold in generalized coordinates. In particular, the directional derivative of φ in the direction $\boldsymbol{l}$ equals

$$\frac{d\varphi}{dl} = \boldsymbol{l}\cdot\nabla\varphi,$$

where $\boldsymbol{l} = l^i\mathbf{e}_i$.

2) The *divergence* of a vector field $\mathbf{A} = \mathbf{A}(x^1, x^2, x^3)$ is defined as the contraction of the (mixed) covariant derivative of $\mathbf{A}$, i.e.,[3]

$$\operatorname{div}\mathbf{A} = A^i_{,i} = \frac{\partial A^i}{\partial x^i} + \begin{Bmatrix} i \\ i\ \ j\end{Bmatrix}A^j. \tag{5.23}$$

The sum $\begin{Bmatrix} i \\ i\ \ j\end{Bmatrix}$ can be expressed in terms of the metric tensor. In fact, using (5.12) and the symmetry of g^{ik}, we have

$$\begin{Bmatrix} i \\ i\ \ j\end{Bmatrix} = g^{ik}[k, ij] = \tfrac{1}{2}g^{ik}([k, ij] + [i, kj]), \tag{5.24}$$

where

$$[k, ij] + [i, kj] = \frac{\partial g_{ik}}{\partial x^j}$$

because of (5.20). Expanding the determinant $G = \det\|g_{ik}\|$ with respect to elements of the ith row, we obtain

$$G = g_{ik}G^{ik} \qquad \text{(no summation over } i\text{),}$$

where G^{ik} is the cofactor of g_{ik} in the determinant G. But G^{ik} is independent of g_{ik}, and hence

$$\frac{\partial G}{\partial g_{ik}} = G^{ik}.$$

Therefore

$$g^{ik} = \frac{G^{ik}}{G} = \frac{1}{G}\frac{\partial G}{\partial g_{ik}}, \tag{5.25}$$

---

[3] Thus div **A** is the first invariant of the tensor $A^i_{,j}$ (see Exercise 9, p. 132).

where we have used (1.52). It follows from (5.23)–(5.25) that

$$\left\{\begin{matrix} i \\ i \quad j \end{matrix}\right\} = \frac{1}{2G}\frac{\partial G}{\partial g_{ik}}\frac{\partial g_{ik}}{\partial x^j} = \frac{1}{2G}\frac{\partial G}{\partial x^j} = \frac{1}{\sqrt{G}}\frac{\partial(\sqrt{G})}{\partial x^j}.$$

Thus, finally,

$$\begin{aligned}\operatorname{div}\mathbf{A} &= \frac{\partial A^i}{\partial x^i} + \frac{A^i}{\sqrt{G}}\frac{\partial(\sqrt{G})}{\partial x^i} = \frac{1}{\sqrt{G}}\frac{\partial}{\partial x^i}(A^i\sqrt{G}) \\ &= \frac{1}{\sqrt{G}}\frac{\partial}{\partial x^i}(g^{ik}A_k\sqrt{G}) = \frac{1}{\sqrt{G}}\sum_{i=1}^{3}\frac{\partial}{\partial x^i}\left(\frac{A^{*i}\sqrt{G}}{\sqrt{g_{ii}}}\right),\end{aligned} \tag{5.26}$$

where the A^{*i} are the physical components of the vector $\mathbf{A}$. In particular,

$$\operatorname{div}\mathbf{A} = \frac{1}{h_1h_2h_3}\sum_{i=1}^{3}\frac{\partial}{\partial x^i}\left(\frac{A_i^*h_1h_2h_3}{h_i}\right)$$

in orthogonal coordinates, where A_i^* can be replaced by A_i if the local basis is orthonormal.

3) By the *Laplacian* of a scalar field φ we mean the quantity

$$\Delta\varphi = \operatorname{div}\operatorname{grad}\varphi.$$

Combining (5.21) and (5.26), we have

$$\Delta\varphi = \frac{1}{\sqrt{G}}\frac{\partial}{\partial x^i}\left(g^{ik}\sqrt{G}\frac{\partial\varphi}{\partial x^k}\right). \tag{5.27}$$

In orthogonal coordinates, (5.27) becomes

$$\Delta\varphi = \frac{1}{h_1h_2h_3}\frac{\partial}{\partial x^i}\left(\frac{h_1h_2h_3}{h_i}\frac{\partial\varphi}{\partial x^i}\right).$$

4) In generalized coordinates, the curl of a vector field $\mathbf{A}$ is defined as the vector product of the operator ∇ and the vector $\mathbf{A}$. Thus

$$\operatorname{curl}\mathbf{A} = \nabla\times\mathbf{A} = \mathbf{e}^j\frac{\partial}{\partial x^j}\times\mathbf{A} = \mathbf{e}^j\times\frac{\partial\mathbf{A}}{\partial x^j} = \mathbf{e}^j\times\mathbf{e}^kA_{k,j}, \tag{5.28}$$

since

$$\frac{\partial\mathbf{A}}{\partial x^j}\cdot\mathbf{e}_k = A_{k,j}$$

[see (5.3)]. But it will be recalled from Secs. 1.6.1 and 1.6.5 that

$$\mathbf{e}^j\times\mathbf{e}^k = \begin{cases} \dfrac{\mathbf{e}_i}{\sqrt{G}} & \text{if } \ i, j, k \text{ is a cyclic permutation of } 1, 2, 3, \\[2ex] -\dfrac{\mathbf{e}_i}{\sqrt{G}} & \text{if } \ i, j, k \text{ is a cyclic permutation of } 2, 1, 3, \\[2ex] 0 & \text{otherwise,} \end{cases}$$

where $G = \det \|g_{ik}\|$. Therefore (5.28) takes the form

$$\operatorname{curl} \mathbf{A} = \sum_{i,j,k} \frac{\mathbf{e}_i}{\sqrt{G}} (A_{k,j} - A_{j,k}),$$

where the indices i, j, k are a cyclic permutation of the numbers 1, 2, 3. Moreover,

$$A_{k,j} = \frac{\partial A_k}{\partial x^j} - \begin{Bmatrix} l \\ k \quad j \end{Bmatrix} A_l,$$

$$A_{j,k} = \frac{\partial A_j}{\partial x^k} - \begin{Bmatrix} l \\ j \quad k \end{Bmatrix} A_l,$$

and hence

$$A_{k,j} - A_{j,k} = \frac{\partial A_k}{\partial x^j} - \frac{\partial A_j}{\partial x^k},$$

since $\begin{Bmatrix} l \\ j \quad k \end{Bmatrix} = \begin{Bmatrix} l \\ k \quad j \end{Bmatrix}$. Thus, finally,

$$\operatorname{curl} \mathbf{A} = \sum_{i,j,k} \frac{\mathbf{e}_i}{\sqrt{G}} \left(\frac{\partial A_k}{\partial x^j} - \frac{\partial A_j}{\partial x^k} \right), \tag{5.29}$$

where i, j, k is a cyclic permutation of 1, 2, 3.

Formula (5.29) leads to the following expressions for the contravariant and physical components of curl **A** (there is no summation over j and k):

$$(\operatorname{curl} \mathbf{A})^i = \frac{1}{\sqrt{G}} \left(\frac{\partial A_k}{\partial x^j} - \frac{\partial A_j}{\partial x^k} \right),$$

$$\begin{aligned} (\operatorname{curl} \mathbf{A})^{*i} &= \frac{\sqrt{g_{ii}}}{\sqrt{G}} \left(\frac{\partial A_k}{\partial x^j} - \frac{\partial A_j}{\partial x^k} \right) \\ &= \frac{\sqrt{g_{ii}}}{\sqrt{G}} \left[\frac{\partial (A_k^* \sqrt{g_{kk}})}{\partial x^j} - \frac{\partial (A_j^* \sqrt{g_{jj}})}{\partial x^k} \right]. \end{aligned} \tag{5.30}$$

In the case of orthogonal coordinates with an orthonormal local basis, (5.30) reduces to

$$(\operatorname{curl} \mathbf{A})_i = \frac{h_i}{h_1 h_2 h_3} \left[\frac{\partial (A_k h_k)}{\partial x^j} - \frac{\partial (A_j h_j)}{\partial x^k} \right] \quad \text{(no summation over } j \text{ and } k).$$

In Sec. 3.7.3 we introduced the unit pseudotensor ε_{jkl}, defined as

$$\varepsilon_{jkl} = (\mathbf{i}_j \times \mathbf{i}_k) \cdot \mathbf{i}_l$$

in a system of rectangular coordinates with orthonormal basis $\mathbf{i}_1, \mathbf{i}_2, \mathbf{i}_3$. In a system of generalized coordinates with local basis $\mathbf{e}_1, \mathbf{e}_2, \mathbf{e}_3$, we replace this definition by

$$\varepsilon_{jkl} = (\mathbf{e}_j \times \mathbf{e}_k) \cdot \mathbf{e}_l,$$

so that

$$\varepsilon_{jkl} = \begin{cases} \sqrt{G} & \text{if } \; j, k, l \text{ is a cyclic permutation of 1, 2, 3,} \\ -\sqrt{G} & \text{if } \; j, k, l \text{ is a cyclic permutation of 2, 1, 3,} \\ 0 & \text{otherwise.} \end{cases}$$

The contravariant components ε^{jkl} then turn out to be

$$\varepsilon^{jkl} = \begin{cases} \dfrac{1}{\sqrt{G}} & \text{if } \; j, k, l \text{ is a cyclic permutation of 1, 2, 3,} \\ -\dfrac{1}{\sqrt{G}} & \text{if } \; j, k, l \text{ is a cyclic permutation of 1, 2, 3,} \\ 0 & \text{otherwise} \end{cases}$$

(why?)

In terms of the unit pseudotensor, the vector product $\mathbf{C} = \mathbf{A} \times \mathbf{B}$ has covariant and contravariant components

$$C_i = \varepsilon_{ijk} A^j B^k,$$
$$C^i = \varepsilon^{ijk} A_j A_k$$

(see Problem 5, p. 40), while curl $\mathbf{A}$ has components

$$(\text{curl } \mathbf{A})^i = \varepsilon^{jki} A_{k,j},$$
$$(\text{curl } \mathbf{A})_i = g_{il} (\text{curl } \mathbf{A})^l = g_{il} \varepsilon^{jkl} A_{k,j}.$$

5.2. Integral Theorems

The integral theorems of vector and tensor analysis are essentially relations between the values of a vector or tensor field inside a volume and the values of the field on the boundary of the volume. In this sense, they are generalizations of the fundamental theorem of calculus

$$\int_a^b \frac{dA(x)}{dx}\, dx = A(b) - A(a),$$

expressing the definite integral of the derivative of a function in terms of the values of the function at the limits of integration (under the assumption that the derivative exists and is continuous).

In Sec. 4.2 we proved two of the most important integral theorems of vector analysis, i.e., Gauss' theorem and Stokes' theorem, which in vector notation take the form

$$\iiint_V \text{div } \mathbf{A}\, dV = \iint_S \mathbf{A} \cdot \mathbf{n}\, dS,$$
$$\iint_S \mathbf{n} \cdot \text{curl } \mathbf{A}\, dS = \oint_L \mathbf{A} \cdot d\mathbf{r}. \tag{5.31}$$

These theorems are of basic importance and are widely used in theoretical physics, particularly in hydrodynamics, elasticity theory and electromagnetic theory. Moreover, as we now show, they can be used to deduce a number of related integral theorems.

5.2.1. Theorems related to Gauss' theorem. Suppose $\mathbf{A} = \mathbf{c}\varphi$ in (5.31), where $\mathbf{c}$ is a fixed but arbitrary vector and φ is a scalar field. Then (5.31) implies

$$\mathbf{c} \cdot \left(\iiint_V \operatorname{grad} \varphi \, dV - \iint_S \varphi \mathbf{n} \, dS \right) = 0,$$

since

$$\operatorname{div} (\mathbf{c}\varphi) = \nabla \cdot (\mathbf{c}\varphi) = \mathbf{c} \cdot \nabla\varphi = \mathbf{c} \cdot \operatorname{grad} \varphi.$$

It follows that[4]

$$\iiint_V \operatorname{grad} \varphi \, dV = \iint_S \varphi \mathbf{n} \, dS, \tag{5.32}$$

a relation we have already encountered [recall (4.30)].

Similarly, substituting $\mathbf{A} = \mathbf{A}' \times \mathbf{c}$ into (5.31), where $\mathbf{A}'$ is another vector field and $\mathbf{c}$ is a constant vector, we find that

$$\mathbf{c} \cdot \iiint_V \operatorname{curl} \mathbf{A}' \, dV = \mathbf{c} \cdot \iint_S (\mathbf{n} \times \mathbf{A}') \, dS,$$

since

$$\operatorname{div} (\mathbf{A}' \times \mathbf{c}) = \nabla \cdot (\mathbf{A}' \times \mathbf{c}) = \mathbf{c} \cdot (\nabla \times \mathbf{A}') = \mathbf{c} \cdot \operatorname{curl} \mathbf{A}',$$

$$(\mathbf{A}' \times \mathbf{c}) \cdot \mathbf{n} = \mathbf{c} \cdot (\mathbf{n} \times \mathbf{A}').$$

Hence, since $\mathbf{c}$ is arbitrary,

$$\iiint_V \operatorname{curl} \mathbf{A} \, dV = \iint_S \mathbf{n} \times \mathbf{A} \, dS, \tag{5.33}$$

after dropping the prime.

Finally, suppose the vector $\mathbf{A}$ in (5.31) has components

$$A_k = T_{ik} c_i,$$

where T_{ik} is a tensor field and $\mathbf{c}$ is an arbitrary vector with components c_k. Then the same argument as before shows that

$$\iiint_V \frac{\partial T_{ik}}{\partial x_k} \, dV = \iint_S T_{ik} n_k \, dS, \tag{5.34}$$

where the n_k are the components of $\mathbf{n}$.

---

[4] Since $\mathbf{c}$ is arbitrary, the fact that the scalar product of $\mathbf{c}$ with another vector vanishes implies that the other vector vanishes.

Remark. Formulas (5.32)–(5.34) all stem from the operator identity

$$\iiint_V \nabla(\cdots)\, dV = \iint_S \mathbf{n}(\cdots)\, dS,$$

where $(\cdots)$ denotes some expression (possibly preceded by a dot or a cross) on which the given operator acts.

Example 1. If the field **A** is such that div **A** vanishes everywhere inside a volume V bounded by a surface S, then (5.31) implies that the flux of **A** through S vanishes:

$$\iint_S \mathbf{A} \cdot \mathbf{n}\, dS = 0. \tag{5.35}$$

Example 2. If div·**A** vanishes everywhere except at some "singular point" M where the divergence is either nonzero or fails to exist, then (5.35) has the same value for every surface S containing M and vanishes for every S which does not contain M (cf. Example 3, p. 159).

Example 3. If $\varphi = \text{const}$, then, just as in Example 2, p. 167, it follows from (5.32) that

$$\iint_S \mathbf{n}\, dS = 0$$

for every closed surface S.

Example 4. If the field **A** is such that curl **A** vanishes everywhere inside a volume V bounded by a surface S, then (5.33) implies

$$\iint_S \mathbf{n} \times \mathbf{A}\, dS = 0.$$

5.2.2. Theorems related to Stokes' theorem. We begin by giving another proof of Stokes' theorem, which is less rigorous but more intuitive than that given in Sec. 4.2.3. Let S be a surface bounded by a closed contour L, and let $\mathbf{A} = \mathbf{A}(\mathbf{r})$ be a vector field which, together with its derivatives $\partial A_i/\partial x_k$, is continuous on $S \cup L$. Partitioning S into N small pieces S_i $(i = 1, 2, \ldots, N)$, let $\mathbf{n}_i$ be the exterior normal to S_i at some point M_i and let L_i be the boundary of S_i traversed in the direction corresponding to $\mathbf{n}_i$ (see Fig. 5.1). Then,

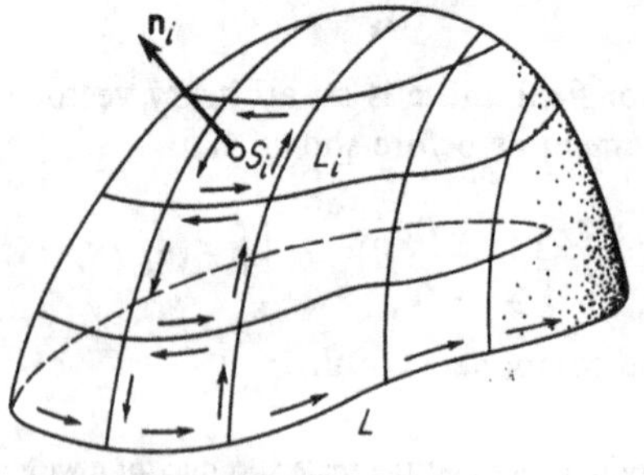

FIG. 5.1. Illustrating an alternative proof of Stokes' theorem.

according to (4.57),

$$(\mathbf{n}_i \cdot \text{curl } \mathbf{A})_{M_i} = \lim_{S_i \to 0} \frac{1}{S_i} \oint_{L_i} \mathbf{A} \cdot d\mathbf{r}, \tag{5.36}$$

where S_i and its boundary L_i shrink to the point M_i, and hence

$$(\mathbf{n}_i \cdot \text{curl } \mathbf{A})_{M_i} S_i = \oint_{L_i} \mathbf{A} \cdot d\mathbf{r} + \varepsilon S_i,$$

where $|\varepsilon|$ can be made arbitrarily small by decreasing the size of S_i. Moreover, given any $\varepsilon > 0$, there is a partition of S into $N = N(\varepsilon)$ parts such that

$$\max_{1 \leqslant i \leqslant N} \varepsilon_i < \varepsilon$$

[the convergence in (5.36) is uniform]. Therefore

$$\left| \sum_{i=1}^{N} (\mathbf{n}_i \cdot \text{curl } \mathbf{A})_{M_i} S_i - \sum_{i=1}^{N} \oint_{L_i} \mathbf{A} \cdot d\mathbf{r} \right| < \varepsilon \sum_{i=1}^{N} S_i = \varepsilon S, \tag{5.37}$$

where $\varepsilon \to 0$ as $N \to \infty$. But

$$\sum_{i=1}^{N} \oint_{L_i} \mathbf{A} \cdot d\mathbf{r} = \oint_{L} \mathbf{A} \cdot d\mathbf{r}, \tag{5.38}$$

since every part of the contours $L_1, \ldots, L_N$ appearing in the sum on the left is traversed twice in opposite directions, except for the parts making up the contour L. Thus

$$\left| \sum_{i=1}^{N} (\mathbf{n}_i \cdot \text{curl } \mathbf{A})_{M_i} S_i - \oint_{L} \mathbf{A} \cdot d\mathbf{r} \right| < \varepsilon S,$$

i.e.,

$$\lim_{N \to \infty} \sum_{i=1}^{N} (\mathbf{n}_i \cdot \text{curl } \mathbf{A})_{M_i} S_i = \oint_{L} \mathbf{A} \cdot d\mathbf{r}.$$

But the limit on the left equals

$$\iint_{S} \mathbf{n} \cdot \text{curl } \mathbf{A} \, dS,$$

by definition, and hence, finally,

$$\iint_{S} \mathbf{n} \cdot \text{curl } \mathbf{A} \, dS = \oint_{L} \mathbf{A} \cdot d\mathbf{r}, \tag{5.39}$$

which is just the vector form of Stokes' theorem [recall (4.60)].

Just as in Sec. 5.2.1, we can now deduce further integral theorems from (5.39) by making appropriate choices of the vector field $\mathbf{A}$. First suppose $\mathbf{A} = \mathbf{c}\varphi$, where $\mathbf{c}$ is a fixed but arbitrary vector and φ is a scalar field. Then (5.39) implies

$$\iint_{S} \mathbf{n} \times \text{grad } \varphi \, dS = \oint_{L} \varphi \, d\mathbf{r}, \tag{5.40}$$

since

$$\mathbf{n} \cdot \operatorname{curl}(\mathbf{c}\varphi) = \mathbf{n} \cdot \nabla \times (\mathbf{c}\varphi) = \mathbf{c} \cdot (\mathbf{n} \times \nabla\varphi) = \mathbf{c} \cdot (\mathbf{n} \times \operatorname{grad} \varphi).$$

Similarly, substituting $\mathbf{A} = \mathbf{A}' \times \mathbf{c}$ into (5.39), where $\mathbf{A}'$ is another vector field and $\mathbf{c}$ is a constant vector, we find that

$$\mathbf{c} \cdot \iint_S (\mathbf{n} \times \nabla) \times \mathbf{A}' \, dS = c \cdot \oint_L d\mathbf{r} \times \mathbf{A}',$$

since

$$\mathbf{n} \cdot \operatorname{curl}(\mathbf{A}' \times \mathbf{c}) = \mathbf{n} \cdot \nabla \times (\mathbf{A}' \times \mathbf{c}) = \mathbf{c} \cdot ((\mathbf{n} \times \nabla) \times \mathbf{A}').$$

Hence, since $\mathbf{c}$ is arbitrary,

$$\iint_S (\mathbf{n} \times \nabla) \times \mathbf{A} \, dS = \oint_L d\mathbf{r} \times \mathbf{A}, \tag{5.41}$$

after dropping the prime.

Remark. Formulas (5.39)–(5.41) all stem from the operator identity

$$\iint_S (\mathbf{n} \times \nabla)(\cdots) \, dS = \oint_L d\mathbf{r}(\cdots),$$

where $(\cdots)$ denotes some expression (possibly preceded by a dot or a cross) on which the given operator acts.

Example 1. If the field $\mathbf{A}$ is such that curl $\mathbf{A}$ vanishes everywhere, then (5.39) implies that the circulation of $\mathbf{A}$ around any closed contour L vanishes.[5]

Example 2. The flux of the curl of a field $\mathbf{A}$ through any closed surface S vanishes. In fact, divide S into two parts S_1 and S_2 bounded by the same closed contour L, and let the direction of traversing L correspond to the exterior normal to S_1. Then

$$\iint_{S_1} \mathbf{n} \cdot \operatorname{curl} \mathbf{A} \, dS = \oint_L \mathbf{A} \cdot d\mathbf{r}, \tag{5.42}$$

while

$$\iint_{S_2} \mathbf{n} \cdot \operatorname{curl} \mathbf{A} \, dS = -\oint_L \mathbf{A} \cdot d\mathbf{r} \tag{5.43}$$

(explain the minus sign). Adding (5.42) and (5.43), we obtain

$$\iint_{S_1} \mathbf{n} \cdot \operatorname{curl} \mathbf{A} \, dS + \iint_{S_2} \mathbf{n} \cdot \operatorname{curl} \mathbf{A} \, dS = 0,$$

or

$$\iint_{S_1 \cup S_2} \mathbf{n} \cdot \operatorname{curl} \mathbf{A} \, dS = \iint_S \mathbf{n} \cdot \operatorname{curl} \mathbf{A} \, dS = 0.$$

---

[5] Provided L can be shrunk continuously to a point without leaving the field (see Sec. 4.2.4).

5.2.3. Green's formulas. Suppose we choose

$$\mathbf{A} = \varphi \nabla \chi$$

in Gauss' theorem (5.31), where φ and ψ are continuous scalar functions with continuous first and second partial derivatives. Then

$$\operatorname{div} \mathbf{A} = \operatorname{div} (\varphi \nabla \psi) = \nabla \cdot (\varphi \nabla \psi) = \varphi \nabla \cdot \nabla \psi + \nabla \varphi \cdot \nabla \psi = \varphi \Delta \psi + \nabla \varphi \cdot \nabla \psi,$$

$$\mathbf{A} \cdot \mathbf{n} = \varphi \mathbf{n} \cdot \nabla \psi = \varphi \frac{\partial \psi}{\partial n},$$

and hence (5.31) becomes

$$\iiint\limits_V (\varphi \Delta \psi + \nabla \varphi \cdot \nabla \psi)\, dV = \iint\limits_S \varphi \frac{\partial \psi}{\partial n}\, dS, \tag{5.44}$$

a result known as *Green's first formula.* Similarly, choosing

$$\mathbf{A} = \varphi \nabla \psi - \psi \nabla \varphi$$

in (5.31), we obtain *Green's second formula*

$$\iiint\limits_V (\varphi \Delta \psi - \psi \Delta \varphi)\, dV = \iint\limits_S \left(\varphi \frac{\partial \psi}{\partial n} - \psi \frac{\partial \varphi}{\partial n}\right) dS. \tag{5.45}$$

Formula (5.45) can also be obtained by interchanging φ and ψ in (5.44) and subtracting the result from (5.44) itself.

Example 1. If $\varphi = \psi$, (5.44) becomes

$$\iiint\limits_V [\varphi \Delta \varphi + (\nabla \varphi)^2]\, dV = \iint\limits_S \varphi \frac{\partial \varphi}{\partial n}\, dS.$$

Example 2. If $\varphi = \text{const}$, (5.44) reduces to

$$\iiint\limits_V \Delta \psi\, dV = \iint\limits_S \frac{\partial \psi}{\partial n}\, dS. \tag{5.46}$$

Formula (5.46) implies the following symbolic representation of the Laplacian operator:

$$\Delta(\cdots) = \lim_{V \to 0} \frac{1}{V} \iint\limits_S \frac{\partial}{\partial n}(\cdots)\, dS.$$

Next we prove an important consequence of Green's second formula:

THEOREM. *Given a volume V bounded by a closed surface S, let $\varphi = \varphi(x_1, x_2, x_3)$ be a continuous scalar field with continuous first and second partial derivatives. Then the value of φ at any interior point M_0 of V is given by*

$$\varphi(M_0) = -\frac{1}{4\pi} \iiint\limits_V \frac{1}{r} \Delta \varphi\, dV - \frac{1}{4\pi} \iint\limits_S \left[\varphi \frac{\partial}{\partial n}\left(\frac{1}{r}\right) - \frac{1}{r} \frac{\partial \varphi}{\partial n}\right] dS. \tag{5.47}$$

Proof. Let M_0 have coordinates x_{10}, x_{20}, x_{30}, and let $\mathbf{r}$ be the radius vector drawn from M_0 to a variable point M, so that

$$r = \sqrt{(x_1 - x_{10})^2 + (x_2 - x_{20})^2 + (x_3 - x_{30})^2}.$$

Surround M_0 by a little sphere of radius ρ with surface ε lying entirely inside S (see Fig. 5.2), and let V' be the volume between the surfaces ε and S. Then, setting $\psi = 1/r$, $V = V'$ in (5.45),[6] we have

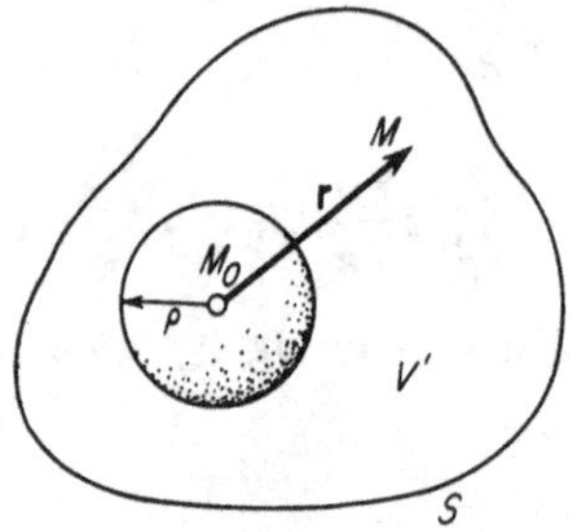

FIG. 5.2. Isolation of a singular point M_0.

$$\iiint_{V'} \left(-\frac{1}{r}\Delta\varphi\right) dV = \iint_S \left[\varphi \frac{\partial}{\partial n}\left(\frac{1}{r}\right) - \frac{1}{r}\frac{\partial \varphi}{\partial n}\right] dS + \iint_\varepsilon \left[\varphi \frac{\partial}{\partial n}\left(\frac{1}{r}\right) - \frac{1}{r}\frac{\partial \varphi}{\partial n}\right] d\varepsilon, \quad (5.48)$$

since, as can easily be verified by direct calculation, $\Delta(1/r) = 0$ except at the "singular point" M_0 where $r = 0$.

The exterior normal to the inner surface of V' is just the exterior normal to the surface ε, and has the same direction as the radius vector $\mathbf{r}$. Therefore the second integral in the right-hand side of (5.48) becomes

$$\iint_\varepsilon \left[\varphi \frac{\partial}{\partial n}\left(\frac{1}{r}\right) - \frac{1}{r}\frac{\partial \varphi}{\partial n}\right] d\varepsilon = \iint_\varepsilon \left[-\varphi \frac{\partial}{\partial r}\left(\frac{1}{r}\right) - \frac{1}{r}\frac{\partial \varphi}{\partial n}\right] d\varepsilon$$
$$= \iint_\varepsilon \left(\frac{\varphi}{r^2} - \frac{1}{r}\frac{\partial \varphi}{\partial n}\right) d\varepsilon.$$

Passing to the limit as $\rho \to 0$, i.e., as the sphere ε shrinks to the point M_0, we have

$$\lim_{\rho \to 0} \iint_\varepsilon \left(\frac{\varphi}{r^2} - \frac{1}{r}\frac{\partial \varphi}{\partial n}\right) d\varepsilon = \lim_{\substack{\rho \to 0 \\ M' \to M_0}} \left[\frac{\varphi(M')}{\rho^2} - \frac{1}{\rho}\left(\frac{\partial \varphi}{\partial n}\right)_{M'}\right] 4\pi\rho^2 = 4\pi\varphi(M_0). \quad (5.49)$$

Letting $\rho \to 0$ in (5.48) [so that V' approaches V] and using (5.49), we find that

$$4\pi\varphi(M_0) = -\iiint_V \frac{1}{r}\Delta\varphi \, dV - \iint_S \left[\varphi \frac{\partial}{\partial n}\left(\frac{1}{r}\right) - \frac{1}{r}\frac{\partial \varphi}{\partial n}\right] dS, \quad (5.50)$$

which is equivalent to (5.47).

---

[6] Like Gauss' theorem itself, Green's formulas hold for volumes bounded by several closed surfaces (recall Remark 4, p. 139).

Remark. An analogous formula holds for a (suitably smooth) vector field $\mathbf{A} = \mathbf{A}(\mathbf{r})$ with components A_1, A_2, A_3. In fact, setting

$$T_{ik} = A_i \frac{\partial}{\partial x_k}\left(\frac{1}{r}\right) - \frac{1}{r}\frac{\partial A_i}{\partial x_k}$$

in (5.34) and repeating the argument leading to (5.50), we find that

$$4\pi A_i(M_0) = -\iiint_V \frac{1}{r}\Delta A_i\, dV - \iint_S \left[A_i \frac{\partial}{\partial x_k}\left(\frac{1}{r}\right) - \frac{1}{r}\frac{\partial A_i}{\partial x_k}\right] n_k\, dS$$

or

$$\mathbf{A}(M_0) = -\frac{1}{4\pi}\iiint_V \frac{1}{r}\Delta\mathbf{A}\, dV - \frac{1}{4\pi}\iint_S \left[\mathbf{A}\frac{\partial}{\partial n}\left(\frac{1}{r}\right) - \frac{1}{r}(\mathbf{n}\cdot\nabla)\mathbf{A}\right] dS \tag{5.51}$$

(concerning expressions for $\Delta\mathbf{A}$, see Exercises 10-12, pp. 183–184).

5.3. Applications to Fluid Dynamics

5.3.1. Equations of fluid motion. Given a moving fluid (liquid or gas) described by a velocity field $\mathbf{v} = \mathbf{v}(\mathbf{r}, t)$, let V be a "material volume," i.e., a volume moving with the fluid and hence always consisting of the same fluid particles. Then V and its surface S are in general functions of time. The total momentum of the volume V is

$$\iiint_V \rho\mathbf{v}\, dV,$$

while the total body force acting on V is

$$\iiint_V \mathbf{f}\, dV, \tag{5.52}$$

where $\mathbf{f}$ is the body force per unit mass.[7] Besides (5.52), V is also subject to internal forces acting across its surface S. Let $\mathbf{p}_n$ be the stress acting on an element of area dS with unit exterior normal $\mathbf{n}$. Then the total force acting on V due to the stress on S is

$$\iint_S \mathbf{p}_n\, dS.$$

It follows from Newton's second law that

$$\frac{d}{dt}\iiint_V \rho\mathbf{v}\, dV = \iiint_V \rho\mathbf{f}\, dV + \iint_S \mathbf{p}_n\, dS. \tag{5.53}$$

[7] For example, in a gravitational field, $\mathbf{f}$ equals the acceleration $\mathbf{g}$ due to gravity.

To transform (5.53) further, we note that

$$\frac{d}{dt}\iiint_V \rho\mathbf{v}\,dV = \iiint_V \rho\frac{d\mathbf{v}}{dt}\,dV. \tag{5.54}$$

This can be seen as follows: Let V_0 be the volume occupied by $V = V(t)$ at time $t = 0$, and let $\mathbf{r} = \mathbf{r}(\boldsymbol{\xi}, t)$ be the position at time t of the fluid particle which had radius vector $\boldsymbol{\xi}$ at time $t = 0$. Then

$$\begin{aligned}\frac{d}{dt}\iiint_V \rho\mathbf{v}\,dV &= \frac{d}{dt}\iiint_V \rho(\mathbf{r}, t)v(\mathbf{r}, t)\,dV \\ &= \frac{d}{dt}\iiint_{V_0} \rho[\mathbf{r}(\boldsymbol{\xi}, t), t]\mathbf{v}[\mathbf{r}(\xi, t), t]\frac{dV}{dV_0}\,dV_0 \\ &= \iiint_{V_0}\left[\frac{d}{dt}(\rho\mathbf{v})\frac{dV}{dV_0} + \rho\mathbf{v}\frac{d}{dt}\left(\frac{dV}{dV_0}\right)\right]dV_0 \\ &= \iiint_{V_0}\left[\frac{d}{dt}(\rho\mathbf{v}) + \rho\mathbf{v}\operatorname{div}\mathbf{v}\right]\frac{dV}{dV_0}\,dV_0,\end{aligned}$$

where we use the fact that[8]

$$\frac{d}{dt}\left(\frac{dV}{dV_0}\right) = \operatorname{div}\mathbf{v}. \tag{5.55}$$

[8] Clearly

$$\frac{dV}{dV_0} = \frac{\partial(x_1, x_2, x_3)}{\partial(\xi_1, \xi_2, \xi_3)} = \begin{vmatrix} \dfrac{\partial x_1}{\partial \xi_1} & \dfrac{\partial x_1}{\partial \xi_2} & \dfrac{\partial x_1}{\partial \xi_3} \\ \dfrac{\partial x_2}{\partial \xi_1} & \dfrac{\partial x_2}{\partial \xi_2} & \dfrac{\partial x_2}{\partial \xi_3} \\ \dfrac{\partial x_3}{\partial \xi_1} & \dfrac{\partial x_3}{\partial \xi_2} & \dfrac{\partial x_3}{\partial \xi_3} \end{vmatrix},$$

and hence

$$\frac{d}{dt}\left(\frac{dV}{dV_0}\right) = \begin{vmatrix} \dfrac{d}{dt}\dfrac{\partial x_1}{\partial \xi_1} & \dfrac{d}{dt}\dfrac{\partial x_1}{\partial \xi_2} & \dfrac{d}{dt}\dfrac{\partial x_1}{\partial \xi_3} \\ \dfrac{\partial x_2}{\partial \xi_1} & \dfrac{\partial x_2}{\partial \xi_2} & \dfrac{\partial x_2}{\partial \xi_3} \\ \dfrac{\partial x_3}{\partial \xi_1} & \dfrac{\partial x_3}{\partial \xi_2} & \dfrac{\partial x_3}{\partial \xi_3} \end{vmatrix} + \text{two similar terms}$$

(involving differentiation of the second and third rows).

After a bit of algebra based on the properties of determinants and the formulas

$$\frac{d}{dt}\frac{\partial x_i}{\partial \xi_j} = \frac{\partial v_i}{\partial \xi_j}, \qquad \frac{\partial v_i}{\partial \xi_j} = \frac{\partial v_i}{\partial x_k}\frac{\partial x_k}{\partial \xi_j},$$

we obtain (5.55). For the details, see R. Aris, *op. cit.*, pp. 83–84.

Therefore

$$\frac{d}{dt}\iiint\limits_V \rho\mathbf{v}\,dV = \iiint\limits_V \left[\frac{d}{dt}(\rho\mathbf{v}) + \rho\mathbf{v}\operatorname{div}\mathbf{v}\right] dV$$

$$= \iiint\limits_V \left[\rho\frac{d\mathbf{v}}{dt} + \mathbf{v}\left(\frac{d\rho}{dt} + \operatorname{div}\mathbf{v}\right)\right] dV = \iiint\limits_V \rho\frac{d\mathbf{v}}{dt}\,dV,$$

in accordance with (5.54), because of the equation of continuity

$$\frac{d\rho}{dt} + \rho\operatorname{div}\mathbf{v} = \frac{\partial\rho}{\partial t} + \mathbf{v}\cdot\nabla\rho + \rho\operatorname{div}\mathbf{v} = \frac{\partial\rho}{\partial t} + \operatorname{div}(\rho\mathbf{v}) = 0$$

[recall (4.50)].[9]

Using (5.54), we can write (5.53) as

$$\iiint\limits_V \rho\frac{d\mathbf{v}}{dt}\,dV = \iiint\limits_V \rho\mathbf{f}\,dV + \iint\limits_S \mathbf{p}_n\,dS$$

$$= \iiint\limits_V \rho\mathbf{f}\,dV + \iint\limits_S \mathbf{p}_k n_k\,dS$$

(recall p. 67), or, in component form,

$$\iiint\limits_V \rho\frac{dv_i}{dt}\,dV = \iiint\limits_V \rho f_i\,dV + \iint\limits_S p_{ik}n_k\,dS, \tag{5.56}$$

where p_{ik} is the stress tensor.[10] To obtain a differential equation describing the motion of the fluid, we use Gauss' theorem in the form (5.34) to transform the surface integral into a volume integral:

$$\iint\limits_S p_{ik}n_k\,dS = \iiint\limits_V \frac{\partial p_{ik}}{\partial x_k}\,dV.$$

Then (5.56) becomes

$$\iiint\limits_V \left(\rho\frac{dv_i}{dt} - \rho f_i - \frac{\partial p_{ik}}{\partial x_k}\right) dV = 0.$$

Since the volume V is arbitrary, we have

$$\rho\frac{dv_i}{dt} = \rho f_i + \frac{\partial p_{ik}}{\partial x_k}, \tag{5.57}$$

---

[9] Note that

$$\frac{d\rho}{dt} = \frac{\partial\rho}{\partial t} + \frac{\partial\rho}{\partial x_i}\frac{dx_i}{dt} = \frac{\partial\rho}{\partial t} + v_i\frac{\partial\rho}{\partial x_i} = \frac{\partial\rho}{\partial t} + \mathbf{v}\cdot\nabla\rho.$$

[10] The tensor p_{ik} is symmetric (see below).

assuming that the integrand is continuous. Here dv_i/dt is the total derivative, which, as shown on p. 166, can be written in the form[11]

$$\frac{dv_i}{dt} = \frac{\partial v_i}{\partial t} + v_k \frac{\partial v_i}{\partial x_k}. \tag{5.58}$$

Thus, finally, combining (5.57) and (5.58), we find that the motion of the fluid is described by the differential equation

$$\rho \frac{\partial v_i}{\partial t} + \rho v_k \frac{\partial v_i}{\partial x_k} = \rho f_i + \frac{\partial p_{ik}}{\partial x_k} \qquad (i = 1, 2, 3). \tag{5.59}$$

Next we make the usual hydrodynamical assumption that the viscous stress tensor p_{ik} is a linear function of the rate of deformation tensor v_{ik}. Then recalling formula (2.61) and footnote 11, p. 96,[12] we have

$$p_{ik} = -p\delta_{ik} + 2\mu v_{ik} + \mu'\delta_{ik}v_{ll}, \tag{5.60}$$

where p is the hydrostatic pressure, μ and μ' are constants of proportionality (called viscosity coefficients), and

$$v_{ll} = \frac{\partial v_l}{\partial x_l} = \operatorname{div} \mathbf{v}$$

[recall (2.22)]. Under certain circumstances,[13] it can be assumed that

$$\mu' + \tfrac{2}{3}\mu = 0,$$

and then (5.60) becomes

$$p_{ik} = -p\delta_{ik} + 2\mu v_{ik} - \tfrac{2}{3}\mu\delta_{ik}v_{ll}. \tag{5.61}$$

If the fluid is incompressible, $\operatorname{div} \mathbf{v} = 0$ and hence

$$p_{ik} = -p\delta_{ik} + 2\mu v_{ik}. \tag{5.62}$$

If the fluid is at rest or if the fluid is "ideal" with no viscosity ($\mu = 0$), we have just

$$p_{ik} = -p\delta_{ik}. \tag{5.63}$$

Substituting (5.61)–(5.63) in turn into (5.59), we deduce the equations of motion for three kinds of fluids:

---

[11] From a formal standpoint, (5.58) is an immediate consequence of the chain rule for partial differentiation:

$$\frac{dv_i}{dt} = \frac{\partial v_i}{\partial t} + \frac{\partial v_i}{\partial x_k}\frac{dx_k}{dt} = \frac{\partial v_i}{\partial t} + v_k \frac{\partial v_i}{\partial x_k}.$$

[12] Cf. also Prob. 1, p. 126.

[13] R. Aris, *op. cit.*, p. 112.

1) *The ideal fluid* ($\mu = 0$). In this case, (5.59) becomes

$$\rho \frac{\partial v_i}{\partial t} + \rho v_k \frac{\partial v_i}{\partial x_k} = \rho f_i - \frac{\partial p}{\partial x_i}$$

or

$$\rho \frac{\partial \mathbf{v}}{\partial t} + \rho(\mathbf{v} \cdot \nabla)\mathbf{v} = \rho \mathbf{f} - \nabla p \tag{5.64}$$

in vector notation.

2) *The viscous incompressible fluid* ($\mu = \text{const} \neq 0$, $\operatorname{div} \mathbf{v} = 0$). We now have the *Navier-Stokes equation*

$$\rho \frac{\partial v_i}{\partial t} + \rho v_k \frac{\partial v_i}{\partial x_k} = \rho f_i - \frac{\partial p}{\partial x_i} + \mu \frac{\partial^2 v_i}{\partial x_k \, \partial x_k},$$

since

$$\frac{\partial}{\partial x_k}(2\mu v_{ik}) = \mu \frac{\partial}{\partial x_k}\left(\frac{\partial v_i}{\partial x_k} + \frac{\partial v_k}{\partial x_i}\right)$$
$$= \mu \frac{\partial^2 v_i}{\partial x_k \, \partial x_k} + \mu \frac{\partial}{\partial x_i}\left(\frac{\partial v_k}{\partial x_k}\right) = \mu \frac{\partial^2 v_i}{\partial x_k \, \partial x_k}.$$

The vector form of the Navier-Stokes equation is

$$\rho \frac{\partial \mathbf{v}}{\partial t} + \rho(\nabla \cdot \mathbf{v})\mathbf{v} = \rho \mathbf{f} - \nabla p + \mu \, \Delta \mathbf{v}.$$

3) *The viscous compressible fluid* ($\mu = \text{const} \neq 0$, $\operatorname{div} \mathbf{v} \neq 0$). In this case, (5.59) becomes

$$\rho \frac{\partial v_i}{\partial t} + \rho v_k \frac{\partial v_i}{\partial x_k} = \rho f_i - \frac{\partial p}{\partial x_i} + \mu \frac{\partial^2 v_i}{\partial x_k \, \partial x_k} + \tfrac{1}{3}\mu \frac{\partial^2 v_k}{\partial x_i \, \partial x_k}, \tag{5.65}$$

or

$$\rho \frac{\partial \mathbf{v}}{\partial t} + \rho(\mathbf{v} \cdot \nabla)\mathbf{v} = \rho \mathbf{f} - \nabla p + \mu \, \Delta \mathbf{v} + \tfrac{1}{3}\mu \operatorname{grad} \operatorname{div} \mathbf{v}. \tag{5.66}$$

Example **(*Archimedes' law*).** The force $\mathbf{F}$ exerted by a fluid on a body of volume V and surface S immersed in the fluid is

$$\mathbf{F} = \iint_S \mathbf{p}_n \, dS = \iint_S \mathbf{p}_k n_k \, dS,$$

with components

$$F_i = \iint_S p_{ik} n_k \, dS.$$

If the fluid is at rest ($\mathbf{v} = 0$), then, according to (5.63) and (5.66),

$$p_{ik} = -p\delta_{ik},$$
$$\nabla p = \rho \mathbf{f} = \rho \mathbf{g}, \tag{5.67}$$

where $\mathbf{g}$ is the acceleration due to gravity (recall footnote 7, p. 203). It follows that

$$F_i = -\iint_S pn_i\, dS$$

or

$$\mathbf{F} = -\iint_S p\mathbf{n}\, dS = -\iiint_V \nabla p\, dV,$$

where (5.32) has been used in the last step. Substituting from (5.67), we find that

$$\mathbf{F} = -\iiint_V \rho\mathbf{g}\, dV = -\mathbf{g}\iiint_V \rho\, dV = -\mathbf{g}m,$$

where m is the mass of fluid displaced by the body. Thus we have proved *Archimedes' law*, i.e., the force exerted by a fluid on a body immersed in the fluid equals the weight $m|\mathbf{g}|$ of the displaced fluid and points in the direction opposite to the force of gravity.

5.3.2. The momentum theorem. Consider a *fixed* volume V immersed in a velocity field $\mathbf{v} = \mathbf{v}(\mathbf{r}, t)$. Then the amount of fluid in V equals

$$\iiint_V \rho\mathbf{v}\, dV.$$

This quantity changes in time, at the rate

$$\frac{\partial}{\partial t}\iiint_V \rho\mathbf{v}\, dt = \iiint_V \frac{\partial}{\partial t}(\rho\mathbf{v})\, dV.$$

with components

$$\frac{\partial}{\partial t}\iiint_V \rho v_i\, dV = \iiint_V \frac{\partial}{\partial t}(\rho v_i)\, dV = \iiint_V \left(v_i\frac{\partial\rho}{\partial t} + \rho\frac{\partial v_i}{\partial t}\right) dV.$$

Suppose there are no body forces, so that $\mathbf{f} = 0$. Then determining $\partial\rho/\partial t$ from the equation of continuity (4.50) and $\rho\partial v_i/\partial t$ from the equation of motion (5.59), we obtain

$$\frac{\partial}{\partial t}\iiint_V \rho v_i\, dV = \iiint_V \left[-v_i\frac{\partial}{\partial x_k}(\rho v_k) - \rho v_k\frac{\partial v_i}{\partial x_k} + \frac{\partial p_{ik}}{\partial x_k}\right] dV$$

$$= -\iiint_V \frac{\partial}{\partial x_k}(\rho v_i v_k - p_{ik})\, dV.$$

Using formula (5.34) to transform the integral on the right, we find that

$$\frac{\partial}{\partial t}\iiint_V \rho v_i\, dt = -\iint_S (\rho v_i v_k - p_{ik})n_k\, dS = -\iint_S \Pi_{ik}n_k\, dS, \quad (5.68)$$

where S is the closed surface bounding V and

$$\Pi_{ik} = \rho v_i v_k - p_{ik}.$$

The left-hand side of (5.68) is the rate of change of the ith component of the momentum of the fluid in the fixed volume V, while the right-hand side is minus the flux of the ith component of the tensor Π_{ik} through the surface S. Thus $\Pi_{ik}n_k$ is the rate at which the ith component of the momentum of the fluid leaves V through the surface element dS whose unit normal has components n_k, while Π_{ik} itself is the rate at which the ith component of the momentum flows through a unit area perpendicular to the x_k-axis. For this reason, Π_{ik} is called the tensor of *momentum flux density*, and in fact, the total momentum flux through any surface S has components

$$\iint_S \Pi_{ik} n_k \, dS.$$

Remark. The momentum flux should not be confused with the flux

$$\iiint_S \rho \mathbf{v} \cdot \mathbf{n} \, dS \tag{5.69}$$

of the vector $\rho\mathbf{v}$ (the momentum of a unit volume of the fluid). As noted on p. 154, (5.69) is the rate at which mass is lost through S. Correspondingly, $\rho\mathbf{v}$ is called the *(mass) flux density* of the fluid.[14]

If the velocity field is stationary, the left-hand side of (5.68) vanishes, and we have the *momentum theorem*

$$\iint_S \Pi_{ik} n_k \, dS = \iint_S (\rho v_i v_k - p_{ik}) n_k \, dS = 0. \tag{5.70}$$

According to this theorem, *in the absence of body forces the flux of the tensor*

$$\Pi_{ik} = \rho v_i v_k - p_{ik}$$

through any closed surface S immersed in a stationary velocity field vanishes.

Let V be any volume immersed in a stationary velocity field, and let S be the surface of V. Then, in the absence of body forces, the momentum theorem allows us to express the total force $\mathbf{F}$ acting on V in terms of the fluid's velocity and density on S. In fact, the ith component of $\mathbf{F}$ is

$$F_i = \iint_S p_{ik} n_k \, dS,$$

---

[14] Note that ρv_k is the rate at which mass flows through a unit area perpendicular to the x_k-axis.

and hence, by (5.70),

$$F_i = \iint_S \rho v_i v_k n_k \, dS,$$

or

$$\mathbf{F} = \iint_S \rho \mathbf{v}(\mathbf{v} \cdot \mathbf{n}) \, dS$$

in vector notation.

Similarly, in the absence of body forces, the momentum theorem can be used to find the force **F** acting on a solid body of arbitrary shape immersed in a stationary velocity field. Suppose the body has surface S. In the fluid we choose a "control surface" S_0, i.e., a closed surface completely surrounding the surface S (see Fig. 5.3). Applying the momentum theorem to the

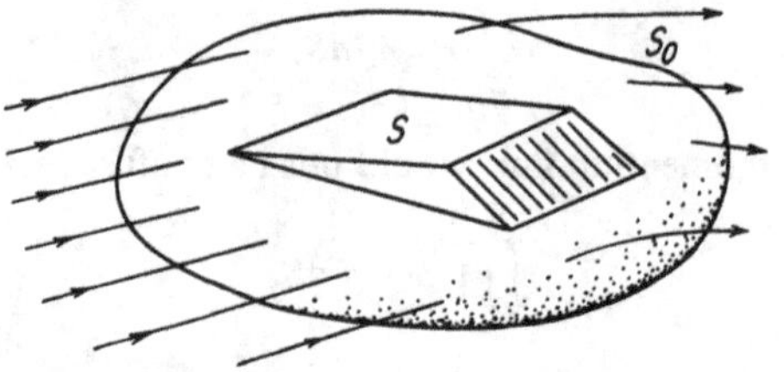

FIG. 5.3. In the absence of viscosity, the force exerted by a moving fluid on a body with surface S can be expressed in terms of the velocity, density and pressure on a suitable "control surface" S_0 surrounding S.

volume between the surfaces S and S_0, we find that

$$\iint_S p_{ik} n_k \, dS + \iint_{S_0} p_{ik} n_k \, dS_0 - \iint_S \rho v_i v_k n_k \, dS - \iint_{S_0} \rho v_i v_k n_k \, dS_0 = 0.$$

The first integral is the ith component of the force exerted by the body on the fluid, and hence its negative is the ith component F_i of the force exerted by the fluid on the body:

$$F_i = -\iint_S p_{ik} n_k \, dS.$$

The third integral vanishes since $\mathbf{v} \cdot \mathbf{n} = v_k n_k = 0$ on the surface S (no fluid flows through the surface of a solid body). It follows that

$$F_i = \iint_{S_0} (p_{ik} - \rho v_i v_k) n_k \, dS_0. \tag{5.71}$$

Thus to find the force acting on a solid body immersed in a stationary flow, we need only take some surface S_0, which can be chosen for its experimental convenience, and then measure the surface stresses and the fluid's velocity and density on S_0.

In the absence of viscosity,

$$p_{ik} = -p\delta_{ik}$$

[recall (5.63)] and (5.71) takes the particularly simple form

$$F_i = -\iint_{S_0} (pn_i + \rho v_i v_k n_k)\, dS_0$$

or

$$\mathbf{F} = -\iint_{S_0} [p\mathbf{n} + \rho\mathbf{v}(\mathbf{v}\cdot\mathbf{n})]\, dS_0 \tag{5.72}$$

in vector notation. Thus, to find the force acting on the body in this case, we need only measure the velocity, density and pressure on the control surface.

5.4. Potential and Irrotational Fields

Suppose a vector field $\mathbf{A} = \mathbf{A}(\mathbf{r})$ is the gradient of a (single-valued) scalar field φ:

$$\mathbf{A} = \text{grad}\ \varphi = \nabla\varphi. \tag{5.73}$$

Then **A** is said to be a *potential field*, and φ is called the (*scalar*) *potential* of the field **A**. Clearly, the potential φ is defined only to within an additive constant. In rectangular coordinates, a potential field **A** has components

$$A_1 = \frac{\partial\varphi}{\partial x_1}, \quad A_2 = \frac{\partial\varphi}{\partial x_2}, \quad A_3 = \frac{\partial\varphi}{\partial x_3}.$$

The great importance of potential fields stems from the fact that they are completely specified by a single scalar, namely the potential.

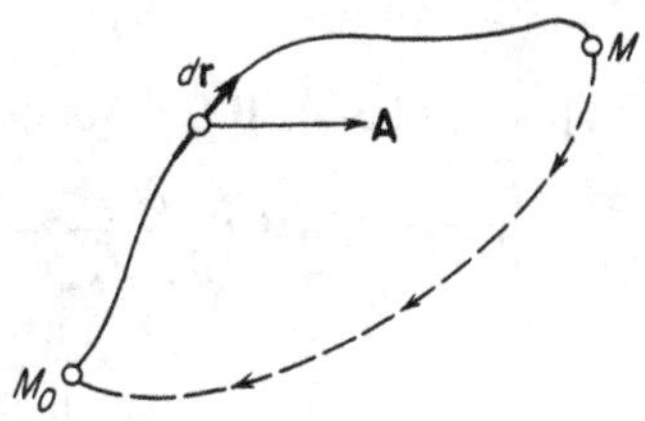

FIG. 5.4. The line integral of a potential field **A** is independent of the path of integration and vanishes if the path of integration is closed.

THEOREM 1. *If a field* **A** *has a single-valued potential, then the value of the line integral*

$$\int_{M_0}^{M} \mathbf{A}\cdot d\mathbf{r}$$

is independent of the path of integration, and depends only on the end points M_0 *and* M *of the path* (*see Fig. 5.4*).

Proof. If φ is the potential of **A**, then

$$\int_{M_0}^{M} \mathbf{A}\cdot d\mathbf{r} = \int_{M_0}^{M} \nabla\varphi\cdot dr = \int_{M_0}^{M} \left(\frac{\partial\varphi}{\partial x_1}dx_1 + \frac{\partial\varphi}{\partial x_2}dx_2 + \frac{\partial\varphi}{\partial x_3}dx_3\right)$$

$$= \int_{M_0}^{M} d\varphi = \varphi(M) - \varphi(M_0).$$

COROLLARY. *If a field* **A** *has a single-valued potential, then the line integral of* **A** *along any closed contour L vanishes.*

Proof. We need only note that

$$\oint_L \mathbf{A} \cdot d\mathbf{r} = \oint_L \nabla\varphi \cdot d\mathbf{r} = \varphi(M_0) - \varphi(M_0) = 0.$$

Remark. If a force field has a single-valued potential, then calculation of the work done by the force simply involves finding the potential difference between the end points of the path. Hydrodynamical problems become much simpler if the velocity field of the moving fluid is potential, since we can then use powerful methods like the theory of functions of a complex variable (in the case of two-dimensional flow).

THEOREM 2. *A necessary and sufficient condition for a vector field* **A** *occupying a simply connected region R to be potential is that* **A** *be irrotational, i.e., that*

$$\text{curl}\,\mathbf{A} = 0.$$

Proof. Suppose **A** is potential, so that $\mathbf{A} = \nabla\varphi$. Then

$$\text{curl}\,\mathbf{A} = \text{curl}\,\nabla\varphi = \begin{vmatrix} \mathbf{i}_1 & \mathbf{i}_2 & \mathbf{i}_3 \\ \dfrac{\partial}{\partial x_1} & \dfrac{\partial}{\partial x_2} & \dfrac{\partial}{\partial x_3} \\ \dfrac{\partial\varphi}{\partial x_1} & \dfrac{\partial\varphi}{\partial x_2} & \dfrac{\partial\varphi}{\partial x_3} \end{vmatrix}$$

[recall (4.58)]. It follows that

$$\text{curl}\,\mathbf{A} = \mathbf{i}_1\left(\frac{\partial^2\varphi}{\partial x_2\,\partial x_3} - \frac{\partial^2\varphi}{\partial x_3\,\partial x_2}\right) + \mathbf{i}_2\left(\frac{\partial^2\varphi}{\partial x_3\,\partial x_1} - \frac{\partial^2\varphi}{\partial x_1\,\partial x_3}\right) + \mathbf{i}_3\left(\frac{\partial^2\varphi}{\partial x_1\,\partial x_2} - \frac{\partial^2\varphi}{\partial x_2\,\partial x_1}\right) = 0,$$

where we assume that φ has continuous first and second partial derivatives. This proves the necessity.

To prove the sufficiency, let L be any closed contour in the region R occupied by the field **A**, and suppose curl $\mathbf{A} = 0$ everywhere in R. Then, since R is simply connected, L is the boundary of some surface S lying entirely in R (see Sec. 4.2.4). Hence, by Stokes' theorem,

$$\oint_L \mathbf{A} \cdot d\mathbf{r} = \iint_S \mathbf{n} \cdot \text{curl}\,\mathbf{A}\, dS = 0,$$

Since L is arbitrary (in R), it follows that given any fixed point M_0, the line integral

$$\int_{M_0}^{M} \mathbf{A} \cdot d\mathbf{r} = \int_{M_0}^{M} A_1\,dx_1 + A_2\,dx_2 + A_3\,dx_3$$

is independent of the path of integration and is a function only of the variable point $M = (x_1, x_2, x_3)$. Denoting this function by $\varphi(x_1, x_2, x_3)$, we have

$$A_i = \frac{\partial \varphi}{\partial x_i} \quad (i = 1, 2, 3). \tag{5.74}$$

In fact,

$$\begin{aligned}\frac{\partial \varphi}{\partial x_1} &= \lim_{\Delta x_1 \to 0} \frac{1}{\Delta x_1} \left\{\int_{M_0}^{(x_1+\Delta x_1, x_2, x_3)} \mathbf{A} \cdot d\mathbf{r} - \int_{M_0}^{(x_1, x_2, x_3)} \mathbf{A} \cdot d\mathbf{r}\right\} \\ &= \lim_{\Delta x_1 \to 0} \frac{1}{\Delta x_1} \int_{(x_1, x_2, x_3)}^{(x_1+\Delta x_1, x_2, x_3)} \mathbf{A} \cdot d\mathbf{r} \qquad (5.75) \\ &= \lim_{\Delta x_1 \to 0} \frac{1}{\Delta x_1} \int_\sigma \mathbf{A} \cdot d\mathbf{r} = \lim_{\Delta x_1 \to 0} \frac{1}{\Delta x_1} \int_\sigma A_1\, dx_1,\end{aligned}$$

where σ is the line segment joining (x_1, x_2, x_3) to $(x_1 + \Delta x_1, x_2, x_3)$. But, by the mean value theorem for integrals,

$$\lim_{\Delta x_1 \to 0} \frac{1}{\Delta x_1} \int_\sigma A_1\, dx_1 = \lim_{\Delta x_1 \to 0} \frac{1}{\Delta x_1} A_1(x_1 + \theta\, \Delta x_1, x_2, x_3)\, \Delta x_1,$$

where $0 < \theta < \Delta x$, i.e.,

$$\lim_{\Delta x_1 \to 0} \frac{1}{\Delta x_1} \int_\sigma A_1\, dx_1 = A_1(x_1, x_2, x_3), \tag{5.76}$$

by the continuity of $\mathbf{A}$ (recall p. 135). Together (5.75) and (5.76) imply (5.74) for $i = 1$, and the proof is similar for $i = 2, 3$. Hence $\mathbf{A}$ is a potential field, as asserted.

5.4.1. Multiple-valued potentials. If the region R occupied by the field $\mathbf{A}$ is multiply connected, then there exists at least one closed contour L which is not the boundary of any surface lying entirely in R (see Sec. 4.2.4). In this case, we can no longer use Stokes' theorem to deduce that

$$\oint_L \mathbf{A} \cdot d\mathbf{r} = 0,$$

and in fact it may turn out that

$$\oint_L \mathbf{A} \cdot d\mathbf{r} = c \neq 0.$$

If so, it is no longer possible to construct a single-valued potential φ such that $\mathbf{A} = \nabla\varphi$, since the integral

$$\int_{M_0}^{M} \mathbf{A} \cdot d\mathbf{r} \tag{5.77}$$

will no longer be path-independent. To see this, let M' be a point of L distinct from M_0 and M', let l be a path joining M_0 to M' and let l' be a path joining M' to M (draw a figure). Then

$$\int_{l+l'} \mathbf{A} \cdot d\mathbf{r} \neq \int_{l+L+l'} \mathbf{A} \cdot d\mathbf{r} = \int_{l+l'} \mathbf{A} \cdot d\mathbf{r} + \oint_L \mathbf{A} \cdot d\mathbf{r} = \int_{l+l'} \mathbf{A} \cdot d\mathbf{r} + c,$$

where, for example, $l + L + l'$ means the path going from M to M' along l, then from M' to M' along L, and finally from M' to M along l'. Thus the value of (5.77) depends on the path joining M_0 to M, i.e., we can talk about a "potential"

$$\varphi = \int_{M_0}^{M} \mathbf{A} \cdot d\mathbf{r}$$

only if φ is allowed to be multiple-valued.

Example. According to the *Biot-Savart law*, a current I (measured in electromagnetic units) flowing in an infinite straight wire along the x_3-axis produces a magnetic field

$$\mathbf{H} = \frac{2I}{r^2} \mathbf{i}_3 \times \mathbf{r} \tag{5.78}$$

(measured in oersteds), where

$$\mathbf{r} = x_1\mathbf{i}_1 + x_2\mathbf{i}_2, \qquad r = \sqrt{x_1^2 + x_2^2}$$

It follows that

$$H_1 = -2I \frac{x_2}{x_1^2 + x_2^2},$$

$$H_2 = 2I \frac{x_1}{x_1^2 + x_2^2},$$

$$H_3 = 0.$$

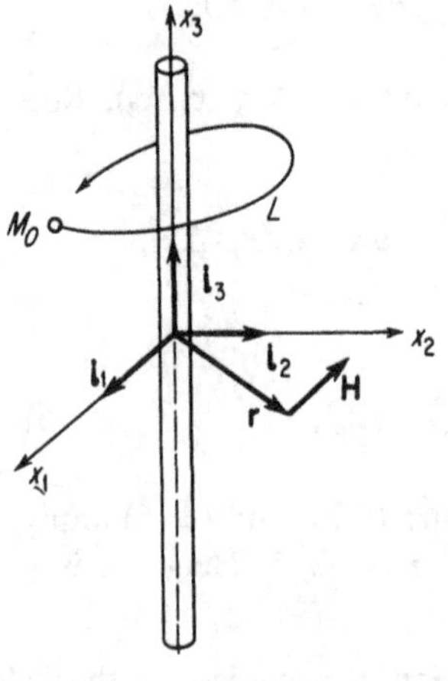

FIG. 5.5. The potential of the magnetic field **H** due to an electric current in an infinite wire is multiple-valued.

The field **H** is defined everywhere except on the x_3-axis where $r = 0$ (see Fig. 5.5). Therefore the field occupies a doubly connected region, i.e., all of space minus the x_3-axis. It is easily verified that curl $\mathbf{H} = 0$ everywhere except on the x_3-axis, where curl **H** is undefined (like **H** itself).

The field **H** can be derived from the multiple-valued potential

$$\varphi = 2I \arc\tan \frac{x_2}{x_1}, \tag{5.79}$$

since

$$\frac{\partial \varphi}{\partial x_1} = -\frac{\dfrac{x_2}{x_1^2}}{1 + \left(\dfrac{x_2}{x_1}\right)^2} = H_1, \quad \frac{\partial \varphi}{\partial x_2} = \frac{\dfrac{1}{x_1}}{1 + \left(\dfrac{x_2}{x_1}\right)^2} = H_2, \quad \frac{\partial \varphi}{\partial x_3} = 0 = H_3.$$

If L is a closed contour surrounding the x_3-axis, then

$$\oint_L \mathbf{H} \cdot d\mathbf{r} = 2I \oint_L \nabla\varphi \cdot d\mathbf{r},$$

where φ is given by (5.79). Therefore

$$\oint_L \mathbf{H} \cdot d\mathbf{r} = 2I \times \text{(the change in polar angle in traversing } L\text{)},$$

i.e.,

$$\oint_L \mathbf{H} \cdot d\mathbf{r} = 2I \cdot 2\pi = 4\pi I,$$

even though curl $\mathbf{H} = 0$ everywhere in the region occupied by the field! In other words, if we start from a point M_0 in the field and go around the wire once in the counterclockwise direction (see Fig. 5.5), we return to M_0 with a new value of the potential exceeding the old value by $4\pi I$.

Remark 1. The gradient of a single valued potential cannot have closed trajectories, since if L were closed, we would have

$$\oint_L \nabla\varphi \cdot d\mathbf{r} = \text{the change in } \varphi \text{ in traversing } L \neq 0.$$

On the other hand, the gradient of a multiple-valued potential can have closed trajectories (see Fig. 5.6), as in the case of the magnetic field (5.78).

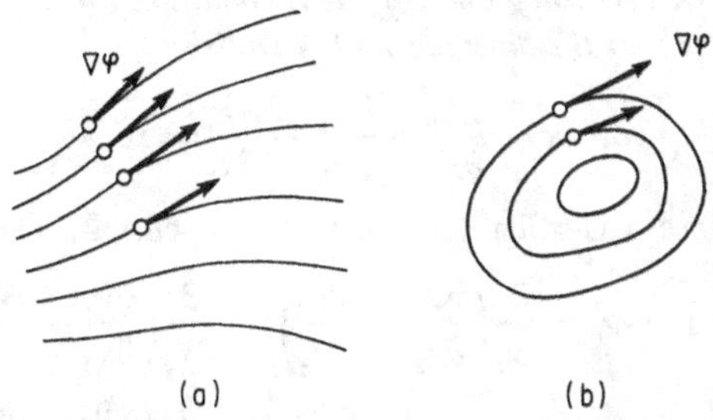

FIG. 5.6. (a) The gradient of a single-valued potential cannot have closed trajectories; (b) The gradient of a multiple-valued potential can have closed trajectories.

Remark 2. A multiple-valued potential can always be made single-valued by enlarging the boundary of the multiply connected region occupied by the field in such a way as to make the region simply connected. The new boundary then excludes closed contours which cannot be shrunk to a point without leaving the region (recall Fig. 4.5, p. 144). Once the region has been made simply connected, it follows from curl $\mathbf{A} = 0$ that $\mathbf{A}$ is a potential field in the ordinary sense, i.e., $\mathbf{A} = \nabla\varphi$ where the potential φ is single-valued.

5.5. Solenoidal Fields

Suppose a vector field $\mathbf{A} = \mathbf{A}(\mathbf{r})$ is the curl of another vector field $\mathbf{\Phi} = \mathbf{\Phi}(\mathbf{r})$:

$$\mathbf{A} = \operatorname{curl} \mathbf{\Phi} = \nabla \times \mathbf{\Phi}.$$

Then $\mathbf{A}$ is said to be a *solenoidal field*, and $\mathbf{\Phi}$ is called the *vector potential* of the field $\mathbf{A}$. Clearly, the vector potential $\mathbf{\Phi}$ is defined only to within the gradient of an arbitrary function f, since $\mathbf{\Phi}' = \mathbf{\Phi} + \operatorname{grad} f$ implies

$$\operatorname{curl} \mathbf{\Phi}' = \operatorname{curl} \mathbf{\Phi} + \operatorname{curl} \operatorname{grad} f = \operatorname{curl} \mathbf{\Phi}$$

and hence

$$\mathbf{A} = \operatorname{curl} \mathbf{\Phi} = \operatorname{curl} \mathbf{\Phi}'.$$

In rectangular coordinates, a solenoidal field $\mathbf{A}$ has components

$$A_1 = \frac{\partial \Phi_3}{\partial x_2} - \frac{\partial \Phi_2}{\partial x_3}, \tag{5.80}$$

$$A_2 = \frac{\partial \Phi_1}{\partial x_3} - \frac{\partial \Phi_3}{\partial x_1}, \tag{5.81}$$

$$A_3 = \frac{\partial \Phi_2}{\partial x_1} - \frac{\partial \Phi_1}{\partial x_2}. \tag{5.82}$$

THEOREM 1. *A necessary and sufficient condition for a vector field* $\mathbf{A}$ *to be solenoidal is that the divergence of* $\mathbf{A}$ *vanish:*

$$\operatorname{div} \mathbf{A} = \frac{\partial A_1}{\partial x_1} + \frac{\partial A_2}{\partial x_2} + \frac{\partial A_3}{\partial x_3} = 0. \tag{5.83}$$

Proof. Suppose $\mathbf{A}$ is solenoidal, so that $\mathbf{A} = \operatorname{curl} \mathbf{\Phi}$. Then

$$\operatorname{div} \mathbf{A} = \operatorname{div} \operatorname{curl} \mathbf{\Phi} = \frac{\partial}{\partial x_1}\left(\frac{\partial \Phi_3}{\partial x_2} - \frac{\partial \Phi_2}{\partial x_3}\right) + \frac{\partial}{\partial x_2}\left(\frac{\partial \Phi_1}{\partial x_3} - \frac{\partial \Phi_3}{\partial x_1}\right) + \frac{\partial}{\partial x_3}\left(\frac{\partial \Phi_2}{\partial x_1} - \frac{\partial \Phi_1}{\partial x_2}\right) = 0,$$

where we assume that the components of $\mathbf{\Phi}$ have continuous first and second partial derivatives. This proves the necessity.

To prove the sufficiency, let Φ_1 be an arbitrary function of x_1:

$$\Phi_1 = f(x_1).$$

Then, integrating (5.82) and (5.81), we obtain

$$\Phi_2 = \int A_3 \, dx_1 + \varphi(x_2, x_3),$$

$$\Phi_3 = -\int A_2 \, dx_1 + \psi(x_2, x_3),$$

where the functions φ and ψ must satisfy a condition implied by (5.80) but are otherwise arbitrary. Substituting these expressions for Φ_2 and Φ_3 into (5.80), we find that

$$A_1 = -\int \frac{\partial A_2}{\partial x_2}\,dx_1 + \frac{\partial \psi}{\partial x_2} - \int \frac{\partial A_3}{\partial x_3}\,dx_1 - \frac{\partial \varphi}{\partial x_3}.$$

But (5.83) implies

$$A_1 = -\int \frac{\partial A_2}{\partial x_2}\,dx_1 - \int \frac{\partial A_3}{\partial x_3}\,dx_1.$$

It follows that

$$\frac{\partial \psi}{\partial x_2} - \frac{\partial \varphi}{\partial x_3} = 0. \tag{5.84}$$

In other words, the vector $\mathbf{\Phi}$ with components

$$f(x_1), \quad \int A_3\,dx_1 + \varphi(x_2, x_3), \quad -\int A_2\,dx_1 + \psi(x_2, x_3),$$

subject to the condition (5.84), is a vector potential for the field $\mathbf{A}$. Hence $\mathbf{A}$ is a solenoidal field, as asserted.

Example. Let $\mathbf{v} = \mathbf{v}(x_1, x_2)$ be the stationary velocity field of an incompressible fluid flowing in the x_1x_2-plane. The trajectories of the field $\mathbf{v}$, called *streamlines*, satisfy the differential equation

$$v_1\,dx_2 - v_2\,dx_1 = 0 \tag{5.85}$$

[cf. (4.33)]. Since the fluid is incompressible,

$$\operatorname{div} \mathbf{v} = \frac{\partial v_1}{\partial x_1} + \frac{\partial v_2}{\partial x_2} = 0,$$

and hence

$$\frac{\partial v_1}{\partial x_1} = -\frac{\partial v_2}{\partial x_2}.$$

But this is just the condition for the left-hand side of (5.85) to be the total differential of some function $\psi(x_1, x_2)$:

$$v_1\,dx_2 - v_2\,dx_1 = \frac{\partial \psi}{\partial x_1}\,dx_1 + \frac{\partial \psi}{\partial x_2}\,dx_2.$$

It follows that

$$v_1 = \frac{\partial \psi}{\partial x_2}, \qquad v_2 = -\frac{\partial \psi}{\partial x_1},$$

where the function ψ, called the *stream function*, is constant along the streamlines. Therefore $\mathbf{v}$ is a solenoidal field with vector potential

$$\mathbf{\Phi} = \mathbf{i}_3\psi.$$

In fact, we have

$$\operatorname{curl} \mathbf{\Phi} = \begin{vmatrix} \mathbf{i}_1 & \mathbf{i}_2 & \mathbf{i}_3 \\ \dfrac{\partial}{\partial x_1} & \dfrac{\partial}{\partial x_2} & \dfrac{\partial}{\partial x_3} \\ 0 & 0 & \psi \end{vmatrix} = \mathbf{i}_1 \frac{\partial \psi}{\partial x_2} - \mathbf{i}_2 \frac{\partial \psi}{\partial x_1} = \mathbf{v}.$$

Note also that

$$\mathbf{v} = \nabla\psi \times \mathbf{i}_3$$

and

$$\operatorname{curl} \mathbf{v} = -\mathbf{i}_3 \, \Delta\psi.$$

Given a closed contour L in a vector field $\mathbf{A}$, the surface formed by the trajectories of $\mathbf{A}$ going through L is called a *vector tube*, and the flux of $\mathbf{A}$ through any cross section of the tube is called the *intensity* of the tube. The last definition relies on

THEOREM 2. *The intensity of a vector tube of a solenoidal field* $\mathbf{A}$ *is constant along the whole tube.*

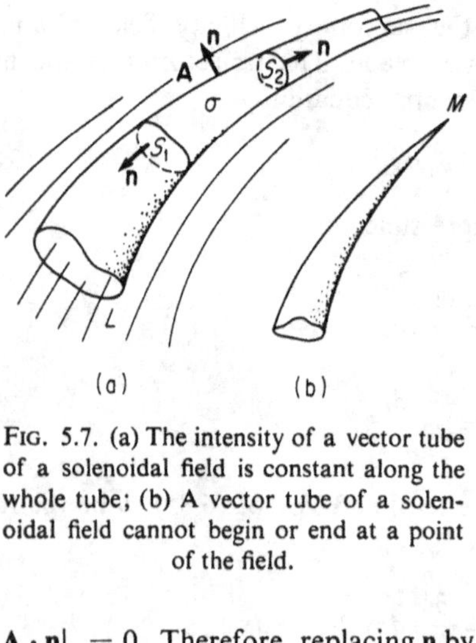

FIG. 5.7. (a) The intensity of a vector tube of a solenoidal field is constant along the whole tube; (b) A vector tube of a solenoidal field cannot begin or end at a point of the field.

Proof. Applying Gauss' theorem to the volume bounded by two cross sections S_1 and S_2 and by the surface σ of the tube between S_1 and S_2 [see Fig. 5.7(a)], we have

$$\iiint_V \operatorname{div} \mathbf{A} \, dV = \iint_\sigma \mathbf{A} \cdot \mathbf{n} \, d\sigma + \iint_{S_1} \mathbf{A} \cdot \mathbf{n} \, dS_1 + \iint_{S_2} \mathbf{A} \cdot \mathbf{n} \, dS_2.$$

The integral on the left vanishes, since $\operatorname{div} \mathbf{A} = 0$. The first integral on the right also vanishes, since $\mathbf{A}$ is perpendicular to $\mathbf{n}$ on the surface of the tube and hence $\mathbf{A} \cdot \mathbf{n}|_\sigma = 0$. Therefore, replacing $\mathbf{n}$ by $-\mathbf{n}$ in the second integral (so that the flux through S_1 is calculated in the same direction as that through S_2), we find that

$$\iint_{S_1} \mathbf{A} \cdot \mathbf{n} \, dS_1 = \iint_{S_2} \mathbf{A} \cdot \mathbf{n} \, dS_2,$$

as asserted.

COROLLARY. *The vector tubes of a solenoidal field can neither begin nor end inside the field. Hence there are only three possibilities:*

1) *The tubes are closed;*
2) *They terminate on the boundary of the region occupied by the field;*
3) *They go off to infinity (if the field is unbounded).*

Proof. Suppose a vector tube terminates at a point M inside the field, as shown in Fig. 5.7(b). Then $|\mathbf{A}|$ must become infinite at M, since the cross section of the tube vanishes at M although its intensity remains constant. But this is impossible, since $\mathbf{A}$ is continuous at every point of the field, as assumed on p. 135. Moreover, the tubes cannot terminate in a cross section of finite area S_0 inside the field, since this would again contradict the continuity of $\mathbf{A}$.

5.6. Laplacian Fields

A vector field $\mathbf{A} = \mathbf{A}(\mathbf{r})$ is said to be *Laplacian* if

$$\operatorname{curl} \mathbf{A} = 0, \quad \operatorname{div} \mathbf{A} = 0$$

at every point of the field. Thus a Laplacian field is both potential and solenoidal. In a simply connected region a Laplacian field is completely determined by a scalar potential φ satisfying *Laplace's equation*

$$\Delta\varphi = 0.$$

In fact, $\operatorname{curl} \mathbf{A} = 0$ implies $\mathbf{A} = \nabla\varphi$ in a simply connected region, and then $0 = \operatorname{div} \mathbf{A} = \operatorname{div} \nabla\varphi = \Delta\varphi$.

5.6.1. Harmonic functions. A continuous function φ with continuous first and second partial derivatives is said to be *harmonic* if it satisfies Laplace's equation $\Delta\varphi = 0$.

Example 1. The functions

$$c, \quad ax_1 + bx_2 + cx_3, \quad x_1^2 - x_3^2, \quad x_1x_2, \quad x_1x_2x_3$$

are all harmonic, and so is

$$(x_1^2 + x_2^2)^{k/2}\left\{a \sin\left(k \arc\tan \frac{x_2}{x_1}\right) + b \cos\left(k \arc\tan \frac{x_2}{x_1}\right)\right\}.$$

Example 2. The function

$$\frac{1}{r} = \frac{1}{\sqrt{x_1^2 + x_2^2 + x_3^2}}$$

is harmonic everywhere except at the origin ($r = 0$). In fact,

$$\frac{\partial}{\partial x_i}\left(\frac{1}{r}\right) = -\frac{x_i}{r^3}, \qquad \frac{\partial^2}{\partial x_i^2}\left(\frac{1}{r}\right) = \frac{3x_i^2 - r^2}{r^5},$$

and hence

$$\Delta\left(\frac{1}{r}\right) = \frac{\partial^2}{\partial x_1^2}\left(\frac{1}{r}\right) + \frac{\partial^2}{\partial x_2^2}\left(\frac{1}{r}\right) + \frac{\partial^2}{\partial x_3^2}\left(\frac{1}{r}\right) = \frac{3x_1^2 + 3x_2^2 + 3x_3^2 - 3r^2}{r^5} = 0.$$

Example 3. The function

$$\ln\sqrt{x_1^2 + x_2^2}$$

(but not $\ln\sqrt{x_1^2 + x_2^2 + x_3^2}$) is also harmonic everywhere except at the origin.

We now prove a number of important properties of harmonic functions:

1) *The integral of the normal derivative of a harmonic function ψ over a closed surface S vanishes if the function is harmonic everywhere in the volume V bounded by S.* This follows at once from the identity (5.46), which reduces to

$$\iint_S \frac{\partial\psi}{\partial n}\, dS = 0$$

if ψ is harmonic ($\Delta\psi = 0$) in V.

2) It follows from (5.45) that *the functions φ and ψ satisfy the relation*

$$\iint_S \varphi\frac{\partial\psi}{\partial n}\, dS = \iint_S \psi\frac{\partial\varphi}{\partial n}\, dS$$

on a closed surface S if the functions are harmonic in the volume V bounded by S.

3) *A function φ harmonic in a volume V can be expressed in terms of the function and its normal derivative on the surface S bounding V.* In fact, if φ is harmonic ($\Delta\varphi = 0$), formula (5.47) reduces to

$$\varphi(M_0) = -\frac{1}{4\pi}\iint_S \left[\varphi\frac{\partial}{\partial n}\left(\frac{1}{r}\right) - \frac{1}{r}\frac{\partial\varphi}{\partial n}\right] dS. \tag{5.86}$$

4) If the surface S in (5.86) is a sphere S_ρ of radius ρ with center at the point M_0, then

$$\varphi(M_0) = \frac{1}{4\pi\rho^2}\iint_{S_\rho} \varphi\, dS_\rho, \tag{5.87}$$

since

$$\iint_{S_\rho} \frac{1}{r}\frac{\partial\varphi}{\partial n}\, dS = \frac{1}{\rho}\iint_{S_\rho} \frac{\partial\varphi}{\partial n}\, dS_\rho = 0$$

by Property 1 and

$$\frac{\partial}{\partial n}\left(\frac{1}{r}\right)\bigg|_{S_\rho} = \frac{\partial}{\partial n}\left(\frac{1}{r}\right)\bigg|_{r=\rho} = -\frac{1}{\rho^2}.$$

According to (5.87), *the value of a harmonic function* φ *at a point* M_0 *is the average of its values on any sphere with center* M_0 *lying in the region where* φ *is harmonic.*

5) *A nonconstant harmonic function* φ *harmonic in a region* R *can have neither a maximum nor a minimum in* R. In fact, suppose $\varphi \not\equiv$ const has a maximum at a point M_0 in R. Then, by the continuity of φ, there is a sphere S_ρ of radius ρ and center M_0 lying entirely in R such that $\varphi(M) \leqslant \varphi(M_0)$ for every point of S_ρ and $\varphi(M) < \varphi(M_0)$ for some points of S_ρ. But then

$$\varphi(M_0) = \frac{1}{4\pi\rho^2}\iint_{S_\rho} \varphi\, dM < \varphi(M_0),$$

which is impossible. Since $-\varphi$ is also harmonic in R, $-\varphi$ cannot have a maximum in R, i.e., φ cannot have a minimum in R.

6) *If* φ *is harmonic in a volume* V *bounded by a surface* S *and if* $\varphi = c =$ const *at every point of* S, *then* $\varphi = c$ *at every point of* V. In fact, setting $\varphi = c$ in the right-hand side of (5.86), we obtain

$$\varphi(M_0) = \frac{c}{4\pi}\iint_S \frac{dS}{r^2},$$

where M_0 is any interior point of V. But

$$\frac{1}{4\pi}\iint_S \frac{dS}{r^2} = 1,$$

as we see at once by substituting the harmonic function $\varphi \equiv 1$ in both sides of (5.86). It follows that $\varphi(M_0) = c$ and hence $\varphi \equiv c$ in V, since M_0 is arbitrary.

7) *Laplace's equation* $\Delta\varphi = 0$ *has a unique solution in a volume* V *taking given values on the surface* S *bounding* V. To see this, let φ_1 and φ_2 be two harmonic functions taking the same values on S. Then $\varphi = \varphi_1 - \varphi_2$ is also harmonic (by the linearity of Laplace's equation) and vanishes identically on S. But then, by Property 6, $\varphi \equiv 0$ in V and hence $\varphi_1 \equiv \varphi_2$ in V.

8) *If* φ *is harmonic in a volume* V *bounded by a surface* S *and if* $\partial\varphi/\partial n = 0$ *at every point of* S, *then* $\varphi =$ const *at every point of* V. In fact, setting $\varphi = \psi$ in (5.44), we obtain

$$\iiint_V [\varphi\,\Delta\varphi + (\nabla\varphi)^2]\, dV = \iint_S \varphi \frac{\partial\varphi}{\partial n}\, dS,$$

which reduces to

$$\iiint_V (\nabla\varphi)^2\, dV = 0,$$

since

$$\Delta\varphi \equiv 0 \text{ in } V, \qquad \frac{\partial\varphi}{\partial n} \equiv 0 \text{ on } S.$$

It follows that $\nabla\varphi \equiv 0$ in V and hence $\varphi \equiv \text{const}$ in V.

9) *If φ_1 and φ_2 are two solutions of Laplace's equation in a volume V whose normal derivatives take the same value on the surface S bounding V, then φ_1 and φ_2 can differ only by a constant.* In fact, $\varphi = \varphi_1 - \varphi_2$ satisfies Laplace's equation in V, while

$$\frac{\partial\varphi}{\partial n} = \frac{\partial\varphi_1}{\partial n} - \frac{\partial\varphi_2}{\partial n}$$

vanishes identically on S. It follows from Property 8 that $\varphi \equiv \text{const}$ in V, i.e., $\varphi_1 \equiv \varphi_2 + \text{const}$ in V.

5.6.2. The Dirichlet and Neumann problems. The problem of solving Laplace's equation

$$\Delta\varphi = 0$$

in a volume V, subject to the boundary condition

$$\varphi|_S = f(x_1, x_2, x_3)$$

on the surface S bounding V, is called the *Dirichlet problem*. According to Property 7 above, such a function φ, if it exists, is unique. On the other hand, the problem of solving Laplace's equation subject to the boundary condition

$$\left.\frac{\partial\varphi}{\partial n}\right|_S = f(x_1, x_2, x_3)$$

is called the *Neumann problem*. According to Property 9, such a function φ, if it exists, is determined only to within an additive constant.

Example. Consider a stationary velocity field $\mathbf{v}$ describing an irrotational flow of an incompressible fluid. Since curl $\mathbf{v} = 0$, we have

$$\mathbf{v} = \nabla\varphi,$$

where the *velocity potential* φ is harmonic since

$$\text{div}\,\mathbf{v} = \text{div}\,(\nabla\varphi) = \Delta\varphi = 0.$$

The normal component of $\mathbf{v}$ must vanish on the surface S of any solid body immersed in the flow, i.e.,

$$v = \mathbf{v} \cdot \mathbf{n} = \mathbf{n} \cdot \nabla\varphi = \frac{\partial\varphi}{\partial n} = 0 \text{ on } S.$$

In the case of two-dimensional flow, we have a vector potential

$$\mathbf{\Phi} = \mathbf{i}_3\psi,$$

where ψ is the stream function and

$$v_1 = \frac{\partial\psi}{\partial x_2}, \qquad v_2 = -\frac{\partial\psi}{\partial x_1}$$

(see the example on pp. 217–218). Since curl $\mathbf{v} = 0$, it follows that

$$\Delta\psi = 0.$$

The function ψ must take a constant value on any rigid contour, since any such contour must be a streamline. Thus the velocity potential φ is the solution of a Neumann problem, while the stream function ψ is the solution of a Dirichlet problem.[15]

5.7. The Fundamental Theorem of Vector Analysis

We now prove the following key

THEOREM (*Fundamental theorem of vector analysis*). *Let* $\mathbf{A} = \mathbf{A}(\mathbf{r})$ *be a continuous vector field with continuous divergence and* curl, *such that* $|\mathbf{A}|$ *falls off at infinity like* $1/r^{1+\varepsilon}$ *while* $|\text{div } \mathbf{A}|$ *and* $|\text{curl } \mathbf{A}|$ *fall off at infinity like* $1/r^{2+\varepsilon}$ *where* $\varepsilon > 0$. *Then* $\mathbf{A}$ *has a unique representation (to within constant vectors) as a sum of a potential field* $\mathbf{A}_1 = \mathbf{A}_1(\mathbf{r})$ *and a solenoidal field* $\mathbf{A}_2 = \mathbf{A}_2(r)$, *i.e.*,

$$\mathbf{A} = \mathbf{A}_1 + \mathbf{A}_2,$$

where

$$\text{curl } \mathbf{A}_1 = 0, \qquad \text{div } \mathbf{A}_2 = 0. \tag{5.88}$$

Proof. First we construct $\mathbf{A}_1$. It follows from (5.88) that

$$\text{div } \mathbf{A} = \text{div } \mathbf{A}_1, \qquad \text{curl } \mathbf{A} = \text{curl } \mathbf{A}_2, \tag{5.89}$$

so that

$$\text{curl } \mathbf{A}_1 = 0, \qquad \text{div } \mathbf{A}_1 = \text{div } \mathbf{A}. \tag{5.90}$$

The first of these equations implies

$$\mathbf{A}_1 = \nabla\varphi + \mathbf{c}_1,$$

[15] In the two-dimensional case, these problems are often solved by the method of conformal mapping.

where $\mathbf{c}_1$ is a constant vector and φ is a single-valued potential (we assume that the region occupied by $\mathbf{A}$ and hence by $\mathbf{A}_1$ is simply connected). The second of the equations (5.90) then gives

$$\operatorname{div}(\nabla\varphi + \mathbf{c}_1) = \operatorname{div}\mathbf{A},$$

or[16]

$$\Delta\varphi = \operatorname{div}\mathbf{A}. \tag{5.91}$$

Thus the constant vector $\mathbf{c}_1$ plays no role in determining the potential φ. To solve (5.91) for φ, we use formula (5.47), where V is all of space (in the limit). The functions φ and $\partial\varphi/\partial n$ both approach zero at infinity, with $\partial\varphi/\partial n$ going to zero faster than $1/r$. It follows that the surface integral

$$\iint_S \left[\varphi\frac{\partial}{\partial n}\left(\frac{1}{r}\right) - \frac{1}{r}\frac{\partial\varphi}{\partial n}\right] dS$$

vanishes in the limit as V becomes all of space, so that (5.47) reduces to

$$\varphi = -\frac{1}{4\pi}\iiint_V \frac{\operatorname{div}\mathbf{A}}{r}\, dV$$

(here div $\mathbf{A}$ is regarded as known). It follows that

$$\mathbf{A}_1 = \nabla\varphi + \mathbf{c}_1 = -\frac{1}{4\pi}\operatorname{grad}\iiint_V \frac{\operatorname{div}\mathbf{A}}{r}\, dV + \mathbf{c}_1. \tag{5.92}$$

Next we construct $\mathbf{A}_2$. According to (5.88) and (5.89),

$$\operatorname{div}\mathbf{A}_2 = 0, \quad \operatorname{curl}\mathbf{A}_2 = \operatorname{curl}\mathbf{A}. \tag{5.93}$$

The first of these equations

$$\mathbf{A}_2 = \operatorname{curl}\mathbf{\Phi} + \mathbf{c}_2, \tag{5.94}$$

where $\mathbf{c}_2$ is a constant vector and $\mathbf{\Phi}$ is a vector potential. Since $\mathbf{\Phi}$ is determined only to within the gradient of an arbitrary function f (recall p. 216), $\mathbf{\Phi}$ can be subjected to the extra condition[17]

$$\operatorname{div}\mathbf{\Phi} = 0. \tag{5.95}$$

Substituting (5.94) into the second of the equations (5.93), we obtain

$$\operatorname{curl}(\operatorname{curl}\mathbf{\Phi} + \mathbf{c}_2) = \operatorname{curl}\mathbf{A}$$

---

[16] An equation like (5.91), of the form $\Delta\varphi = f(\mathbf{r})$, is called *Poisson's equation.* If $f(\mathbf{r}) \equiv 0$, Poisson's equation reduces to Laplace's equation.

[17] In fact, if div $\mathbf{\Phi} = \chi \neq 0$, we need only replace $\mathbf{\Phi}$ by $\mathbf{\Phi}$ + grad f, where f satisfies the Poisson equation $\Delta f = -\chi$, since then

$$\operatorname{div}(\mathbf{\Phi} + \operatorname{grad} f) = \operatorname{div}\mathbf{\Phi} + \Delta f = \chi - \chi = 0.$$

(the constant vector $\mathbf{c}_2$ plays no role in determining the vector potential $\mathbf{\Phi}$). Thus

$$\text{curl curl } \mathbf{\Phi} = \text{curl } \mathbf{A},$$

which implies

$$\Delta\mathbf{\Phi} = -\text{curl } \mathbf{A} \tag{5.96}$$

because of (4.96) and (5.96). Just as we used (5.47) to solve (5.91) for φ, we can use (5.51) to solve for $\mathbf{\Phi}$, obtaining[18]

$$\mathbf{\Phi} = \frac{1}{4\pi}\iiint_V \frac{\text{curl } \mathbf{A}}{r}\, dV,$$

and hence

$$\mathbf{A}_2 = \frac{1}{4\pi}\text{curl}\iiint_V \frac{\text{curl } \mathbf{A}}{r}\, dV + \mathbf{c}_2. \tag{5.97}$$

Combining (5.92) and (5.97), we finally have the representation

$$\mathbf{A} = \mathbf{A}_1 + \mathbf{A}_2 = -\frac{1}{4\pi}\text{grad}\iiint_V \frac{\text{div } \mathbf{A}}{r}\, dV + \frac{1}{4\pi}\text{curl}\iiint_V \frac{\text{curl } \mathbf{A}}{r}\, dV, \tag{5.98}$$

where the constants $\mathbf{c}_1$ and $\mathbf{c}_2$ have been dropped since $\mathbf{A}$ is assumed to vanish at infinity. By construction, $\mathbf{A}_1$ is a potential field (curl $\mathbf{A}_1 = 0$), while $\mathbf{A}_2$ is a solenoidal field (div $\mathbf{A}_2 = 0$).

We must still prove the uniqueness of the representation (5.98). Suppose $\mathbf{A}$ has another representation $\mathbf{A} = \mathbf{A}_1' + \mathbf{A}_2'$ where curl $\mathbf{A}_1' = 0$, div $\mathbf{A}_2' = 0$. Then, since div $\mathbf{A}$ = div $\mathbf{A}_1$ = div $\mathbf{A}_1'$, we have

$$\text{div } (\mathbf{A}_1 - \mathbf{A}_1') = 0, \qquad \text{curl } (\mathbf{A}_1 - \mathbf{A}_1') = 0.$$

Similarly curl $\mathbf{A}$ = curl $\mathbf{A}_2$ = curl $\mathbf{A}_2'$, and hence

$$\text{div } (\mathbf{A}_2 - \mathbf{A}_2') = 0, \qquad \text{curl } (\mathbf{A}_2 - \mathbf{A}_2') = 0.$$

But then the same argument leading from (5.90) and (5.93) to (5.92) and (5.97) shows that

$$\mathbf{A}_1 - \mathbf{A}_1' = \mathbf{c}_1, \qquad \mathbf{A}_2 - \mathbf{A}_2' = \mathbf{c}_2,$$

where $\mathbf{c}_1$ and $\mathbf{c}_2$ are constant vectors. Thus the representation $\mathbf{A} = \mathbf{A}_1 + \mathbf{A}_2$ is unique to within constant vectors, and the proof of the theorem is complete.

---

[18] The surface integral vanishes for the same reason as before.

Remark. Roughly speaking, the sources and sinks of the given field **A** all appear in $\mathbf{A}_1$ (since div $\mathbf{A} =$ div $\mathbf{A}_1$), while all the "vorticity" of **A** appears in $\mathbf{A}_2$ (since curl $\mathbf{A} =$ curl $\mathbf{A}_2$).

5.8. Applications to Electromagnetic Theory

5.8.1. Maxwell's equations. Just as nonrelativistic mechanical phenomena are described by Newton's equations, electromagnetic phenomena are described by a set of four equations called *Maxwell's equations.* An electromagnetic field is characterized by two vector fields (both time-dependent in general), the *electric field* $\mathbf{E} = \mathbf{E}(\mathbf{r}, t)$ and the *magnetic field* $\mathbf{H} = \mathbf{H}(\mathbf{r}, t)$. The charges and currents producing the electromagnetic field are themselves characterized by two fields (one scalar, the other vector), the *charge density* $\rho = \rho(\mathbf{r}, t)$ and the *current density* $\mathbf{j} = \mathbf{j}(\mathbf{r}, t)$.[19]

We begin by considering electromagnetic phenomena *in vacuum.* Then, regardless of the charge and current distributions, it is an experimental fact that the electric and magnetic fields are related by *Faraday's law of induction*

$$\frac{1}{c}\frac{\partial}{\partial t}\iint_S \mathbf{H}\cdot\mathbf{n}\,dS = -\oint_L \mathbf{E}\cdot d\mathbf{r}, \tag{5.99}$$

where $c = 3 \times 10^{10}$ cm/sec is the velocity of light and S is any surface bounded by a closed contour L. According to (5.99), the rate of change of the flux of the magnetic field through a surface "supported" by a contour L equals minus the *electromotive force* around L, i.e., the circulation of the electric field around L. Moreover, the magnetic field always satisfies the relation

$$\iint_\Sigma \mathbf{H}\cdot\mathbf{n}\,d\Sigma = 0, \tag{5.100}$$

where Σ is an arbitrary *closed* surface. In other words, the flux of the magnetic field through any closed surface vanishes.

Applying Stokes' theorem (5.39) to (5.99), we obtain

$$\frac{1}{c}\frac{\partial}{\partial t}\iint_S \mathbf{H}\cdot\mathbf{n}\,dS = -\iint_S \mathbf{n}\cdot\text{curl}\,\mathbf{E}\,dS$$

or

$$\iint_S \left(\frac{1}{c}\frac{\partial \mathbf{H}}{\partial t} + \text{curl}\,\mathbf{E}\right)\cdot\mathbf{n}\,dS = 0.$$

---

[19] We will use Gaussian units, in which electric quantities (like **E** and ρ) are measured in electrostatic units, while magnetic quantities (like **H** and **j**) are measured in electromagnetic units.

But S is arbitrary and hence

$$\frac{\partial \mathbf{H}}{\partial t} = -c \operatorname{curl} \mathbf{E}. \tag{5.101}$$

Similarly, applying Gauss' theorem (5.31) to (5.100), we find that

$$\iiint_V \operatorname{div} \mathbf{H}\, dV = 0,$$

where V is the volume bounded by Σ, and hence

$$\operatorname{div} \mathbf{H} = 0, \tag{5.102}$$

since V is arbitrary. Equations (5.101) and (5.102) are two of the four Maxwell equations (the homogeneous pair). They are the differential forms of equations (5.99) and (5.100).[20]

The relation between the fields $\mathbf{E}$ and $\mathbf{H}$ and the charge and current densities ρ and $\mathbf{j}$ is given by the following two experimental laws:

$$\frac{\partial}{\partial t}\iint_S \mathbf{E} \cdot \mathbf{n}\, dS = c\oint_L \mathbf{H} \cdot d\mathbf{r} - 4\pi\iint_S \mathbf{j} \cdot \mathbf{n}\, dS, \tag{5.103}$$

$$\iint_\Sigma \mathbf{E} \cdot \mathbf{n}\, d\Sigma = 4\pi\iiint_V \rho\, dV. \tag{5.104}$$

Here S is again any closed surface bounded by a closed contour L, and Σ is any closed surface bounding a volume V. According to (5.103), the rate of change of the flux of the electric field through a surface S "supported" by a contour L equals c times the *magnetomotive force* (i.e., the circulation of the magnetic field around L) minus 4π times the flux of the current density through S. Similarly, (5.104) states that the flux of the electric field through any closed surface equals 4π times the charge enclosed by the surface.

To get the other two Maxwell equations, we apply Stokes' theorem to (5.103) and Gauss' theorem to (5.104), obtaining

$$\frac{\partial}{\partial t}\iint_S \mathbf{E} \cdot \mathbf{n}\, dS = c\iint_S \mathbf{n} \cdot \operatorname{curl} \mathbf{H}\, dS - 4\pi\iint_S \mathbf{j} \cdot \mathbf{n}\, dS$$

and

$$\iiint_V \operatorname{div} \mathbf{E}\, dV = \iiint_V \rho\, dV.$$

It follows that

$$\frac{\partial \mathbf{E}}{\partial t} = c \operatorname{curl} \mathbf{H} - 4\pi\mathbf{j}, \tag{5.105}$$

$$\operatorname{div} \mathbf{E} = 4\pi\rho, \tag{5.106}$$

[20] Alternatively, we might have regarded (5.101) and (5.102) as given and then deduced (5.99) and (5.100), called Maxwell's equations in integral form.

since S and V are arbitrary. Equations (5.105) and (5.106) are the remaining two Maxwell equations (the inhomogeneous pair).

So far we have only considered electromagnetic phenomena in vacuum. In the presence of a medium with dielectric and magnetic properties, two more vector fields are needed to complete the description of the electromagnetic field, the *polarization* $\mathbf{P} = \mathbf{P}(\mathbf{r}, t)$, equal to the electric moment per unit volume, and the *magnetization* $\mathbf{M} = \mathbf{M}(\mathbf{r}, t)$, equal to the magnetic moment per unit volume. Alternatively, the effect of polarization and magnetization can be taken into account by defining two new vector fields, the *electric displacement*

$$\mathbf{D} = \mathbf{E} + 4\pi\mathbf{P}$$

and the *magnetic induction*

$$\mathbf{B} = \mathbf{H} + 4\pi\mathbf{M}.$$

It then turns out that Maxwell's equations (5.101)–(5.102), (5.105)–(5.106) in empty space (vacuum) are replaced by

$$\frac{\partial \mathbf{B}}{\partial t} = -c \operatorname{curl} \mathbf{E}, \tag{5.107}$$

$$\operatorname{div} \mathbf{B} = 0, \tag{5.108}$$

$$\frac{\partial \mathbf{D}}{\partial t} = c \operatorname{curl} \mathbf{H} - 4\pi\mathbf{j}, \tag{5.109}$$

$$\operatorname{div} \mathbf{D} = 4\pi\rho \tag{5.110}$$

in a medium with dielectric and magnetic properties.

Remark. The relations between the various fields $\mathbf{E}$, $\mathbf{D}$, $\mathbf{B}$, $\mathbf{H}$ and $\mathbf{j}$ that hold in media other than empty space are called *constitutive relations*. For example,

$$\mathbf{D} = \varepsilon\mathbf{E}, \quad \mathbf{B} = \mu\mathbf{H}$$

in a medium of dielectric constant ε and magnetic permeability μ, while in a medium of conductivity σ we have *Ohm's law*

$$\mathbf{j} = \sigma\mathbf{E}.$$

Constitutive relations of a more complicated kind are often encountered, in particular in anisotropic media where, for example, ε may be a tensor (recall Example 2, p. 110). To solve an electromagnetic problem involving media other than empty space, Maxwell's equations (5.107)–(5.110) must be supplemented by the relevant constitutive relations.

5.8.2. The scalar and vector potentials. In electromagnetic theory, a very important role is played by two auxiliary functions, the *scalar potential* $\varphi = \varphi(\mathbf{r}, t)$ and the *vector potential* $\mathbf{A} = \mathbf{A}(\mathbf{r}, t)$. Confining ourselves to the case of empty space (so that $\mathbf{D} = \mathbf{E}$, $\mathbf{B} = \mathbf{H}$), we begin by considering the homogeneous Maxwell equations

$$\frac{\partial \mathbf{H}}{\partial t} + c \operatorname{curl} \mathbf{E} = 0, \tag{5.111}$$

$$\operatorname{div} \mathbf{H} = 0. \tag{5.112}$$

Since the divergence of the curl of any vector vanishes, we can satisfy (5.112) by setting

$$\mathbf{H} = \operatorname{curl} \mathbf{A}. \tag{5.113}$$

Then, substituting (5.113) into (5.111), we obtain

$$\operatorname{curl}\left(\frac{1}{c}\frac{\partial \mathbf{A}}{\partial t} + \mathbf{E}\right) = 0. \tag{5.114}$$

Since the curl of the gradient of any function vanishes, we can satisfy (5.114) by setting

$$\mathbf{E} + \frac{1}{c}\frac{\partial \mathbf{A}}{\partial t} = -\nabla\varphi$$

or

$$\mathbf{E} = -\frac{1}{c}\frac{\partial \mathbf{A}}{\partial t} - \nabla\varphi, \tag{5.115}$$

where φ is an arbitrary function of $\mathbf{r}$ and t. It should be noted that replacing $\mathbf{A}$ by

$$\mathbf{A} + \nabla f \tag{5.116}$$

and φ by

$$\varphi - \frac{1}{c}\frac{\partial f}{\partial t}, \tag{5.117}$$

where f is an arbitrary function of $\mathbf{r}$ and t, has no effect on the fields $\mathbf{E}$ and $\mathbf{H}$, since

$$\begin{aligned} -\frac{1}{c}\frac{\partial(\mathbf{A} + \nabla f)}{\partial t} - \nabla\left(\varphi - \frac{1}{c}\frac{\partial f}{\partial t}\right) &= -\frac{1}{c}\frac{\partial \mathbf{A}}{\partial t} - \frac{1}{c}\nabla\frac{\partial f}{\partial t} - \nabla\varphi + \frac{1}{c}\nabla\frac{\partial f}{\partial t} \\ &= -\frac{1}{c}\frac{\partial \mathbf{A}}{\partial t} - \nabla\varphi = \mathbf{E}, \end{aligned}$$

while

$$\operatorname{curl}(\mathbf{A} + \nabla f) = \operatorname{curl}\mathbf{A} + \operatorname{curl}\operatorname{grad} f = \operatorname{curl}\mathbf{A} = \mathbf{H}.$$

Next we substitute (5.113) and (5.115) into the inhomogeneous Maxwell equations

$$\frac{\partial \mathbf{E}}{\partial t} = c \operatorname{curl} \mathbf{H} - 4\pi\mathbf{j},$$

$$\operatorname{div} \mathbf{E} = 4\pi\rho.$$

This gives the following equations satisfied by the scalar and vector potentials:

$$\frac{\partial}{\partial t}\left(-\frac{1}{c}\frac{\partial \mathbf{A}}{\partial t} - \nabla\varphi\right) = c \operatorname{curl} \operatorname{curl} \mathbf{A} - 4\pi\mathbf{j}, \tag{5.118}$$

$$\operatorname{div}\left(-\frac{1}{c}\frac{\partial \mathbf{A}}{\partial t} - \nabla\varphi\right) = 4\pi\rho. \tag{5.119}$$

To simplify (5.118) and (5.119) further, we impose the *Lorentz condition*

$$\operatorname{div} \mathbf{A} + \frac{1}{c}\frac{\partial\varphi}{\partial t} = 0$$

on the potentials **A** and φ. This entails no loss of generality, since we need only choose the arbitrary function f in (5.116) and (5.117) to satisfy the equation

$$\operatorname{div}(\mathbf{A} + \nabla f) + \frac{1}{c}\frac{\partial}{\partial t}\left(\varphi - \frac{1}{c}\frac{\partial f}{\partial t}\right) = 0$$

or[21]

$$\Delta f - \frac{1}{c^2}\frac{\partial^2 f}{\partial t^2} = -\operatorname{div} \mathbf{A} - \frac{1}{c}\frac{\partial\varphi}{\partial t}. \tag{5.120}$$

Using (4.96) and the Lorentz condition, we transform (5.118) and (5.119) into

$$\Delta\mathbf{A} - \frac{1}{c^2}\frac{\partial^2\mathbf{A}}{\partial t^2} = -\frac{4\pi}{c}\mathbf{j}, \tag{5.121}$$

and

$$\Delta\varphi - \frac{1}{c^2}\frac{\partial^2\varphi}{\partial t^2} = -4\pi\rho. \tag{5.122}$$

An equation of the form (5.120)–(5.122) is called an *inhomogeneous wave equation*. In terms of the *D'Alembertian operator*

$$\Box \equiv \Delta - \frac{1}{c^2}\frac{\partial^2}{\partial t^2},$$

(5.121) and (5.122) take the form

$$\Box\mathbf{A} = -\frac{4\pi}{c}\mathbf{j}, \tag{5.123}$$

$$\Box\varphi = -4\pi\varphi.$$

5.8.3. Energy of the electromagnetic field. Poynting's vector. First we consider the case of empty space. Adding the scalar product of (5.101) with

---

[21] The solution of equations (5.120)–(5.122) is discussed in Prob. 10, p. 242.

H to the scalar product of (5.105) with **E**, we obtain

$$\mathbf{E}\cdot\frac{\partial\mathbf{E}}{\partial t}+\mathbf{H}\cdot\frac{\partial\mathbf{H}}{\partial t}=c\mathbf{E}\cdot\operatorname{curl}\mathbf{H}-4\pi\mathbf{j}\cdot\mathbf{E}-c\mathbf{H}\cdot\operatorname{curl}\mathbf{E}. \tag{5.124}$$

Noting that

$$\operatorname{div}(\mathbf{E}\times\mathbf{H})=\mathbf{H}\cdot\operatorname{curl}\mathbf{E}-\mathbf{E}\cdot\operatorname{curl}\mathbf{H},$$

we can write (5.124) in the form

$$\frac{\partial}{\partial t}\frac{E^2+H^2}{8\pi}=-\operatorname{div}\left[\frac{c}{4\pi}(\mathbf{E}\times\mathbf{H})\right]-\mathbf{j}\cdot\mathbf{E}.$$

After integrating over an arbitrary volume V, this becomes

$$\frac{\partial}{\partial t}\iiint_V\frac{E^2+H^2}{8\pi}\,dV=-\iiint_V\operatorname{div}\mathbf{P}\,dV-\iiint_V\mathbf{j}\cdot\mathbf{E}\,dV, \tag{5.125}$$

where

$$\mathbf{P}=\frac{c}{4\pi}(\mathbf{E}\times\mathbf{H}). \tag{5.126}$$

Using Gauss' theorem, we can write (5.125) in the form

$$\frac{\partial}{\partial t}\iiint_V\frac{E^2+H^2}{8\pi}\,dV=-\iint_S\mathbf{P}\cdot\mathbf{n}\,dS-\iiint_V\mathbf{j}\cdot\mathbf{E}\,dV, \tag{5.127}$$

where S is the surface bounding the volume V. If $\mathbf{P}=0$ and $\mathbf{j}=0$, the integral in the left-hand side of (5.127) is a constant. It represents the energy of the electromagnetic field, distributed with density[22]

$$\frac{E^2+H^2}{8\pi}. \tag{5.128}$$

In the general case where $\mathbf{P}\neq 0$ and $\mathbf{j}\neq 0$, (5.127) states that the rate of change of the energy of the electromagnetic field inside V equals the flux of the vector **P** across S *into* V minus the rate at which work is done by the electric field on the moving charges inside V.[23] Thus if conservation of energy

---

[22] See e.g., J. B. Marion, *Classical Electromagnetic Radiation*, Academic Press, Inc., New York (1965), p. 118.

[23] If $\mathbf{v}=\mathbf{v}(\mathbf{r},t)$ is the velocity field of the moving charges, then $\mathbf{j}=\rho\mathbf{v}$ and

$$\iiint_V\mathbf{j}\cdot\mathbf{E}\,dV=\iiint_V\rho\mathbf{v}\cdot\mathbf{E}\,dV.$$

The quantity $\rho\mathbf{E}\,dV$ is the force exerted by the electric field **E** on the charge element $\rho\,dV$. Hence the scalar product of **v** and $\rho\mathbf{E}\,dV$ is the rate at which **E** does work on the charge element, and the rate at which **E** does work on all the moving charges in V is given by

$$\iiint_V\rho\mathbf{v}\cdot\mathbf{E}\,dV.$$

This is also the rate at which energy is dissipated in Joule heat (J. B. Marion, *op. cit.*, p. 119).

is to hold, the vector **P**, called *Poynting's vector*, must represent the density of flow of energy in the electromagnetic field (attributable to the phenomenon of radiation).

In the case of a medium of dielectric constant ε, magnetic permeability μ and conductivity σ, we have

$$\mathbf{D} = \varepsilon \mathbf{E}, \quad \mathbf{B} = \mu \mathbf{H}, \quad \mathbf{j} = \sigma \mathbf{E},$$

and (5.127) is replaced by

$$\frac{\partial}{\partial t} \iiint\limits_V \frac{\varepsilon E^2 + \mu H^2}{8\pi} \, dV = -\iint\limits_S \mathbf{P} \cdot \mathbf{n} \, dS - \iiint\limits_V \sigma E^2 \, dV.$$

The energy density of the electromagnetic field is now

$$\frac{\varepsilon E^2 + \mu H^2}{8\pi},$$

which reduces to (5.128) is $\varepsilon = 1$, $\mu = 1$, i.e., in vacuum. The flux of electromagnetic energy into V is given by the same integral as before, involving the Poynting vector (5.126), and the last term again represents the rate at which energy is dissipated in Joule heat.

SOLVED PROBLEMS

Problem 1. Prove that if ψ is a harmonic function ($\Delta\psi = 0$), then the vector $\mathbf{r}\psi$ satisfies the *biharmonic equation*

$$\Delta\Delta(\mathbf{r}\psi) = \frac{\partial^2}{\partial x_i \, \partial x_i} \Delta(\mathbf{r}\psi) = \frac{\partial^4(\mathbf{r}\psi)}{\partial x_i \, \partial x_i \, \partial x_k \, \partial x_k} = 0.$$

Solution. Clearly

$$\frac{\partial}{\partial x_k}(\mathbf{r}\psi) = \frac{\partial}{\partial x_k}(\mathbf{i}_i x_i \psi) = \mathbf{i}_k \psi + \mathbf{r} \frac{\partial \psi}{\partial x_k},$$

and hence

$$\frac{\partial^2}{\partial x_k \, \partial x_k}(\mathbf{r}\psi) = \mathbf{i}_k \frac{\partial \psi}{\partial x_k} + \mathbf{i}_k \frac{\partial \psi}{\partial x_k} + \mathbf{r} \frac{\partial^2 \psi}{\partial x_k \, \partial x_k}$$

$$= 2\mathbf{i}_k \frac{\partial \psi}{\partial x_k} + \mathbf{r} \frac{\partial^2 \psi}{\partial x_k \, \partial x_k}$$

or

$$\Delta(\mathbf{r}\psi) = 2\, \nabla\psi + \mathbf{r}\Delta\psi.$$

Therefore

$$\Delta(\mathbf{r}\psi) = 2\, \nabla\psi,$$

since $\Delta\psi = 0$ by hypothesis. But the operators Δ and ∇ commute (why?), and hence

$$\Delta\Delta(\mathbf{r}\psi) = 2\,\Delta\nabla\psi = 2\nabla\Delta\psi = 0,$$

as asserted.

Problem 2. Prove that if a velocity field $\mathbf{v}$ is potential ($\mathbf{v} = \nabla\varphi$), then the acceleration field $d\mathbf{v}/dt$ is also potential.

Solution. Setting $\mathbf{A} = \mathbf{B} = \mathbf{v}$ in formula (4.93), we obtain

$$\nabla(v^2) = 2(\mathbf{v}\cdot\nabla)\mathbf{v} + 2\mathbf{v} \times \operatorname{curl}\mathbf{v}.$$

Therefore the acceleration field

$$\frac{d\mathbf{v}}{dt} = \frac{\partial\mathbf{v}}{\partial t} + (\mathbf{v}\cdot\nabla)\mathbf{v}$$

can be written in the form

$$\frac{d\mathbf{v}}{dt} = \frac{\partial\mathbf{v}}{\partial t} + \nabla\left(\frac{v^2}{2}\right) - \mathbf{v} \times \operatorname{curl}\mathbf{v}. \tag{5.129}$$

But

$$\mathbf{v} = \nabla\varphi, \quad \operatorname{curl}\mathbf{v} = \operatorname{curl}\nabla\varphi = 0,$$

and hence

$$\frac{d\mathbf{v}}{dt} = \frac{\partial}{\partial t}(\nabla\varphi) + \nabla\left(\frac{v^2}{2}\right) = \nabla\left(\frac{\partial\varphi}{\partial t} + \frac{v^2}{2}\right).$$

Therefore $d\mathbf{v}/dt$ has a potential Φ, equal to

$$\Phi = \frac{\partial\varphi}{\partial t} + \frac{v^2}{2}.$$

Problem 3. Integrate the equation of fluid motion (5.64) if the flow is irrotational ($\operatorname{curl}\mathbf{v} = 0$) and *barotropic* [$\rho = \rho(p)$], while the body forces are derivable from a potential ($\mathbf{f} = -\nabla\Omega$).

Solution. Using (5.129) and the fact that $\operatorname{curl}\mathbf{v} = 0$, we can write (5.64) in the form

$$\frac{\partial\mathbf{v}}{\partial t} + \nabla\left(\frac{v^2}{2}\right) = \mathbf{f} - \frac{\nabla p}{\rho}. \tag{5.130}$$

But $\operatorname{curl}\mathbf{v} = 0$ implies $\mathbf{v} = \nabla\varphi$, while

$$\frac{\nabla p}{\rho} = \nabla\int\frac{dp}{\rho(p)}$$

(the flow is barotropic). Therefore (5.130) becomes

$$\nabla\left(\frac{\partial\varphi}{\partial t} + \frac{v^2}{2} + \Omega + \int\frac{dp}{\rho(p)}\right) = 0,$$

with solution

$$\frac{\partial \varphi}{\partial t} + \frac{v^2}{2} + \Omega + \int \frac{dp}{\rho(p)} = \psi(t), \tag{5.131}$$

where $\psi(t)$ is a function of time only. In the case of stationary motion of an incompressible fluid (ρ = const) subject to no body forces, (5.131) reduces to *Bernoulli's law*

$$\frac{\rho v^2}{2} + p = \text{const.} \tag{5.132}$$

***Problem 4* (*The Kutta-Joukowski theorem*).** Let L be a rigid plane contour immersed in a stationary flow $\mathbf{v}$, where $\mathbf{v}$ is incompressible, irrotational (curl $\mathbf{v} = 0$) and plane (i.e., dependent on only two coordinates, say x_1 and x_2). Suppose the velocity of the fluid far from L has the constant value $\mathbf{v}_\infty$, and suppose there are no body forces ($\mathbf{f} = 0$). Suppose finally that the "perturbational velocity" $\mathbf{v}' = \mathbf{v} - \mathbf{v}_\infty$ due to the presence of L falls off like $1/r$, where r is the distance from the origin O chosen to lie inside the contour L (see Fig. 5.8).

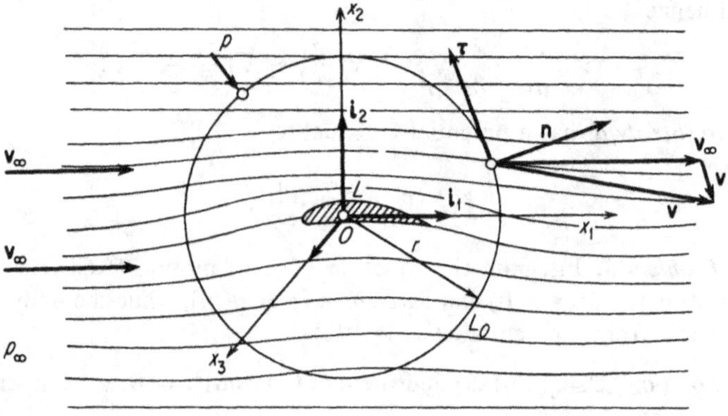

FIG. 5.8. Illustrating the Kutta-Joukowski theorem on the force acting on a rigid contour immersed in a stationary flow of an ideal fluid.

Prove that the force $\mathbf{F}$ exerted by the moving fluid on the contour L is

$$\mathbf{F} = \rho \Gamma \mathbf{v}_\infty \times \mathbf{i}_3,$$

where ρ is the density of the fluid, $\mathbf{i}_3$ is a unit vector along the x_3-axis, and

$$\Gamma = \oint_L \mathbf{v} \cdot d\mathbf{r}$$

is the circulation of the velocity field around L. In other words, show that the lift experienced by the contour L is perpendicular to $\mathbf{v}_\infty$ and is of magnitude $\rho v_\infty \Gamma$, a result known as the *Kutta-Joukowski theorem.*

Solution. Applying the momentum theorem (5.72) to the volume of fluid contained in a cylinder of unit height with faces in the shape of the plane region between the contour L and a circle L_0 of radius r surrounding L (see Fig. 5.8), we find that the force $\mathbf{F}$ acting on a unit length of an infinite right cylinder intersecting the x_1x_2-plane in the contour L equals[24]

$$\mathbf{F} = -\oint_{L_0} p\mathbf{n}\, ds - \oint_{L_0} \rho\mathbf{v}(\mathbf{v}\cdot\mathbf{n})\, ds, \tag{5.133}$$

where ds is an element of arc length along L_0, p is the pressure and ρ is the density. Since Bernoulli's law applies here, we can substitute (5.132) into (5.133), obtaining

$$\mathbf{F} = \frac{\rho}{2}\oint_{L_0} v^2\mathbf{n}\, ds - \rho\oint_{L_0} \mathbf{v}(\mathbf{v}\cdot\mathbf{n})\, ds$$

(ρ is constant), where we use the fact that

$$\oint_{L_0} \mathbf{n}\, ds = \oint_{L_0} \left(-\mathbf{i}_1\frac{dy}{ds} + \mathbf{i}_2\frac{dx}{ds}\right) ds = \oint_{L_0} (-\mathbf{i}_1\, dy + \mathbf{i}_2\, dx) = 0. \tag{5.134}$$

Then, replacing $\mathbf{v}$ by $\mathbf{v}_\infty + \mathbf{v}'$, where $\mathbf{v}_\infty$ is the velocity of the fluid at infinity and $\mathbf{v}'$ is the perturbational velocity due to the presence of L, we have

$$\mathbf{F} = \frac{\rho v_\infty^2}{2}\oint_{L_0} \mathbf{n}\, ds + \rho\oint_{L_0} (\mathbf{v}_\infty\cdot\mathbf{v}')\mathbf{n}\, ds + \frac{\rho}{2}\oint_{L_0} v'^2\,\mathbf{n}\, ds - \rho\mathbf{v}_\infty\oint_{L_0} (\mathbf{v}\cdot\mathbf{n})\, ds$$
$$- \rho\oint_{L_0} \mathbf{v}'(\mathbf{v}_\infty\cdot\mathbf{n})\, ds - \rho\oint_{L_0} \mathbf{v}'(\mathbf{v}'\cdot\mathbf{n})\, ds.$$

The first integral vanishes because of (5.134), while the fourth integral vanishes because of the incompressibility of the fluid (explain this further). The third and sixth integrals are both of order $1/r$, and their sum will accordingly be denoted by $O(1/r)$. It follows that

$$\mathbf{F} = \oint_{L_0} \rho[(\mathbf{v}_\infty\cdot\mathbf{v}')\mathbf{n} - \mathbf{v}'(\mathbf{v}_\infty\cdot\mathbf{n})]\, ds + O\left(\frac{1}{r}\right),$$

and hence

$$\mathbf{F} = \rho\oint_{L_0} \mathbf{v}_\infty \times (\mathbf{n}\times\mathbf{v}')\, ds + O\left(\frac{1}{r}\right)$$

---

[24] The integrals over the faces of the cylinder vanish since $\mathbf{v} \perp \mathbf{n}$ and hence $\mathbf{v}\cdot\mathbf{n} = 0$ on the faces, while the value of $\iint p\mathbf{n}\, dS$ on one face is the negative of its value on the other face.

by the formula (1.30) for the vector triple product. But

$$\oint_{L_0} \mathbf{v}_\infty \times (\mathbf{n} \times \mathbf{v}_\infty)\, ds = \mathbf{v}_\infty \times \left(\oint_{L_0} \mathbf{n}\, ds\right) \times \mathbf{v}_\infty = 0,$$

and therefore

$$\begin{aligned} \mathbf{F} &= \rho \oint_{L_0} \mathbf{v}_\infty \times (\mathbf{n} \times \mathbf{v})\, ds + O\left(\frac{1}{r}\right) \\ &= \rho \mathbf{v}_\infty \times \oint_{L_0} (\mathbf{n} \times \mathbf{v})\, ds + O\left(\frac{1}{r}\right). \end{aligned} \tag{5.135}$$

Now let $\boldsymbol{\tau}$ be the unit tangent vector to L_0, so that

$$\mathbf{n} = \boldsymbol{\tau} \times \mathbf{i}_3$$

(see Fig. 5.8), and

$$\mathbf{n} \times \mathbf{v} = (\boldsymbol{\tau} \times \mathbf{i}_3) \times \mathbf{v} = \mathbf{i}_3(\boldsymbol{\tau} \cdot \mathbf{v}) - \boldsymbol{\tau}(\mathbf{i}_3 \cdot \mathbf{v}) = \mathbf{i}_3(\mathbf{v} \cdot \boldsymbol{\tau}).$$

Thus (5.135) becomes

$$\mathbf{F} = \rho \mathbf{v}_\infty \times \mathbf{i}_3 \oint_{L_0} \mathbf{v} \cdot d\mathbf{r} + O\left(\frac{1}{r}\right),$$

where $d\mathbf{r} = \boldsymbol{\tau} ds$, or

$$\mathbf{F} = \rho \Gamma \mathbf{v}_\infty \times \mathbf{i}_3 + O\left(\frac{1}{r}\right), \tag{5.136}$$

where

$$\Gamma = \oint_{L_0} \mathbf{v} \cdot d\mathbf{r}$$

is the circulation of $\mathbf{v}$ around L_0. Since the radius of the circle L_0 is arbitrary, we can take the limit as $r \to \infty$ in (5.136), obtaining

$$\mathbf{F} = \rho\left(\lim_{r \to \infty} \Gamma\right) \mathbf{v}_\infty \times \mathbf{i}_3.$$

But, since the motion is irrotational, the circulation around any circle L_0 surrounding L is the same as the circulation around the contour L itself, and hence

$$\lim_{r \to \infty} \Gamma = \Gamma = \oint_L \mathbf{v} \cdot d\mathbf{r}.$$

It follows that

$$\mathbf{F} = \rho \Gamma \mathbf{v}_\infty \times \mathbf{i}_3,$$

where

$$\Gamma = \oint_L \mathbf{v} \cdot d\mathbf{r}.$$

This completes the proof of the Kutta-Joukowski theorem.

Problem 5. The equation for equilibrium of a fluid is

$$\rho \mathbf{f} = \nabla p \tag{5.137}$$

[set $\mathbf{v} = 0$ in (5.66)], where $\mathbf{f}$ is the body force per unit mass. Prove that equilibrium of a fluid is possible only in a force field such that the lines of force (the trajectories) of **f** are orthogonal to the trajectories of curl **f**.

Solution. Taking the curl of both sides of (5.137), we obtain

$$\text{curl}\,(\rho\mathbf{f}) = \nabla\rho \times \mathbf{f} + \rho\,\text{curl}\,\mathbf{f} = 0. \tag{5.138}$$

Then taking the scalar product of (5.138) with **f**, we find that

$$\mathbf{f}\cdot\text{curl}\,\mathbf{f} = 0.$$

Therefore **f** must be orthogonal to curl **f** at every point of the field. This is always the case in a plane force field. Obviously, equilibrium is also possible in a potential field (curl $\mathbf{f} = 0$). The equipotential surfaces of such a field are also surfaces of equal density (since $\nabla\rho \times \mathbf{f} = 0$) and surfaces of equal pressure (since $\nabla p \times \mathbf{f} = 0$).

Problem 6. Consider two parallel infinite plates a distance h apart, where the top plate moves with velocity u_1 and the bottom plate with velocity u_2. Then the stationary velocity field of an incompressible fluid of viscosity μ moving between the plates is of the form

$$\mathbf{v} = v_1\mathbf{i}_1 = \left[\frac{1}{2\mu}\frac{dp}{dx_1}(x_2^2 - hx_2) + \frac{u_1 - u_2}{h}x_2 + u_2\right]\mathbf{i}_1, \tag{5.139}$$

where x_1 lies in the lower plate and points in the common direction of motion of the two plates, x_2 is perpendicular to the plates, and dp/dx_1 is a constant pressure gradient. Find the circulation Γ of the velocity around a circle L_R of radius R perpendicular to the flow with its center halfway between the plates.

First solution. By direct calculation, we have

$$\Gamma = \oint_{L_R} v_1\,dx_1 = \frac{1}{2\mu}\frac{dp}{dx_1}\oint_{L_R}(x_2^2 - hx_2)\,dx_1 + \frac{u_1 - u_2}{h}\oint_{L_R} x_2\,dx_1 + u_2\oint_{L_R} dx_1.$$

In terms of the polar angle φ,

$$x_1 = R\cos\varphi, \qquad x_2 = \frac{h}{2} + R\sin\varphi,$$

and hence

$$\oint_{L_R} dx_1 = -R\int_0^{2\pi}\sin\varphi\,d\varphi = 0,$$

$$\oint_{L_R} x_2\,dx_1 = -R\int_0^{2\pi}\left(\frac{h}{2} + R\sin\varphi\right)\sin\varphi\,d\varphi = -\pi R^2,$$

$$\oint_{L_R} x_2^2\,dx_1 = -R\int_0^{2\pi}\left(\frac{h}{2} + R\sin\varphi\right)^2\sin\varphi\,d\varphi = -R^2h\pi.$$

It follows that

$$\Gamma = -\frac{u_1 - u_2}{h}\pi R^2.$$

Second solution. Applying Stokes' theorem to the disk S bounded by L_R, we find that

$$\begin{aligned}\Gamma &= \iint_S (\text{curl } \mathbf{v})_3 \, dS = -\iint_S \frac{\partial v_1}{\partial x_2} \, dS \\ &= -\iint_S \left[\frac{1}{2\mu}\frac{dp}{dx_1}(2x_2 - h) + \frac{u_1 - u_2}{h}\right] dS \\ &= -\int_0^R \int_0^{2\pi} \left[\frac{1}{2\mu}\frac{dp}{dx_1}(h + 2r\sin\varphi - h) + \frac{u_1 - u_2}{h}\right] r \, dr \, d\varphi \\ &= -\frac{u_1 - u_2}{h}\pi R^2.\end{aligned}$$

***Problem* 7.** Express the kinetic energy of a stationary irrotational flow of an incompressible fluid of density ρ occupying a volume V in terms of an integral over the surface S bounding V. Prove that if the fluid is at rest on the surface S, then it is at rest throughout the volume V.

Solution. Let $\mathbf{v}$ be the velocity field. Since curl $\mathbf{v} = 0$, we have $\mathbf{v} = \nabla\varphi$, where φ is the velocity potential. Thus the kinetic energy is

$$T = \frac{\rho}{2}\iiint_V \frac{v^2}{2} \, dV = \frac{\rho}{2}\iiint_V (\nabla\varphi)^2 \, dV.$$

Using the incompressibility condition

$$\text{div } \mathbf{v} = \text{div } (\nabla\varphi) = \Delta\varphi = 0$$

and the formula

$$\text{div } (\varphi\mathbf{A}) = \varphi \text{ div } \mathbf{A} + \mathbf{A} \cdot \nabla\varphi,$$

we find that

$$T = \frac{\rho}{2}\iiint_V \text{div } (\varphi \nabla\varphi) \, dV - \frac{\rho}{2}\iiint_V \varphi \text{ div } (\nabla\varphi) = \frac{\rho}{2}\iiint_V \text{div } (\varphi \nabla\varphi) \, dV$$

(set $\mathbf{A} = \nabla\varphi$). Using Gauss' theorem to transform the volume integral into a surface integral, we obtain

$$T = \frac{\rho}{2}\iint_S \varphi(\mathbf{n} \cdot \nabla\varphi) \, dS = \frac{\rho}{2}\iint_S \varphi \frac{\partial\varphi}{\partial n} \, dS.$$

If the fluid is at rest on S, then $\mathbf{v}|_S = 0$ and hence

$$\left.\frac{\partial\varphi}{\partial n}\right|_S = \mathbf{v} \cdot \mathbf{n}|_S = 0,$$

which implies $T = 0$, i.e., the fluid is at rest throughout V.

Problem 8. Let $\mathbf{v} = \mathbf{v}(\mathbf{r}, t)$ be the velocity field of a moving fluid, and let $\mathbf{A} = \mathbf{A}(\mathbf{r}, t)$ be another vector field occupying the same region of space.[25] Find a necessary condition for conservation of the trajectories of $\mathbf{A}$, i.e., a condition for the trajectories of $\mathbf{A}$ to always consist of the same fluid particles.

Solution. Let MN be a trajectory of $\mathbf{A}$ at time t, and suppose MN is deformed into $M'N'$ at time t' while remaining a trajectory. Let a and d be two neighboring particles of MN, which go into two neighboring particles b and c of $M'N'$ (see Fig. 5.9). Then $\mathbf{A} \times \delta\mathbf{r} = 0$ at time t, while $\mathbf{A}' \times \delta\mathbf{r}' = 0$ at time $t' = t + \Delta t$, where the meaning of $\mathbf{A}$, $\mathbf{A}'$, $\delta\mathbf{r}$ and $\delta\mathbf{r}'$ is shown

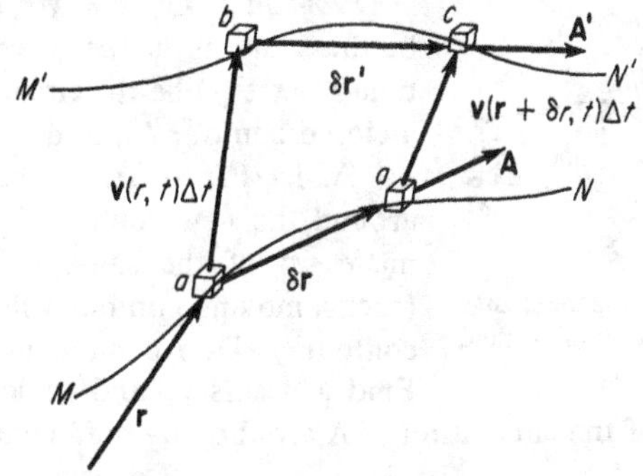

FIG. 5.9. If the trajectories of the field A are conserved, then the particles forming the trajectory *MN* at time *t* move in such a way as to again form a trajectory *M'N'* at any other time *t'*.

in the figure. From the quadrilateral $abcd$ we find that

$$\delta\mathbf{r}' = \delta\mathbf{r} + \mathbf{v}(\mathbf{r} + \delta\mathbf{r}, t)\,\Delta t - \mathbf{v}(\mathbf{r}, t)\,\Delta t = \delta\mathbf{r} + \frac{\partial \mathbf{v}}{\partial x_k}\,\delta x_k\,\Delta t$$
$$= \delta\mathbf{r} + (\delta\mathbf{r} \cdot \nabla)\mathbf{v}\,\Delta t,$$

if only linear terms are retained. Therefore

$$\left(\mathbf{A} + \frac{d\mathbf{A}}{dt}\Delta t\right) \times [\delta\mathbf{r} + (\delta\mathbf{r} \cdot \nabla)\mathbf{v}\,\Delta t] = 0,$$

since $M'N'$ is a trajectory. Expanding the vector product, recalling that $\mathbf{A} \times \delta\mathbf{r} = 0$ and retaining only linear terms, we have

$$\frac{d\mathbf{A}}{dt} \times \delta\mathbf{r} + [\mathbf{A} \times (\delta\mathbf{r} \cdot \nabla)\mathbf{v}] = 0 \tag{5.140}$$

or

$$\left[\frac{d\mathbf{A}}{dt} - (\mathbf{A} \cdot \nabla)\mathbf{v}\right] \times \mathbf{A} = 0,$$

[25] For example, $\mathbf{A}$ might be the acceleration field $d\mathbf{v}/dt$ or the "vorticity field" curl $\mathbf{v}$.

since $\delta\mathbf{r}$ is parallel to $\mathbf{A}$. This is the required necessary condition for conservation of trajectories of $\mathbf{A}$.

Remark. It can be shown[26] that

$$\frac{d\mathbf{A}}{dt} - (\mathbf{A}\cdot\nabla)\mathbf{v} + \mathbf{A}\operatorname{div}\mathbf{v} = 0 \tag{5.141}$$

is a necessary and sufficient condition for conservation both of trajectories of $\mathbf{A}$ and of the intensity of vector tubes of $\mathbf{A}$. Note that (5.141) implies (5.140).

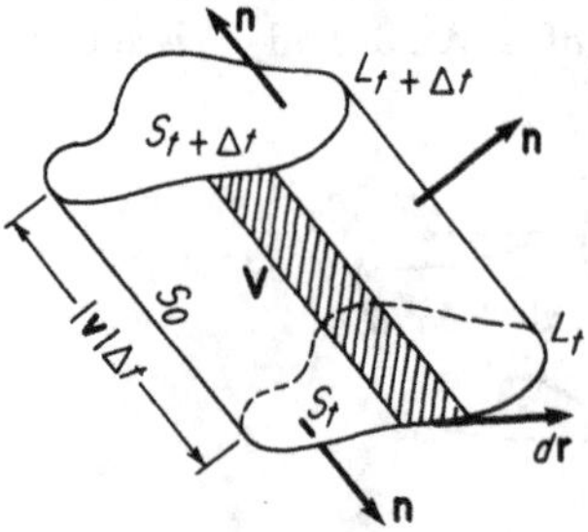

FIG. 5.10. Illustrating the change of circulation around a fluid contour.

***Problem* 9.** Let $\mathbf{v} = \mathbf{v}(\mathbf{r}, t)$ and $\mathbf{A} = \mathbf{A}(\mathbf{r}, t)$ be the same as in the preceding problem. At time t, let $\Gamma(t)$ be the circulation of $\mathbf{A}$ around a closed contour L_t, and at subsequent times $t + \Delta t$ let $\Gamma(t + \Delta t)$ be the circulation of $\mathbf{A}$ around the new contour $L_{t+\Delta t}$ (see Fig. 5.10) made up of the same fluid particles as L_t (such a moving contour will be called a "fluid contour"). Derive an expression for $d\Gamma/dt$. Find a necessary and sufficient condition for conservation of the circulation of $\mathbf{A}$ around any fluid contour.

Solution. Since

$$\Gamma(t) = \oint_{L_t} A(\mathbf{r}, t)\cdot d\mathbf{r},$$

we have

$$\begin{aligned}\frac{d\Gamma}{dt} &= \lim_{\Delta t\to 0}\frac{1}{\Delta t}\left\{\oint_{L_{t+\Delta t}} \mathbf{A}(\mathbf{r}, t + \Delta t)\cdot d\mathbf{r} - \oint_{L_t} \mathbf{A}(\mathbf{r})\cdot d\mathbf{r}\right\}\\ &= \lim_{\Delta t\to 0}\frac{1}{\Delta t}\left\{\oint_{L_{t+\Delta t}} [\mathbf{A}(\mathbf{r}, t + \Delta t) - \mathbf{A}(\mathbf{r}, t)]\cdot d\mathbf{r}\right.\\ &\qquad\qquad \left. + \oint_{L_{t+\Delta t}} \mathbf{A}(\mathbf{r}, t)\cdot d\mathbf{r} - \oint_{L_t} \mathbf{A}(\mathbf{r}, t)\cdot d\mathbf{r}\right\}\\ &= \oint_{L_t}\frac{\partial\mathbf{A}}{\partial t}\cdot d\mathbf{r} + \lim_{\Delta t\to 0}\frac{1}{\Delta t}\left\{\oint_{L_{t+\Delta t}} \mathbf{A}(\mathbf{r}, t)\cdot d\mathbf{r} - \oint_{L_t} \mathbf{A}(\mathbf{r}, t)\cdot d\mathbf{r}\right\}.\end{aligned} \tag{5.142}$$

To calculate the last limit on the right in (5.142), we consider the flux of the vector curl $\mathbf{A}$ through the closed surface formed by two "caps" S_t, $S_{t+\Delta t}$ supported by the contours L_t, $L_{t+\Delta t}$ and the lateral surface S_0 made up of the

[26] See N. E. Kochin, I. A. Kibel and N. V. Roze, *Theoretical Hydrodynamics* (translated and edited by D. Boyanovitch and J. R. M. Radok), John Wiley and Sons, Inc., New York (1964), Sec. 5.5.

segments of the trajectories of the fluid particles "swept out" in time Δt (see Fig. 5.10). Since curl **A** is a solenoidal vector, it follows from Gauss' theorem that

$$\iint_{S_{t+\Delta t}} \mathbf{n} \cdot \operatorname{curl} \mathbf{A}\, dS + \iint_{S_t} \mathbf{n} \cdot \operatorname{curl} \mathbf{A}\, dS + \iint_{S_0} \mathbf{n} \cdot \operatorname{curl} \mathbf{A}\, dS = 0,$$

where dS denotes a generic surface element. Applying Stokes' theorem to the first two integrals on the left and bearing in mind that the interior normal to S_t points in the direction of advance of a right-handed screw turned along L, we obtain

$$\oint_{L_{t+\Delta t}} \mathbf{A}(\mathbf{r}, t) \cdot d\mathbf{r} - \oint_{L_t} \mathbf{A}(\mathbf{r}, t) \cdot d\mathbf{r} = -\iint_{S_0} \mathbf{n} \cdot \operatorname{curl} \mathbf{A}\, dS.$$

But

$$\mathbf{n}\, dS = d\mathbf{r} \times \mathbf{v}\, \Delta t$$

on the lateral surface S_0. It follows that

$$\oint_{L_{t+\Delta t}} \mathbf{A}(\mathbf{r}, t) \cdot d\mathbf{r} - \oint_{L_t} \mathbf{A}(\mathbf{r}, t) \cdot d\mathbf{r} = \Delta t \oint_{L_t} (\operatorname{curl} \mathbf{A} \times \mathbf{v}) \cdot d\mathbf{r}, \tag{5.143}$$

where we have used formula (1.27).

Next, substituting (5.143) into (5.142), we find that

$$\frac{d\Gamma}{dt} = \oint_{L_t} \left\{\frac{\partial \mathbf{A}}{\partial t} + (\operatorname{curl} \mathbf{A} \times \mathbf{v})\right\} \cdot d\mathbf{r} \tag{5.144}$$

or

$$\frac{d\Gamma}{dt} = \iint_{S_t} \left\{\frac{\partial}{\partial t} \operatorname{curl} \mathbf{A} + \operatorname{curl}(\operatorname{curl} \mathbf{A} \times \mathbf{v})\right\} \cdot \mathbf{n}\, dS,$$

after applying Stokes' theorem again. Writing $\mathbf{B} = \operatorname{curl} \mathbf{A}$ and then using Prob. 8, p. 180 to expand curl $(\mathbf{B} \times \mathbf{v})$, we finally obtain

$$\frac{d\Gamma}{dt} = \iint_{S_t} \left\{\frac{d\mathbf{B}}{dt} - (\mathbf{B} \cdot \nabla)\mathbf{v} + \mathbf{B} \operatorname{div} \mathbf{v}\right\} \cdot \mathbf{n}\, dS.$$

Since S_t (like L_t) is arbitrary, it follows that $d\Gamma/dt = 0$ and hence $\Gamma = \text{const}$ if and only if

$$\frac{d\mathbf{B}}{dt} - (\mathbf{B} \cdot \nabla)\mathbf{v} + \mathbf{B} \operatorname{div} \mathbf{v} = 0.$$

Recalling the remark on p. 240, we conclude that a necessary and sufficient condition for conservation of the circulation of a vector field $\mathbf{A} = \mathbf{A}(\mathbf{r}, t)$ around an arbitrary fluid contour is that the trajectories of curl **A** and the intensity of vector tubes of curl **A** both be conserved. In particular, (5.144) implies *Thomson's theorem*, which states that the time derivative of the circulation of the velocity field $\mathbf{v}$ around any fluid contour equals the circulation of the acceleration field $d\mathbf{v}/dt$ around the same contour. In fact, setting $\mathbf{A} = \mathbf{v}$ in (5.144) and noting that

$$\nabla\left(\frac{v^2}{2}\right) = (\mathbf{v} \cdot \nabla)\mathbf{v} + \mathbf{v} \times \operatorname{curl} \mathbf{v}$$

(recall Prob. 2, p. 233), we have

$$\frac{d\Gamma}{dt} = \oint_{L_t} \left\{\frac{\partial \mathbf{v}}{\partial t} + (\mathbf{v} \cdot \nabla)\mathbf{v} - \nabla\left(\frac{v^2}{2}\right)\right\} \cdot d\mathbf{r} = \oint_{L_t} \frac{d\mathbf{v}}{dt} \cdot d\mathbf{r},$$

since

$$\oint_{L_t} \nabla\left(\frac{v^2}{2}\right) \cdot d\mathbf{r} = 0,$$

being the integral of a single-valued function around a closed contour.

Remark. Given a vector field $\mathbf{A} = \mathbf{A}(\mathbf{r}, t)$, a trajectory of the field curl $\mathbf{A}$ is called a *vortex line* of $\mathbf{A}$, and a surface formed by the vortex lines going through some closed contour is called a *vortex tube*. Thus we have shown that a necessary and sufficient condition for conservation of the circulation of $\mathbf{A}$ around an arbitrary fluid contour is that the vortex lines of $\mathbf{A}$ and the intensity of vortex tubes of $\mathbf{A}$ both be conserved.

Problem 10. Show that if r is the distance between the points (x_1, x_2, x_3) and (ξ_1, ξ_2, ξ_3), then the function

$$f(x_1, x_2, x_3) = \frac{g\left(\xi_1, \xi_2, \xi_3, t - \dfrac{r}{c}\right)}{r},$$

where g is twice differentiable but otherwise arbitrary, satisfies the equation

$$\Box f = 0,$$

where $\Box$ is the D'Alembertian operator

$$\Box \equiv \frac{\partial^2}{\partial x_1^2} + \frac{\partial^2}{\partial x_2^2} + \frac{\partial^2}{\partial x_3^2} - \frac{1}{c^2}\frac{\partial^2}{\partial t^2}.$$

Solution. Calculating Δf, we have

$$\Delta f = \frac{1}{r}\nabla g + g\nabla\frac{1}{r}.$$

Using (4.74) and denoting differentiation with respect to the argument $t - \dfrac{r}{c}$ by a prime, we obtain

$$f = \frac{1}{r}\left(-\frac{g'}{c}\nabla r\right) - \frac{g}{r^2}\nabla r = -\left(\frac{g'}{cr} + \frac{g}{r^2}\right)\nabla r = -\left(\frac{g'}{cr} + \frac{g}{r^2}\right)\frac{\mathbf{r}}{r}$$

(recall Exercise 2a, p. 182). It follows that

$$\Delta f = \operatorname{div} \nabla f = -\operatorname{div}\left[\mathbf{r}\left(\frac{g'}{cr^2} + \frac{g}{cr^3}\right)\right]$$

$$= -\left(\frac{g'}{cr^2} + \frac{g}{r^3}\right) \operatorname{div} \mathbf{r} - \mathbf{r} \cdot \nabla\left(\frac{g'}{cr^2} + \frac{g}{r^3}\right)$$

$$= -3\left(\frac{g'}{cr^2} + \frac{g}{r^3}\right) - \mathbf{r} \cdot \frac{\mathbf{r}}{r}\left(-\frac{g''}{c^2r^2} - \frac{2g'}{cr^3} - \frac{g'}{cr^3} - \frac{3g}{r^4}\right) = \frac{g''}{c^2r}$$

(recall Exercise 3a, p. 182). But

$$g'' = \frac{\partial^2 g}{\partial\left(t - \frac{c}{r}\right)^2} = \frac{\partial^2 g}{\partial t^2} = r\frac{\partial^2 f}{\partial t^2},$$

and hence

$$\Delta f = \frac{1}{c^2}\frac{\partial^2 f}{\partial t^2},$$

i.e.,

$$\Box f = 0,$$

as required.

Remark. Consider the equation

$$\Box f = -4\pi g(x_1, x_2, x_3, t), \tag{5.145}$$

where g vanishes outside of some finite volume V. Then by an argument like that used to solve Poisson's equation in the proof of the fundamental theorem of vector analysis (see Sec. 5.7), it can be shown that the solution of (5.145) can be written in the form

$$f(x_1, x_2, x_3, t) = \iiint_V \frac{g\left(\xi_1, \xi_2, \xi_3, t - \frac{r}{c}\right)}{r}\, d\xi_1\, d\xi_2\, d\xi_3,$$

where

$$r = \sqrt{(x_1 - \xi_1)^2 + (x_2 - \xi_2)^2 + (x_3 - \xi_3)^2}.$$

In particular, the solutions of the equations (5.123) satisfied by the vector and scalar electromagnetic potentials can be written as

$$\mathbf{A}(x_1, x_2, x_3, t) = \frac{1}{c}\iiint_V \frac{\mathbf{j}\left(\xi_1, \xi_2, \xi_3, t - \frac{r}{c}\right)}{r}\, d\xi_1\, d\xi_2\, d\xi_3, \tag{5.146}$$

$$\varphi(x_1, x_2, x_3, t) = \iiint_V \frac{\rho\left(\xi_1, \xi_2, \xi_3, t - \frac{r}{c}\right)}{r}\, d\xi_1\, d\xi_2\, d\xi_3. \tag{5.147}$$

These solutions show that the current and charge at the point (ξ_1, ξ_2, ξ_3) influence the potentials at the point (x_1, x_2, x_3) with a time delay r/c equal to the time it takes a signal travelling with the velocity of light to traverse the distance r between the two points. For this reason, (5.146) and (5.147) are often called *retarded potentials.*

Problem 11. Show that electromagnetic disturbances propagate with velocity

$$\frac{c}{\sqrt{\mu\varepsilon}}$$

in a medium of dielectric constant ε and magnetic permeability μ, provided there are no free charges or currents ($\rho = 0, \mathbf{j} = 0$).

Solution. Taking the curl of the third Maxwell equation (5.109) with $\mathbf{D} = \varepsilon\mathbf{E}$, we have

$$\text{curl curl } \mathbf{H} = \frac{\varepsilon}{c} \text{ curl } \frac{\partial \mathbf{E}}{\partial t} = \frac{\varepsilon}{c} \frac{\partial}{\partial t} \text{ curl } \mathbf{E}.$$

Together with the first and second Maxwell equations (5.107) and (5.108), this implies

$$\Delta \mathbf{H} = \frac{\mu\varepsilon}{c^2} \frac{\partial^2 \mathbf{H}}{\partial t^2}, \tag{5.148}$$

where we use (4.96) and the constitutive relation $\mathbf{B} = \mu\mathbf{H}$. In the same way, it is easily verified that

$$\Delta \mathbf{E} = \frac{\mu\varepsilon}{c^2} \frac{\partial^2 \mathbf{E}}{\partial t^2}. \tag{5.149}$$

It follows that the initial values

$$\mathbf{E}\big|_{t=t_0} = \mathbf{E}_0(r), \qquad \mathbf{H}\big|_{t=t_0} = \mathbf{H}_0(r)$$

of the electromagnetic field at time $t = t_0$ propagate with velocity $c/\sqrt{\mu\varepsilon}$, as asserted. In other words, (5.148) and (5.149) describe the propagation of *electromagnetic waves*, and hence are called *wave equations.*

Problem 12. Show that the function

$$\mathbf{H} = \mathbf{H}_0 e^{i(\omega t - \mathbf{k}\cdot\mathbf{r})} \qquad (\mathbf{H}_0 = \textbf{const}) \tag{5.150}$$

satisfies equation (5.148) if

$$k = |\mathbf{k}| = \frac{\omega\sqrt{\mu\varepsilon}}{c}.$$

Solution. Clearly, we have

$$\frac{\partial^2 \mathbf{H}}{\partial t^2} = (i\omega)^2 \mathbf{H}_0 e^{i(\omega t - \mathbf{k}\cdot\mathbf{r})} = -\omega^2 \mathbf{H}_0 e^{i(\omega t - \mathbf{k}\cdot\mathbf{r})} = -\omega^2 \mathbf{H} \tag{5.151}$$

and

$$\Delta\mathbf{H} = \mathbf{H}_0\,\Delta e^{i(\omega t - \mathbf{k}\cdot\mathbf{r})} = \mathbf{H}_0 e^{i\omega t}\,\Delta e^{-i\mathbf{k}\cdot\mathbf{r}}.$$

Moreover the jth component of $\nabla e^{-i(\mathbf{k}\cdot\mathbf{r})}$ is

$$\frac{\partial}{\partial x_j} e^{-i(k_1x_1+k_2x_2+k_3x_3)} = -ik_j e^{-i\mathbf{k}\cdot\mathbf{r}},$$

so that

$$\nabla e^{-i\mathbf{k}\cdot\mathbf{r}} = -i\mathbf{k}e^{-i\mathbf{k}\cdot\mathbf{r}},$$

and similarly

$$\Delta e^{-i\mathbf{k}\cdot\mathbf{r}} = \nabla\cdot\nabla e^{-i\mathbf{k}\cdot\mathbf{r}} = (i\mathbf{k}\cdot i\mathbf{k})e^{-i\mathbf{k}\cdot\mathbf{r}} = -k^2 e^{-i\mathbf{k}\cdot\mathbf{r}}.$$

Therefore

$$\Delta\mathbf{H} = \mathbf{H}_0 e^{i\omega t}\,\Delta e^{-i\mathbf{k}\cdot\mathbf{r}} = -k^2\mathbf{H}_0 e^{i(\omega t-\mathbf{k}\cdot\mathbf{r})} = -k^2\mathbf{H}. \qquad (5.152)$$

It follows from (5.151) and (5.152) that

$$\Delta\mathbf{H} - \frac{\mu\varepsilon}{c^2}\frac{\partial^2\mathbf{H}}{\partial t^2} = \left(-k^2 + \frac{\mu\varepsilon}{c^2}\,\omega^2\right)\mathbf{H},$$

and hence (5.148) is satisfied if

$$k = |\mathbf{k}| = \frac{\omega}{c}\sqrt{\mu\varepsilon}.$$

Remark. The vector $\mathbf{k}$ is called the *wave vector*, and its magnitude k can be expressed in terms of the *wavelength* λ, i.e., the distance (along $\mathbf{k}$) between consecutive points where $\mathbf{H}$ takes identical values at any fixed time. In other words, λ satisfies the relation

$$e^{i(\mathbf{k}\cdot\mathbf{r})} = e^{i\left[\mathbf{k}\cdot\left(\mathbf{r}+\lambda\frac{\mathbf{k}}{k}\right)\right]} = e^{i\mathbf{k}\cdot\mathbf{r}}e^{i\lambda k},$$

which implies

$$\lambda k = 2\pi$$

or

$$\lambda = \frac{2\pi}{k}.$$

The solution (5.150) of equation (5.148) and the similar solution

$$\mathbf{E} = \mathbf{E}_0 e^{i(\omega t-\mathbf{k}\cdot\mathbf{r})}$$

of equation (5.149) describe a *plane monochromatic electromagnetic wave.*[27] In fact, the wave front (the surface of constant phase) belongs to the family of planes

$$\mathbf{k}\cdot\mathbf{r} = \text{const.}$$

Since these planes are perpendicular to the vector $\mathbf{k}$ (see Prob. 9, Case 4,

---

[27] The adjective "monochromatic" refers to the fact that the (angular) frequency ω is constant.

p. 44), a plane wave always propagates in the direction of the wave vector **k**. Using the Fourier integral, we can represent an arbitrary electromagnetic wave as a superposition of plane monochromatic waves with different frequencies ω and wave vectors **k**.

Problem 13. Describe the propagation of electromagnetic waves in a medium of dielectric constant ε, magnetic permeability μ and conductivity σ.

Solution. Inside a conducting medium we have

$$\rho = 0$$

(no free charge) and

$$\mathbf{j} = \sigma \mathbf{E}$$

(Ohm's law). Therefore Maxwell's equations (5.107)–(5.110) can be written as

$$\operatorname{curl} \mathbf{E} = -\frac{\mu}{c}\frac{\partial \mathbf{H}}{\partial t},$$

$$\operatorname{div} \mathbf{E} = 0,$$

$$\operatorname{curl} \mathbf{H} = \frac{\varepsilon}{c}\frac{\partial \mathbf{E}}{\partial t} + \frac{4\pi\sigma}{c}\mathbf{E},$$

$$\operatorname{div} \mathbf{H} = 0.$$

It follows from these equations that[28]

$$\frac{\mu\varepsilon}{c^2}\frac{\partial^2 \mathbf{H}}{\partial t^2} - \Delta \mathbf{H} + \frac{4\pi\mu\sigma}{c^2}\frac{\partial \mathbf{H}}{\partial t} = 0, \tag{5.153}$$

and similarly

$$\frac{\mu\varepsilon}{c^2}\frac{\partial^2 \mathbf{E}}{\partial t^2} - \Delta \mathbf{E} + \frac{4\pi\mu\sigma}{c^2}\frac{\partial \mathbf{E}}{\partial t} = 0. \tag{5.154}$$

Suppose we look for a solution of (5.153) of the form

$$\mathbf{H} = \mathbf{H}_0 e^{i(\omega t - \mathbf{k}\cdot\mathbf{r})}. \tag{5.155}$$

Substituting (5.155) into (5.153), we find that **k** must satisfy the equation

$$-\mathbf{k}\cdot\mathbf{k} - \frac{4\pi\mu\sigma}{c^2}i\omega + \frac{\mu\varepsilon}{c^2}\omega^2 = 0. \tag{5.156}$$

It follows that **k** must be complex, i.e.,

$$\mathbf{k} = \mathbf{k}_1 + i\mathbf{k}_2. \tag{5.157}$$

---

[28] Take the curl of the third equation, and then use the first and second equations.

Substituting (5.157) into (5.156), we obtain

$$k_1^2 - k_2^2 = \frac{\mu\varepsilon}{c^2}\omega^2, \qquad k_1 k_2 = -\frac{2\pi\mu\sigma\omega}{c^2 \cos\theta}, \tag{5.158}$$

where $\cos\theta = \cos(\mathbf{k}_1, \mathbf{k}_2)$ and $\pi/2 < \theta \leqslant \pi$ (since $k_1 k_2 > 0$). Solving (5.158) for k_1 and k_2, we find that

$$k_1 = \frac{\omega}{c}\sqrt{\frac{\mu\varepsilon}{2}}\left[\sqrt{1 + \left(\frac{4\pi\sigma}{\varepsilon\omega\cos\theta}\right)^2} + 1\right]^{1/2},$$

$$k_2 = \frac{\omega}{c}\sqrt{\frac{\mu\varepsilon}{2}}\left[\sqrt{1 + \left(\frac{4\pi\sigma}{\varepsilon\omega\cos\theta}\right)^2} - 1\right]^{1/2}.$$

Combining (5.155) and (5.157), we find that

$$\mathbf{H} = \mathbf{H}_0 e^{\mathbf{k}_2\cdot\mathbf{r}} e^{i(\omega t - \mathbf{k}_1\cdot\mathbf{r})}, \tag{5.159}$$

and similarly

$$\mathbf{E} = \mathbf{E}_0 e^{\mathbf{k}_2\cdot\mathbf{r}} e^{i(\omega t - \mathbf{k}_1\cdot\mathbf{r})}, \tag{5.160}$$

since equations (5.153) and (5.154) are of exactly the same form.

EXERCISES

1. Prove that the rules for covariant differentiation of sums and products of tensors coincide with the corresponding rules for ordinary differentiation of tensors. For example, if A_{ik} and B_{ik} are second-order tensors, prove that

$$(A_{ik} + B_{ik})_{,l} = A_{ik,l} + B_{ik,l},$$

$$(A_{ik}B_{mn})_{,l} = A_{ik,l}B_{mn} + A_{ik}B_{mn,l}.$$

2. Show that

$$\frac{\partial g_{ij}}{\partial x^k} - \frac{\partial g_{jk}}{\partial x^i} = [i, jk] - [k, ij].$$

3. Prove that in orthogonal coordinates $\begin{Bmatrix} i \\ j\,k \end{Bmatrix} = 0$ if the indices i, j, k are distinct, while

$$\begin{Bmatrix} i \\ i\,i \end{Bmatrix} = \frac{\partial}{\partial x^i}\ln h_i, \qquad \begin{Bmatrix} i \\ i\,j \end{Bmatrix} = \frac{\partial}{\partial x^j}\ln h_i, \qquad \begin{Bmatrix} i \\ j\,j \end{Bmatrix} = -\frac{h_j}{h_i^2}\frac{\partial h_j}{\partial x^i}$$

in terms of the metric coefficients h_i (no summation over repeated indices $i \neq j$).

4. Show that

$$\frac{\partial}{\partial x^k}(g_{ij}A^iB^j) = A_{i,k}B^i + A^iB_{i,k}.$$

5. Prove that the covariant derivative of the magnitude of a vector $\mathbf{A}$ equals

$$|\mathbf{A}|_{,i} = \frac{1}{|\mathbf{A}|} A_{j,i}A^j.$$

6. Prove that the analogues of equations (5.59), (5.61) and (5.65) in generalized coordinates x^1, x^2, x^3 are

$$\rho\frac{\partial v^i}{\partial t} + \rho v^k v^i_{,k} = \rho f^i + p^{ik}_{,k},$$

$$p^{ik} = -pg^{ik} + 2\mu v^{ik} - \tfrac{2}{3}\mu g^{ik} v^l_{,l},$$

$$\rho\frac{\partial v^i}{\partial t} + \rho v^k v^i_{,k} = \rho f^i - g^{ik}\frac{\partial p}{\partial x^k} + \mu g^{km}(v^i_{,m})_{,k} + \tfrac{1}{3}\mu g^{ik}\frac{\partial}{\partial x^k}(v^l_{,l}).$$

Write the last of these equations in terms of the covariant components of the velocity field.

Hint. The contravariant components of the rate of deformation tensor are

$$v^{ik} = g^{il}g^{km}v_{lm} = g^{il}g^{km}(v_{l,m} + v_{m,l}).$$

7. Prove that each of the fields

$$\mathbf{A} = 3x_2^4x_3^2\mathbf{i}_1 + 4x_1^3x_3^2\mathbf{i}_2 - 3x_1^2x_2^2\mathbf{i}_3,$$

$$\mathbf{B} = \nabla\varphi \times \nabla\psi \qquad (\varphi \text{ and } \psi \text{ scalar functions})$$

is solenoidal.

8. Find the general form of the function $f(r)$ such that the field $\mathbf{A} = f(r)\mathbf{r}$ is solenoidal.

Ans. $f(r) = c/r^3$ where c is an arbitrary constant.

9. Prove that each of the fields

$$\mathbf{A} = (6x_1x_2 + x_3^3)\mathbf{i}_1 + (3x_1^2 - x_3)\mathbf{i}_2 + (3x_1x_3^2 - x_2)\mathbf{i}_3,$$

$$\mathbf{B} = \varphi\nabla\varphi \qquad (\varphi \text{ a scalar function})$$

is irrotational. Find the potential of the field $\mathbf{A}$.

Ans. $\mathbf{A} = (3x_1^2x_2 + x_1x_3^3 - x_2x_3 + \text{const}).$

10. Prove that if the fields $\mathbf{A}$ and $\mathbf{B}$ are irrotational, then the field $\mathbf{A} \times \mathbf{B}$ is solenoidal.

11. Let V be a finite volume bounded by a surface S. Suppose we have found a vector field $\mathbf{A} = \mathbf{A}(\mathbf{r})$ in V, with prescribed divergence and curl, satisfying the boundary condition

$$\mathbf{A}\cdot\mathbf{n}|_S = f(\mathbf{r}).$$

Prove that $\mathbf{A}$ is unique. Is f completely arbitrary?

12. Deduce (5.139) from the Navier-Stokes equation.

13. Let $\mathbf{v} = \mathbf{v}(\mathbf{r})$ be the stationary velocity of a moving incompressible fluid. Prove that the flux of curl $\mathbf{v}$ is the same through every cross section of a vortex tube (of $\mathbf{v}$). Prove that the intensity of a vortex tube (see p. 218) equals the circulation of $\mathbf{v}$ around any closed contour bounding a cross section of the tube.

14. Let $\mathbf{v} = \mathbf{v}(\mathbf{r}, t)$ be the velocity field of a moving fluid. Prove that the circulation

$$\oint_{L_t} \mathbf{v} \cdot d\mathbf{r}$$

around any closed fluid contour L_t (see p. 240) is independent of the time, provided the body forces have a potential and the fluid is barotropic (i.e., the density of the fluid is a function of the pressure).

Hint. Use Thomson's theorem.

15. Under the same conditions as in the preceding problem, prove *Helmholtz's theorem* which states that the vortex lines (of $\mathbf{v}$) and the intensity of vortex tubes are both conserved (in time).

Hint. Use the remark on p. 240.

16. Show that the waves (5.159) and (5.160) are attenuated, i.e., that $|\mathbf{H}|$ and $|\mathbf{E}|$ decrease as $|\mathbf{r}|$ increases (for sufficiently large $|\mathbf{r}|$ and sufficiently narrow wave fronts).

BIBLIOGRAPHY

Aris, R., *Vectors, Tensors, and the Basic Equations of Fluid Mechanics*, Prentice-Hall, Inc., Englewood Cliffs, N.J. (1962).

Brand, L., *Vector and Tensor Analysis*, John Wiley and Sons, Inc., New York (1947).

Coburn, N., *Vector and Tensor Analysis*, The Macmillan Co., New York (1955).

Craig, H. V., *Vector and Tensor Analysis*, McGraw-Hill Book Co., New York (1943).

Hay, G. E., *Vector and Tensor Analysis*, Dover Publications, Inc., New York (1953).

Jeffreys, H., *Cartesian Tensors*, Cambridge University Press, London (1961).

Lass, H., *Vector and Tensor Analysis*, McGraw-Hill Book Co., New York (1950).

Marion, J. B., *Principles of Vector Analysis*, Academic Press, Inc, New York (1965).

Newell, H. E., Jr., *Vector Analysis*, McGraw-Hill Book Co., New York (1955).

Phillips, H. B., *Vector Analysis*, John Wiley and Sons, New York (1933).

Sokolnikoff, I. S., *Tensor Analysis: Theory and Applications to Geometry and Mechanics of Continua*, second edition, John Wiley and Sons, New York (1964).

Spain, B., *Tensor Calculus*, third edition, John Wiley and Sons, New York (1960).

Taylor, J. H., *Vector Analysis with an Introduction to Tensor Analysis*, Prentice-Hall, Inc., Englewood Cliffs, N.J. (1939).

Temple, G. F. J., *Cartesian Tensors: An Introduction*, John Wiley and Sons, New York (1960).

Wills, A. P., *Vector Analysis with an Introduction to Tensor Analysis*, Prentice-Hall, Inc., Englewood Cliffs, N.J. (1931).

Wrede, R. C., *Introduction to Vector and Tensor Analysis*, John Wiley and Sons, New York (1963).

INDEX

V

W

Z

A CATALOG OF SELECTED

DOVER BOOKS

IN SCIENCE AND MATHEMATICS

Astronomy

BURNHAM'S CELESTIAL HANDBOOK, Robert Burnham, Jr. Thorough guide to the stars beyond our solar system. Exhaustive treatment. Alphabetical by constellation: Andromeda to Cetus in Vol. 1; Chamaeleon to Orion in Vol. 2; and Pavo to Vulpecula in Vol. 3. Hundreds of illustrations. Index in Vol. 3. 2,000pp. 6⅛ x 9¼.
Vol. I: 0-486-23567-X
Vol. II: 0-486-23568-8
Vol. III: 0-486-23673-0

EXPLORING THE MOON THROUGH BINOCULARS AND SMALL TELESCOPES, Ernest H. Cherrington, Jr. Informative, profusely illustrated guide to locating and identifying craters, rills, seas, mountains, other lunar features. Newly revised and updated with special section of new photos. Over 100 photos and diagrams. 240pp. 8¼ x 11. 0-486-24491-1

THE EXTRATERRESTRIAL LIFE DEBATE, 1750–1900, Michael J. Crowe. First detailed, scholarly study in English of the many ideas that developed from 1750 to 1900 regarding the existence of intelligent extraterrestrial life. Examines ideas of Kant, Herschel, Voltaire, Percival Lowell, many other scientists and thinkers. 16 illustrations. 704pp. 5⅜ x 8½. 0-486-40675-X

THEORIES OF THE WORLD FROM ANTIQUITY TO THE COPERNICAN REVOLUTION, Michael J. Crowe. Newly revised edition of an accessible, enlightening book recreates the change from an earth-centered to a sun-centered conception of the solar system. 242pp. 5⅜ x 8½. 0-486-41444-2

A HISTORY OF ASTRONOMY, A. Pannekoek. Well-balanced, carefully reasoned study covers such topics as Ptolemaic theory, work of Copernicus, Kepler, Newton, Eddington's work on stars, much more. Illustrated. References. 521pp. 5⅜ x 8½.
0-486-65994-1

A COMPLETE MANUAL OF AMATEUR ASTRONOMY: TOOLS AND TECHNIQUES FOR ASTRONOMICAL OBSERVATIONS, P. Clay Sherrod with Thomas L. Koed. Concise, highly readable book discusses: selecting, setting up and maintaining a telescope; amateur studies of the sun; lunar topography and occultations; observations of Mars, Jupiter, Saturn, the minor planets and the stars; an introduction to photoelectric photometry; more. 1981 ed. 124 figures. 25 halftones. 37 tables. 335pp. 6½ x 9¼. 0-486-40675-X

AMATEUR ASTRONOMER'S HANDBOOK, J. B. Sidgwick. Timeless, comprehensive coverage of telescopes, mirrors, lenses, mountings, telescope drives, micrometers, spectroscopes, more. 189 illustrations. 576pp. 5⅝ x 8¼. (Available in U.S. only.)
0-486-24034-7

STARS AND RELATIVITY, Ya. B. Zel'dovich and I. D. Novikov. Vol. 1 of *Relativistic Astrophysics* by famed Russian scientists. General relativity, properties of matter under astrophysical conditions, stars, and stellar systems. Deep physical insights, clear presentation. 1971 edition. References. 544pp. 5⅝ x 8¼. 0-486-69424-0

Engineering

DE RE METALLICA, Georgius Agricola. The famous Hoover translation of greatest treatise on technological chemistry, engineering, geology, mining of early modern times (1556). All 289 original woodcuts. 638pp. 6¾ x 11. 0-486-60006-8

FUNDAMENTALS OF ASTRODYNAMICS, Roger Bate et al. Modern approach developed by U.S. Air Force Academy. Designed as a first course. Problems, exercises. Numerous illustrations. 455pp. 5⅜ x 8½. 0-486-60061-0

DYNAMICS OF FLUIDS IN POROUS MEDIA, Jacob Bear. For advanced students of ground water hydrology, soil mechanics and physics, drainage and irrigation engineering and more. 335 illustrations. Exercises, with answers. 784pp. 6⅛ x 9¼. 0-486-65675-6

THEORY OF VISCOELASTICITY (Second Edition), Richard M. Christensen. Complete consistent description of the linear theory of the viscoelastic behavior of materials. Problem-solving techniques discussed. 1982 edition. 29 figures. xiv+364pp. 6⅛ x 9¼. 0-486-42880-X

MECHANICS, J. P. Den Hartog. A classic introductory text or refresher. Hundreds of applications and design problems illuminate fundamentals of trusses, loaded beams and cables, etc. 334 answered problems. 462pp. 5⅜ x 8½. 0-486-60754-2

MECHANICAL VIBRATIONS, J. P. Den Hartog. Classic textbook offers lucid explanations and illustrative models, applying theories of vibrations to a variety of practical industrial engineering problems. Numerous figures. 233 problems, solutions. Appendix. Index. Preface. 436pp. 5⅜ x 8½. 0-486-64785-4

STRENGTH OF MATERIALS, J. P. Den Hartog. Full, clear treatment of basic material (tension, torsion, bending, etc.) plus advanced material on engineering methods, applications. 350 answered problems. 323pp. 5⅜ x 8½. 0-486-60755-0

A HISTORY OF MECHANICS, René Dugas. Monumental study of mechanical principles from antiquity to quantum mechanics. Contributions of ancient Greeks, Galileo, Leonardo, Kepler, Lagrange, many others. 671pp. 5⅜ x 8½. 0-486-65632-2

STABILITY THEORY AND ITS APPLICATIONS TO STRUCTURAL MECHANICS, Clive L. Dym. Self-contained text focuses on Koiter postbuckling analyses, with mathematical notions of stability of motion. Basing minimum energy principles for static stability upon dynamic concepts of stability of motion, it develops asymptotic buckling and postbuckling analyses from potential energy considerations, with applications to columns, plates, and arches. 1974 ed. 208pp. 5⅜ x 8½. 0-486-42541-X

METAL FATIGUE, N. E. Frost, K. J. Marsh, and L. P. Pook. Definitive, clearly written, and well-illustrated volume addresses all aspects of the subject, from the historical development of understanding metal fatigue to vital concepts of the cyclic stress that causes a crack to grow. Includes 7 appendixes. 544pp. 5⅜ x 8½. 0-486-40927-9

ANIMALS: 1,419 Copyright-Free Illustrations of Mammals, Birds, Fish, Insects, etc., Jim Harter (ed.). Clear wood engravings present, in extremely lifelike poses, over 1,000 species of animals. One of the most extensive pictorial sourcebooks of its kind. Captions. Index. 284pp. 9 x 12. 0-486-23766-4

1001 QUESTIONS ANSWERED ABOUT THE SEASHORE, N. J. Berrill and Jacquelyn Berrill. Queries answered about dolphins, sea snails, sponges, starfish, fishes, shore birds, many others. Covers appearance, breeding, growth, feeding, much more. 305pp. 5¼ x 8¼. 0-486-23366-9

ATTRACTING BIRDS TO YOUR YARD, William J. Weber. Easy-to-follow guide offers advice on how to attract the greatest diversity of birds: birdhouses, feeders, water and waterers, much more. 96pp. 5³⁄₁₆ x 8¼. 0-486-28927-3

MEDICINAL AND OTHER USES OF NORTH AMERICAN PLANTS: A Historical Survey with Special Reference to the Eastern Indian Tribes, Charlotte Erichsen-Brown. Chronological historical citations document 500 years of usage of plants, trees, shrubs native to eastern Canada, northeastern U.S. Also complete identifying information. 343 illustrations. 544pp. 6½ x 9¼. 0-486-25951-X

STORYBOOK MAZES, Dave Phillips. 23 stories and mazes on two-page spreads: Wizard of Oz, Treasure Island, Robin Hood, etc. Solutions. 64pp. 8¼ x 11. 0-486-23628-5

AMERICAN NEGRO SONGS: 230 Folk Songs and Spirituals, Religious and Secular, John W. Work. This authoritative study traces the African influences of songs sung and played by black Americans at work, in church, and as entertainment. The author discusses the lyric significance of such songs as "Swing Low, Sweet Chariot," "John Henry," and others and offers the words and music for 230 songs. Bibliography. Index of Song Titles. 272pp. 6½ x 9¼. 0-486-40271-1

MOVIE-STAR PORTRAITS OF THE FORTIES, John Kobal (ed.). 163 glamor, studio photos of 106 stars of the 1940s: Rita Hayworth, Ava Gardner, Marlon Brando, Clark Gable, many more. 176pp. 8⅜ x 11¼. 0-486-23546-7

YEKL and THE IMPORTED BRIDEGROOM AND OTHER STORIES OF YIDDISH NEW YORK, Abraham Cahan. Film Hester Street based on *Yekl* (1896). Novel, other stories among first about Jewish immigrants on N.Y.'s East Side. 240pp. 5⅜ x 8½. 0-486-22427-9

SELECTED POEMS, Walt Whitman. Generous sampling from *Leaves of Grass*. Twenty-four poems include "I Hear America Singing," "Song of the Open Road," "I Sing the Body Electric," "When Lilacs Last in the Dooryard Bloom'd," "O Captain! My Captain!"–all reprinted from an authoritative edition. Lists of titles and first lines. 128pp. 5³⁄₁₆ x 8¼. 0-486-26878-0

SONGS OF EXPERIENCE: Facsimile Reproduction with 26 Plates in Full Color, William Blake. 26 full-color plates from a rare 1826 edition. Includes "The Tyger," "London," "Holy Thursday," and other poems. Printed text of poems. 48pp. 5¼ x 7. 0-486-24636-1

THE BEST TALES OF HOFFMANN, E. T. A. Hoffmann. 10 of Hoffmann's most important stories: "Nutcracker and the King of Mice," "The Golden Flowerpot," etc. 458pp. 5⅜ x 8½. 0-486-21793-0

THE BOOK OF TEA, Kakuzo Okakura. Minor classic of the Orient: entertaining, charming explanation, interpretation of traditional Japanese culture in terms of tea ceremony. 94pp. 5⅜ x 8½. 0-486-20070-1

Mathematics

FUNCTIONAL ANALYSIS (Second Corrected Edition), George Bachman and Lawrence Narici. Excellent treatment of subject geared toward students with background in linear algebra, advanced calculus, physics and engineering. Text covers introduction to inner-product spaces, normed, metric spaces, and topological spaces; complete orthonormal sets, the Hahn-Banach Theorem and its consequences, and many other related subjects. 1966 ed. 544pp. 6⅛ x 9¼. 0-486-40251-7

ASYMPTOTIC EXPANSIONS OF INTEGRALS, Norman Bleistein & Richard A. Handelsman. Best introduction to important field with applications in a variety of scientific disciplines. New preface. Problems. Diagrams. Tables. Bibliography. Index. 448pp. 5⅜ x 8½. 0-486-65082-0

VECTOR AND TENSOR ANALYSIS WITH APPLICATIONS, A. I. Borisenko and I. E. Tarapov. Concise introduction. Worked-out problems, solutions, exercises. 257pp. 5⅜ x 8¼. 0-486-63833-2

AN INTRODUCTION TO ORDINARY DIFFERENTIAL EQUATIONS, Earl A. Coddington. A thorough and systematic first course in elementary differential equations for undergraduates in mathematics and science, with many exercises and problems (with answers). Index. 304pp. 5⅜ x 8½. 0-486-65942-9

FOURIER SERIES AND ORTHOGONAL FUNCTIONS, Harry F. Davis. An incisive text combining theory and practical example to introduce Fourier series, orthogonal functions and applications of the Fourier method to boundary-value problems. 570 exercises. Answers and notes. 416pp. 5⅜ x 8½. 0-486-65973-9

COMPUTABILITY AND UNSOLVABILITY, Martin Davis. Classic graduate-level introduction to theory of computability, usually referred to as theory of recurrent functions. New preface and appendix. 288pp. 5⅜ x 8½. 0-486-61471-9

ASYMPTOTIC METHODS IN ANALYSIS, N. G. de Bruijn. An inexpensive, comprehensive guide to asymptotic methods–the pioneering work that teaches by explaining worked examples in detail. Index. 224pp. 5⅜ x 8½ 0-486-64221-6

APPLIED COMPLEX VARIABLES, John W. Dettman. Step-by-step coverage of fundamentals of analytic function theory–plus lucid exposition of five important applications: Potential Theory; Ordinary Differential Equations; Fourier Transforms; Laplace Transforms; Asymptotic Expansions. 66 figures. Exercises at chapter ends. 512pp. 5⅜ x 8½. 0-486-64670-X

INTRODUCTION TO LINEAR ALGEBRA AND DIFFERENTIAL EQUATIONS, John W. Dettman. Excellent text covers complex numbers, determinants, orthonormal bases, Laplace transforms, much more. Exercises with solutions. Undergraduate level. 416pp. 5⅜ x 8½. 0-486-65191-6

RIEMANN'S ZETA FUNCTION, H. M. Edwards. Superb, high-level study of landmark 1859 publication entitled "On the Number of Primes Less Than a Given Magnitude" traces developments in mathematical theory that it inspired. xiv+315pp. 5⅜ x 8½. 0-486-41740-9

CALCULUS OF VARIATIONS WITH APPLICATIONS, George M. Ewing. Applications-oriented introduction to variational theory develops insight and promotes understanding of specialized books, research papers. Suitable for advanced undergraduate/graduate students as primary, supplementary text. 352pp. 5⅜ x 8½. 0-486-64856-7

COMPLEX VARIABLES, Francis J. Flanigan. Unusual approach, delaying complex algebra till harmonic functions have been analyzed from real variable viewpoint. Includes problems with answers. 364pp. 5⅜ x 8½. 0-486-61388-7

AN INTRODUCTION TO THE CALCULUS OF VARIATIONS, Charles Fox. Graduate-level text covers variations of an integral, isoperimetrical problems, least action, special relativity, approximations, more. References. 279pp. 5⅜ x 8½. 0-486-65499-0

COUNTEREXAMPLES IN ANALYSIS, Bernard R. Gelbaum and John M. H. Olmsted. These counterexamples deal mostly with the part of analysis known as "real variables." The first half covers the real number system, and the second half encompasses higher dimensions. 1962 edition. xxiv+198pp. 5⅜ x 8½. 0-486-42875-3

CATASTROPHE THEORY FOR SCIENTISTS AND ENGINEERS, Robert Gilmore. Advanced-level treatment describes mathematics of theory grounded in the work of Poincaré, R. Thom, other mathematicians. Also important applications to problems in mathematics, physics, chemistry and engineering. 1981 edition. References. 28 tables. 397 black-and-white illustrations. xvii + 666pp. 6⅛ x 9¼. 0-486-67539-4

INTRODUCTION TO DIFFERENCE EQUATIONS, Samuel Goldberg. Exceptionally clear exposition of important discipline with applications to sociology, psychology, economics. Many illustrative examples; over 250 problems. 260pp. 5⅜ x 8½. 0-486-65084-7

NUMERICAL METHODS FOR SCIENTISTS AND ENGINEERS, Richard Hamming. Classic text stresses frequency approach in coverage of algorithms, polynomial approximation, Fourier approximation, exponential approximation, other topics. Revised and enlarged 2nd edition. 721pp. 5⅜ x 8½. 0-486-65241-6

INTRODUCTION TO NUMERICAL ANALYSIS (2nd Edition), F. B. Hildebrand. Classic, fundamental treatment covers computation, approximation, interpolation, numerical differentiation and integration, other topics. 150 new problems. 669pp. 5⅜ x 8½. 0-486-65363-3

THREE PEARLS OF NUMBER THEORY, A. Y. Khinchin. Three compelling puzzles require proof of a basic law governing the world of numbers. Challenges concern van der Waerden's theorem, the Landau-Schnirelmann hypothesis and Mann's theorem, and a solution to Waring's problem. Solutions included. 64pp. 5⅜ x 8½. 0-486-40026-3

THE PHILOSOPHY OF MATHEMATICS: AN INTRODUCTORY ESSAY, Stephan Körner. Surveys the views of Plato, Aristotle, Leibniz & Kant concerning propositions and theories of applied and pure mathematics. Introduction. Two appendices. Index. 198pp. 5⅜ x 8½. 0-486-25048-2

INTRODUCTORY REAL ANALYSIS, A.N. Kolmogorov, S. V. Fomin. Translated by Richard A. Silverman. Self-contained, evenly paced introduction to real and functional analysis. Some 350 problems. 403pp. 5⅜ x 8½. 0-486-61226-0

APPLIED ANALYSIS, Cornelius Lanczos. Classic work on analysis and design of finite processes for approximating solution of analytical problems. Algebraic equations, matrices, harmonic analysis, quadrature methods, much more. 559pp. 5⅜ x 8½. 0-486-65656-X

AN INTRODUCTION TO ALGEBRAIC STRUCTURES, Joseph Landin. Superb self-contained text covers "abstract algebra": sets and numbers, theory of groups, theory of rings, much more. Numerous well-chosen examples, exercises. 247pp. 5⅜ x 8½. 0-486-65940-2

QUALITATIVE THEORY OF DIFFERENTIAL EQUATIONS, V. V. Nemytskii and V.V. Stepanov. Classic graduate-level text by two prominent Soviet mathematicians covers classical differential equations as well as topological dynamics and ergodic theory. Bibliographies. 523pp. 5⅜ x 8½. 0-486-65954-2

THEORY OF MATRICES, Sam Perlis. Outstanding text covering rank, nonsingularity and inverses in connection with the development of canonical matrices under the relation of equivalence, and without the intervention of determinants. Includes exercises. 237pp. 5⅜ x 8½. 0-486-66810-X

INTRODUCTION TO ANALYSIS, Maxwell Rosenlicht. Unusually clear, accessible coverage of set theory, real number system, metric spaces, continuous functions, Riemann integration, multiple integrals, more. Wide range of problems. Undergraduate level. Bibliography. 254pp. 5⅜ x 8½. 0-486-65038-3

MODERN NONLINEAR EQUATIONS, Thomas L. Saaty. Emphasizes practical solution of problems; covers seven types of equations. ". . . a welcome contribution to the existing literature...."–*Math Reviews.* 490pp. 5⅜ x 8½. 0-486-64232-1

MATRICES AND LINEAR ALGEBRA, Hans Schneider and George Phillip Barker. Basic textbook covers theory of matrices and its applications to systems of linear equations and related topics such as determinants, eigenvalues and differential equations. Numerous exercises. 432pp. 5⅜ x 8½. 0-486-66014-1

LINEAR ALGEBRA, Georgi E. Shilov. Determinants, linear spaces, matrix algebras, similar topics. For advanced undergraduates, graduates. Silverman translation. 387pp. 5⅜ x 8½. 0-486-63518-X

ELEMENTS OF REAL ANALYSIS, David A. Sprecher. Classic text covers fundamental concepts, real number system, point sets, functions of a real variable, Fourier series, much more. Over 500 exercises. 352pp. 5⅜ x 8½. 0-486-65385-4

SET THEORY AND LOGIC, Robert R. Stoll. Lucid introduction to unified theory of mathematical concepts. Set theory and logic seen as tools for conceptual understanding of real number system. 496pp. 5⅜ x 8¼. 0-486-63829-4

Math–Decision Theory, Statistics, Probability

ELEMENTARY DECISION THEORY, Herman Chernoff and Lincoln E. Moses. Clear introduction to statistics and statistical theory covers data processing, probability and random variables, testing hypotheses, much more. Exercises. 364pp. 5⅜ x 8½. 0-486-65218-1

STATISTICS MANUAL, Edwin L. Crow et al. Comprehensive, practical collection of classical and modern methods prepared by U.S. Naval Ordnance Test Station. Stress on use. Basics of statistics assumed. 288pp. 5⅜ x 8½. 0-486-60599-X

SOME THEORY OF SAMPLING, William Edwards Deming. Analysis of the problems, theory and design of sampling techniques for social scientists, industrial managers and others who find statistics important at work. 61 tables. 90 figures. xvii +602pp. 5⅜ x 8½. 0-486-64684-X

LINEAR PROGRAMMING AND ECONOMIC ANALYSIS, Robert Dorfman, Paul A. Samuelson and Robert M. Solow. First comprehensive treatment of linear programming in standard economic analysis. Game theory, modern welfare economics, Leontief input-output, more. 525pp. 5⅜ x 8½. 0-486-65491-5

PROBABILITY: AN INTRODUCTION, Samuel Goldberg. Excellent basic text covers set theory, probability theory for finite sample spaces, binomial theorem, much more. 360 problems. Bibliographies. 322pp. 5⅜ x 8½. 0-486-65252-1

GAMES AND DECISIONS: INTRODUCTION AND CRITICAL SURVEY, R. Duncan Luce and Howard Raiffa. Superb nontechnical introduction to game theory, primarily applied to social sciences. Utility theory, zero-sum games, n-person games, decision-making, much more. Bibliography. 509pp. 5⅜ x 8½. 0-486-65943-7

INTRODUCTION TO THE THEORY OF GAMES, J. C. C. McKinsey. This comprehensive overview of the mathematical theory of games illustrates applications to situations involving conflicts of interest, including economic, social, political, and military contexts. Appropriate for advanced undergraduate and graduate courses; advanced calculus a prerequisite. 1952 ed. x+372pp. 5⅜ x 8½. 0-486-42811-7

FIFTY CHALLENGING PROBLEMS IN PROBABILITY WITH SOLUTIONS, Frederick Mosteller. Remarkable puzzlers, graded in difficulty, illustrate elementary and advanced aspects of probability. Detailed solutions. 88pp. 5⅜ x 8½. 65355-2

PROBABILITY THEORY: A CONCISE COURSE, Y. A. Rozanov. Highly readable, self-contained introduction covers combination of events, dependent events, Bernoulli trials, etc. 148pp. 5⅜ x 8¼. 0-486-63544-9

STATISTICAL METHOD FROM THE VIEWPOINT OF QUALITY CONTROL, Walter A. Shewhart. Important text explains regulation of variables, uses of statistical control to achieve quality control in industry, agriculture, other areas. 192pp. 5⅜ x 8½. 0-486-65232-7

Math–Geometry and Topology

ELEMENTARY CONCEPTS OF TOPOLOGY, Paul Alexandroff. Elegant, intuitive approach to topology from set-theoretic topology to Betti groups; how concepts of topology are useful in math and physics. 25 figures. 57pp. 5⅜ x 8½. 0-486-60747-X

COMBINATORIAL TOPOLOGY, P. S. Alexandrov. Clearly written, well-organized, three-part text begins by dealing with certain classic problems without using the formal techniques of homology theory and advances to the central concept, the Betti groups. Numerous detailed examples. 654pp. 5⅜ x 8¼. 0-486-40179-0

EXPERIMENTS IN TOPOLOGY, Stephen Barr. Classic, lively explanation of one of the byways of mathematics. Klein bottles, Moebius strips, projective planes, map coloring, problem of the Koenigsberg bridges, much more, described with clarity and wit. 43 figures. 210pp. 5⅜ x 8½. 0-486-25933-1

THE GEOMETRY OF RENÉ DESCARTES, René Descartes. The great work founded analytical geometry. Original French text, Descartes's own diagrams, together with definitive Smith-Latham translation. 244pp. 5⅜ x 8½. 0-486-60068-8

EUCLIDEAN GEOMETRY AND TRANSFORMATIONS, Clayton W. Dodge. This introduction to Euclidean geometry emphasizes transformations, particularly isometries and similarities. Suitable for undergraduate courses, it includes numerous examples, many with detailed answers. 1972 ed. viii+296pp. 6⅛ x 9¼. 0-486-43476-1

PRACTICAL CONIC SECTIONS: THE GEOMETRIC PROPERTIES OF ELLIPSES, PARABOLAS AND HYPERBOLAS, J. W. Downs. This text shows how to create ellipses, parabolas, and hyperbolas. It also presents historical background on their ancient origins and describes the reflective properties and roles of curves in design applications. 1993 ed. 98 figures. xii+100pp. 6½ x 9¼. 0-486-42876-1

THE THIRTEEN BOOKS OF EUCLID'S ELEMENTS, translated with introduction and commentary by Sir Thomas L. Heath. Definitive edition. Textual and linguistic notes, mathematical analysis. 2,500 years of critical commentary. Unabridged. 1,414pp. 5⅜ x 8½. Three-vol. set.
Vol. I: 0-486-60088-2 Vol. II: 0-486-60089-0 Vol. III: 0-486-60090-4

SPACE AND GEOMETRY: IN THE LIGHT OF PHYSIOLOGICAL, PSYCHOLOGICAL AND PHYSICAL INQUIRY, Ernst Mach. Three essays by an eminent philosopher and scientist explore the nature, origin, and development of our concepts of space, with a distinctness and precision suitable for undergraduate students and other readers. 1906 ed. vi+148pp. 5⅜ x 8½. 0-486-43909-7

GEOMETRY OF COMPLEX NUMBERS, Hans Schwerdtfeger. Illuminating, widely praised book on analytic geometry of circles, the Moebius transformation, and two-dimensional non-Euclidean geometries. 200pp. 5⅝ x 8¼. 0-486-63830-8

DIFFERENTIAL GEOMETRY, Heinrich W. Guggenheimer. Local differential geometry as an application of advanced calculus and linear algebra. Curvature, transformation groups, surfaces, more. Exercises. 62 figures. 378pp. 5⅜ x 8½. 0-486-63433-7

History of Math

THE WORKS OF ARCHIMEDES, Archimedes (T. L. Heath, ed.). Topics include the famous problems of the ratio of the areas of a cylinder and an inscribed sphere; the measurement of a circle; the properties of conoids, spheroids, and spirals; and the quadrature of the parabola. Informative introduction. clxxxvi+326pp. 5⅜ x 8½. 0-486-42084-1

A SHORT ACCOUNT OF THE HISTORY OF MATHEMATICS, W. W. Rouse Ball. One of clearest, most authoritative surveys from the Egyptians and Phoenicians through 19th-century figures such as Grassman, Galois, Riemann. Fourth edition. 522pp. 5⅜ x 8½. 0-486-20630-0

THE HISTORY OF THE CALCULUS AND ITS CONCEPTUAL DEVELOPMENT, Carl B. Boyer. Origins in antiquity, medieval contributions, work of Newton, Leibniz, rigorous formulation. Treatment is verbal. 346pp. 5⅜ x 8½. 0-486-60509-4

THE HISTORICAL ROOTS OF ELEMENTARY MATHEMATICS, Lucas N. H. Bunt, Phillip S. Jones, and Jack D. Bedient. Fundamental underpinnings of modern arithmetic, algebra, geometry and number systems derived from ancient civilizations. 320pp. 5⅜ x 8½. 0-486-25563-8

A HISTORY OF MATHEMATICAL NOTATIONS, Florian Cajori. This classic study notes the first appearance of a mathematical symbol and its origin, the competition it encountered, its spread among writers in different countries, its rise to popularity, its eventual decline or ultimate survival. Original 1929 two-volume edition presented here in one volume. xxviii+820pp. 5⅜ x 8½. 0-486-67766-4

GAMES, GODS & GAMBLING: A HISTORY OF PROBABILITY AND STATISTICAL IDEAS, F. N. David. Episodes from the lives of Galileo, Fermat, Pascal, and others illustrate this fascinating account of the roots of mathematics. Features thought-provoking references to classics, archaeology, biography, poetry. 1962 edition. 304pp. 5⅜ x 8½. (Available in U.S. only.) 0-486-40023-9

OF MEN AND NUMBERS: THE STORY OF THE GREAT MATHEMATICIANS, Jane Muir. Fascinating accounts of the lives and accomplishments of history's greatest mathematical minds–Pythagoras, Descartes, Euler, Pascal, Cantor, many more. Anecdotal, illuminating. 30 diagrams. Bibliography. 256pp. 5⅜ x 8½. 0-486-28973-7

HISTORY OF MATHEMATICS, David E. Smith. Nontechnical survey from ancient Greece and Orient to late 19th century; evolution of arithmetic, geometry, trigonometry, calculating devices, algebra, the calculus. 362 illustrations. 1,355pp. 5⅜ x 8½. Two-vol. set. Vol. I: 0-486-20429-4 Vol. II: 0-486-20430-8

A CONCISE HISTORY OF MATHEMATICS, Dirk J. Struik. The best brief history of mathematics. Stresses origins and covers every major figure from ancient Near East to 19th century. 41 illustrations. 195pp. 5⅜ x 8½. 0-486-60255-9

Physics

OPTICAL RESONANCE AND TWO-LEVEL ATOMS, L. Allen and J. H. Eberly. Clear, comprehensive introduction to basic principles behind all quantum optical resonance phenomena. 53 illustrations. Preface. Index. 256pp. 5⅜ x 8½. 0-486-65533-4

QUANTUM THEORY, David Bohm. This advanced undergraduate-level text presents the quantum theory in terms of qualitative and imaginative concepts, followed by specific applications worked out in mathematical detail. Preface. Index. 655pp. 5⅜ x 8½. 0-486-65969-0

ATOMIC PHYSICS (8th EDITION), Max Born. Nobel laureate's lucid treatment of kinetic theory of gases, elementary particles, nuclear atom, wave-corpuscles, atomic structure and spectral lines, much more. Over 40 appendices, bibliography. 495pp. 5⅜ x 8½. 0-486-65984-4

A SOPHISTICATE'S PRIMER OF RELATIVITY, P. W. Bridgman. Geared toward readers already acquainted with special relativity, this book transcends the view of theory as a working tool to answer natural questions: What is a frame of reference? What is a "law of nature"? What is the role of the "observer"? Extensive treatment, written in terms accessible to those without a scientific background. 1983 ed. xlviii+172pp. 5⅜ x 8½. 0-486-42549-5

AN INTRODUCTION TO HAMILTONIAN OPTICS, H. A. Buchdahl. Detailed account of the Hamiltonian treatment of aberration theory in geometrical optics. Many classes of optical systems defined in terms of the symmetries they possess. Problems with detailed solutions. 1970 edition. xv + 360pp. 5⅜ x 8½. 0-486-67597-1

PRIMER OF QUANTUM MECHANICS, Marvin Chester. Introductory text examines the classical quantum bead on a track: its state and representations; operator eigenvalues; harmonic oscillator and bound bead in a symmetric force field; and bead in a spherical shell. Other topics include spin, matrices, and the structure of quantum mechanics; the simplest atom; indistinguishable particles; and stationary-state perturbation theory. 1992 ed. xiv+314pp. 6⅛ x 9¼. 0-486-42878-8

LECTURES ON QUANTUM MECHANICS, Paul A. M. Dirac. Four concise, brilliant lectures on mathematical methods in quantum mechanics from Nobel Prize-winning quantum pioneer build on idea of visualizing quantum theory through the use of classical mechanics. 96pp. 5⅜ x 8½. 0-486-41713-1

THIRTY YEARS THAT SHOOK PHYSICS: THE STORY OF QUANTUM THEORY, George Gamow. Lucid, accessible introduction to influential theory of energy and matter. Careful explanations of Dirac's anti-particles, Bohr's model of the atom, much more. 12 plates. Numerous drawings. 240pp. 5⅜ x 8½. 0-486-24895-X

ELECTRONIC STRUCTURE AND THE PROPERTIES OF SOLIDS: THE PHYSICS OF THE CHEMICAL BOND, Walter A. Harrison. Innovative text offers basic understanding of the electronic structure of covalent and ionic solids, simple metals, transition metals and their compounds. Problems. 1980 edition. 582pp. 6⅛ x 9¼. 0-486-66021-4

HYDRODYNAMIC AND HYDROMAGNETIC STABILITY, S. Chandrasekhar. Lucid examination of the Rayleigh-Benard problem; clear coverage of the theory of instabilities causing convection. 704pp. 5⅜ x 8¼. 0-486-64071-X

INVESTIGATIONS ON THE THEORY OF THE BROWNIAN MOVEMENT, Albert Einstein. Five papers (1905–8) investigating dynamics of Brownian motion and evolving elementary theory. Notes by R. Fürth. 122pp. 5⅜ x 8½. 0-486-60304-0

THE PHYSICS OF WAVES, William C. Elmore and Mark A. Heald. Unique overview of classical wave theory. Acoustics, optics, electromagnetic radiation, more. Ideal as classroom text or for self-study. Problems. 477pp. 5⅜ x 8½. 0-486-64926-1

GRAVITY, George Gamow. Distinguished physicist and teacher takes reader-friendly look at three scientists whose work unlocked many of the mysteries behind the laws of physics: Galileo, Newton, and Einstein. Most of the book focuses on Newton's ideas, with a concluding chapter on post-Einsteinian speculations concerning the relationship between gravity and other physical phenomena. 160pp. 5⅜ x 8½. 0-486-42563-0

PHYSICAL PRINCIPLES OF THE QUANTUM THEORY, Werner Heisenberg. Nobel Laureate discusses quantum theory, uncertainty, wave mechanics, work of Dirac, Schroedinger, Compton, Wilson, Einstein, etc. 184pp. 5⅜ x 8½. 0-486-60113-7

ATOMIC SPECTRA AND ATOMIC STRUCTURE, Gerhard Herzberg. One of best introductions; especially for specialist in other fields. Treatment is physical rather than mathematical. 80 illustrations. 257pp. 5⅜ x 8½. 0-486-60115-3

AN INTRODUCTION TO STATISTICAL THERMODYNAMICS, Terrell L. Hill. Excellent basic text offers wide-ranging coverage of quantum statistical mechanics, systems of interacting molecules, quantum statistics, more. 523pp. 5⅜ x 8½. 0-486-65242-4

THEORETICAL PHYSICS, Georg Joos, with Ira M. Freeman. Classic overview covers essential math, mechanics, electromagnetic theory, thermodynamics, quantum mechanics, nuclear physics, other topics. First paperback edition. xxiii + 885pp. 5⅜ x 8½. 0-486-65227-0

PROBLEMS AND SOLUTIONS IN QUANTUM CHEMISTRY AND PHYSICS, Charles S. Johnson, Jr. and Lee G. Pedersen. Unusually varied problems, detailed solutions in coverage of quantum mechanics, wave mechanics, angular momentum, molecular spectroscopy, more. 280 problems plus 139 supplementary exercises. 430pp. 6½ x 9¼. 0-486-65236-X

THEORETICAL SOLID STATE PHYSICS, Vol. 1: Perfect Lattices in Equilibrium; Vol. II: Non-Equilibrium and Disorder, William Jones and Norman H. March. Monumental reference work covers fundamental theory of equilibrium properties of perfect crystalline solids, non-equilibrium properties, defects and disordered systems. Appendices. Problems. Preface. Diagrams. Index. Bibliography. Total of 1,301pp. 5⅜ x 8½. Two volumes. Vol. I: 0-486-65015-4 Vol. II: 0-486-65016-2

WHAT IS RELATIVITY? L. D. Landau and G. B. Rumer. Written by a Nobel Prize physicist and his distinguished colleague, this compelling book explains the special theory of relativity to readers with no scientific background, using such familiar objects as trains, rulers, and clocks. 1960 ed. vi+72pp. 5⅜ x 8½. 0-486-42806-0

CATALOG OF DOVER BOOKS

CATALOG OF DOVER BOOKS